Werner Ebeling (1936) studied physics at the Universities of Rostock and Moscow; Diploma in Physics (1959), Dr. rer. nat. (1963) and Dr. habil. (1968) under Prof. Hans Falkenhagen in Rostock. He has served as Professor at the Rostock University, as visiting Professor at the Universities Brussels, Minnesota, Paris, Riga, Stuttgart, Torun, Vera Cruz and is now Professor of Theoretical Physics at the Humboldt University Berlin. He is working in the field of statistical physics, plasma theory, nonlinear dynamics, and the theory of selforganization and evolution.

Andreas Forster (1961) studied physics at the Lomonosov University of Moscow. First academic degree (Dipl.-Phys.) in 1986 for a work on fluctuations in reaction-diffusion systems under Dr. A. S. Mikhailov. Since 1986 he is working in the Laboratory for Statistical Thermodynamics and Theoretical Biophysics at the Humboldt University of Berlin. In 1991 thesis work (Dr. rer. nat.) in plasma physics under Prof. W. Ebeling. His main research interests are thermodynamics and kinetics of nonideal plasmas.

Rainer Radtke (1945) studied physics at the Humboldt University of Berlin. Diploma in Physics (1971) and Dr. rer. nat. (1974) in the field of non-Debye plasmas at the Central Institute of Electron Physics, Berlin. From 1973 to 1991 work on transport and radiation properties of dense, high-power pulse plasmas in the Laboratory of Dense Plasmas of this institute. Since 1992 he deals with spectroscopy of boundary plasmas in fusion facilities at the Max-Planck Institute for Plasma Physics.

Die Deutsche Bibliothek - CIP-Einheitsaufnahme

Physics of nonideal plasmas ed. by Werner Ebeling ...
Stuttgart; Leipzig: Teubner, 1992
 (Teubner-Texte zur Physik; Bd. 26)
  ISBN 978-3-322-99737-1   ISBN 978-3-322-99736-4 (eBook)
  DOI 10.1007/978-3-322-99736-4

NE: Ebeling, Werner [Hrsg.]; GT

TEUBNER-TEXTE zur Physik · Band 26
ISSN 0233-0911

Gesamtherstellung: Druckerei „G. W. Leibniz" GmbH, Gräfenhainichen

TEUBNER-TEXTE zur Physik · Band 26

Herausgeber/Editors: Werner Ebeling, Berlin

Wolfgang Meiling, Dresden

Armin Uhlmann, Leipzig

Bernd Wilhelmi, Jena

Physics of Nonideal Plasmas

Edited by
Werner Ebeling, Andreas Förster, Rainer Radtke

B. G. Teubner Verlagsgesellschaft
Stuttgart · Leipzig 1992

This book is based upon the invited lectures and some selected contributed papers of the VI. International Workshop on Physics of Nonideal Plasmas (PNP VI), Gosen (Germany), 1991. The main topics covered are molecular dynamics, Monte Carlo results, thermodynamics, phase transitions, kinetics of transitions, collective modes, transport properties, radiation and spectroscopy, astrophysics, laser and ion fusion. Leading experts from Austria, Belgium, France, Germany, Japan, the Netherlands, Poland, Russia, Switzerland, and the United States review the progress in theory as well as in diagnostics and measurement achieved during the last years. Additional papers discuss applications of strongly coupled plasma physics for, e.g., radiation sources or the determination of the structure of the giant planets and the sun.

Dieses Buch basiert auf den eingeladenen Hauptvorträgen und weiteren ausgewählten Beiträgen der 6. Internationalen Arbeitstagung für Physik Nichtidealer Plasmen (PNP VI), Gosen (Deutschland), 1991. Folgende Themengebiete werden behandelt: Molekulare Dynamik und Monte Carlo Resultate, Thermodynamik und Phasenübergänge, Kinetik von Übergängen, Kollektive Anregungen und Transporteigenschaften, Strahlung und Spektroskopie, Astrophysik und Laser- und Ionenfusion. Führende Experten aus Belgien, Deutschland, Frankreich, Japan, den Niederlanden, Polen, Österreich, Rußland, der Schweiz und den Vereinigten Staaten geben eine Übersicht zu den Fortschritten in der Theorie wie auch in den Meßmethoden und der Diagnostik, die in den letzten Jahren erreicht wurden. Weitere Artikel diskutieren Anwendungen der Physik stark gekoppelter Plasmen z.B. für Strahlungsquellen oder die Bestimmung der Struktur der großen Planeten und der Sonne.

Ce livre se fonde sur les discours principaux invités et d'autres exposés sélectionnés du 6. Congrès International pour la Physique des Plasmas Non-idéals (PNP VI), à Gosen (Allemagne), en 1991. Les sujets suivants sont traités: la dynamique moléculaire et les résultats Monte Carlo, la thermodynamique, le changement de phase, la cinétique des transitions, les excitations collectives, les propriétés du transport, le rayonnement et la spectroscopie, l'astrophysique, la fusion du laser et des ions. Spécialistes les plus renommés d'Allemagne, Autriche, Belgique, France, Pays Bas, Pologne, Russie, Suisse et États-Unis offrent une vue d'ensemble du progrés dans la théorie, le diagnostic et les méthodes de mesure, achevés dans les derniers ans. Il y a aussi d'autres articles dans lesquels les applications de la physique des plasmas fortement couplés sont discutées, par exemple la détermination de la structure des grandes planetes et du soleil.

В эту книгу включены материалы основных докладов, прочитанных на 6-ой Рабочей вестрече по физике неидеальной плазмы (PNP VI) в Гозене (Германия) в 1991 г., а также некоторые другие сообщения, представленные на этой конференции. Главным образом освещается следующий круг вопросов: молекулярная динамика, моделирование методом Монте Карло, термодинамика, фазовые переходы, кинетика переходов, коллективные состояния, явления переноса, спектроскопия, астрофизика, лазерный и ионный термоядерный синтез. Ведущие специалисты из Австрии, Бельгии, Германии, Нидерландов, Польши, России, Соединенных Штатов, Франции, Швейцарии и Японии дают обзор новейших результатов как в области теории, так и в диагностике и методах измерения, полученных за последнее время. В книге также обсуждаются прикладные аспекты физики сильно связанной плазмы, в частности, создание источников излучения и определение структуры больших планет и солнца.

TABLE OF CONTENTS

Chapter 4: Collective modes and transport properties

Chapter 5: Radiation and spectroscopy

Chapter 6: Astrophysical Problems

Chapter 7: Laser and Ion Fusion

The VI. International Workshop on Physics of Nonideal Plasmas (PNP VI) took place from November 18th to 21th, 1991, in Gosen (Germany) at the Science Communication & Conference Centre of the Humboldt University of Berlin. The workshop was organized by the Institute for Theoretical Physics of the Humboldt University and by the Central Institute of Electron Physics, Berlin, with financial support given by the Deutsche Forschungsgemeinschaft, Bonn, and the Robert-Bosch-Stiftung, Stuttgart. The workshop was attended by more than 100 scientists from 14 countries who presented about 120 papers, including 18 invited lectures.

The series of PNP workshops, which started in 1980, provides a biennial forum for both experimental and theoretical research in the field of nonideal plasmas. These meetings are organized alternately by the Central Institute of Electron Physics, Berlin, and/or by one of the universities of Berlin, Greifswald, and Rostock. They took place in Matzlow-Garwitz (1980), Wustrow (1982 and 1988), Biesenthal (1984), and Greifswald (1986). Since the beginning, the workshop has been concerned mainly with fundamental studies of the thermodynamic, transport, and radiative properties of nonideal plasmas. These fields were also covered at PNP VI in Gosen, but new topics such as high-pressure laser plasmas, dense astrophysical plasmas, molecular dynamics and Monte-Carlo results, and the kinetics of transitions have completed the programme. In particular, several papers addressed the role of nonideal plasmas for radiation sources, for inertial confinement fusion, for helio-seismology, and for the determination of the structure of the giant planets.

With the growing interest in nonideal plasmas the number of participants was also growing from workshop to workshop with the result that at the last meeting a critical number was exceeded allowing no longer to present all contributions orally. Therefore, we organized four poster sessions parallel to the afternoon sessions for the presentation of about 50% of all contributions. The present book is based upon the lectures and some selected contributed papers given by leading experts from plasma physics laboratories in Amsterdam, Belgrade, Berlin, Bochum, Boston, Brussels, Chernogolovka, College Park, Darmstadt, Dolgoprudny, Eindhoven, Gainesville, Graz, Greifswald, Heidelberg, Kiel, Lausanne, Livermore, Los Alamos, Los

Angeles, Lyon, Moscow, New York, Odessa, Orsay, Paris, Rostock, St. Peterburg, Tokyo, Troitsk, Tuscon, Vienna, and Warsaw. Although the book must be a small selection of the many papers which were presented, we think that it gives a representative survey of the workshop and shows the progress in theory as well as in diagnostics and measurement achieved during the last years.

The field of nonideal plasmas has developed rapidly over the past decade. As indicated by the papers at this workshop, it is now characterized by a better understanding of the fundamental processes, although essential properties are still not measured as systematically as under ideal conditions. Also, much work remains to be done on the development of theoretical tools until a quantitative understanding and control of nonideal plasmas is achieved.

We want to recommend the reader to visit the next PNP meeting, which will be hosted by the Rostock University in 1993, and see which problems will have been solved during the next two years, at least partially. In this connection we should mention also the series of Conferences on Strongly Coupled Plasma Physics (Orleans-la-Source, 1977; Les-Houches, 1982; Santa Cruz, 1986; Lake Yamanaka, 1989) which will be continued in 1992 at Rochester.

Finally, we would like to express our thanks to the members of the International Programme Committee, to the local organizing committee, and to the B.G. Teubner Verlagsgesellschaft mbH Stuttgart Leipzig, in particular to our reader Jürgen Weiß, for their support and cooperation.

Berlin, March 1992   W. Ebeling, A. Förster, and R. Radtke

SCIENTIFIC AND LOCAL ORGANIZING COMMITTEE

Scientific Committee

Berni J. Alder (Livermore)
Claude Deutsch (Orsay)
Hugh E. DeWitt (Livermore)
Vladimir E. Fortov (Moscow)
Klaus Günther (Berlin)
Friedrich Hensel (Marburg)
Setsuo Ichimaru (Tokyo)
Gabor J. Kalman (Boston)
Yuri L. Klimontovich (Moscow)
Wolf-Dietrich Kraeft (Greifswald)
Dietrich Kremp (Rostock)
Jürgen Meyer-ter-Vehn (Garching)
Gerd Röpke (Rostock)
Kurt Suchy (Düsseldorf)
Mario P. Tosi (Trieste)

Local Organizing Committee

Werner Ebeling (Chairman)
Andreas Förster (Scientific Secretary)
Bernd Groß
Heiko Lehmann
Ines Leike
Ulf Leonhardt
Rainer Radtke (Chairman)

NUMERICAL SIMULATION OF COULOMBIC FREEZING

Hugh E. DeWitt, Lawrence Livermore National Lab
Livermore, CA 94550

Wayne L. Slattery, Los Alamos National Laboratory
Los Alamos, NM 87545

Juxing Yang, Levich Institute, City College of CUNY
New York, NY 10031

Abstract

The fluid to crystalline solid first order phase transition of the classical one component plasma (OCP) has been studied by Monte Carlo simulation in three dimensions for temperatures below the thermodynamic freezing temperature ($\Gamma = Z^2 e^2/akT = 180$, a = Wigner-Zeitz radius). With N = 686 we found freezing from a metastable supercooled fluid into microcrystals for values of Γ ranging from 250 to 700. In one case, $\Gamma = 500$, the system froze into a perfect bcc lattice from a random start. With more particles the system froze into two or more crystals one bcc and one fcc and smaller examples of hcp. The lattice planes were examined and various kinds of crystal defects could be observed. We developed a program for determining the local environment of each particle as bcc, fcc, hcp, or fluid in order to identify microcrystals in the system at any stage of the freezing process. Generally freezing proceeds rapidly when any single microcrystal attains a sufficient size or roughly 60 to 70 particles. The observations of freezing seem to agree with classical nucleation theory. No separate glass phase (non-crystalline) was seen.

Introduction

The OCP in three dimensions is a system of classical point charges moving in a fixed uniform background. In nature it is a rough approximation of the conditions in a white dwarf star in which one has fully ionized nuclei such as carbon, oxygen, and some smaller portions of elements as heavy as iron all moving in a nearly uniform background provided by relativistically degenerate electrons [1]. In the outer portion of a white dwarf one has a strongly coupled Coulomb fluid ($\Gamma > 1$) but the interior is sufficiently dense and thus highly correlated that it freezes into a solid. In real astrophysical applications one must also consider the complications due to possible phase separation of the heavier nuclei, small screening effects from the electrons, and possibly significant quantum effects on the solid.   In the work reported here we will consider only the mathematical limit with point charges, a rigid background (fixed volume) and no quantum effects. This ideal system has been extensively studied by Monte Carlo simulation which is well adapted to this case in which the phase transition is isochoric and sufficiently long computer runs give the Coulomb interaction energy as obtained from the canonical ensemble [2].

The strongly coupled OCP fluid ($\Gamma > 1$) has an energy equation of state of the general form:

$$u(\Gamma) \ = \ U/NkT \ = \ A\,\Gamma \ + \ B\,\Gamma^s \ + \ C \ \text{---} \ , \quad 1 < \Gamma < \Gamma_m \sim 180 \tag{1}$$

where A is the 'fluid Madelung constant' with a value close to -.9, and the exponent s is approximately 1/3 from the best fit to Monte Carlo data [3]. When the Coulomb fluid freezes, the lattice energy equation of state has the form:

$$u(\Gamma) \ = \ U/NkT \ = \ A_m\Gamma \ + \ 3/2 \ + \ C_1/\Gamma \ + C_2/\Gamma^2 + \ \text{---} \ , \quad \Gamma > \Gamma_m \tag{2}$$

where

A_m = -.895929 ,	$C1$ = 10.843 ,	Γ_m = 178,	bcc
= -.895873,	= 12.348,	= 192,	fcc
= -.895838,	= - - - ,	~ 200,	hcp

Since the Madelung constants, A_m, are extremely close for the three cubic lattices the freezing process will compete strongly with each other. The known first order anharmonic corrections are a small portion of the total energy [4]. The bcc lattice has the lowest free energy and thus is the most stable. The estimated values for Γ_m, the thermodynamic freezing parameters, are obtained from the crossing of the fluid and solid free energies. The fluid energy is also quite close to the lattice energy. At the fluid-solid transition the change of energy is less than 1%.

Since the earlier complete study of the fluid equation of state was done with MC simulations involving N = 686 particles (a bcc number), we used this same number for the freezing studies [3]. As with the fluid energy runs, we started the system from random positions, i.e. no correlations. This is equivalent to an abrupt change from T = ∞ down to the desired physical temperature corresponding to Γ. For Γ values in the fluid region the energy drops rapidly to thermal equilibrium value in a period of about a million configurations. For Γ values in the solid region, i.e. $\Gamma > \Gamma_m$, the system first equilibrates into a supercooled fluid state and remains there for many tens of millions of configurations with the energy dropping slightly. During this long supercooled phase undergoes heterophase fluctuations [5] in which microscopic crystalline clusters form and come apart. Eventually the energy drops steeply and the system goes into a lattice. Unlike the MC studyof Ogata and Ichimaru with N = 432 particles [6] we do not use any intermediate quenches, since going directly from a random configuration to the final value of gamma allows the system to explore the full range of phase space possibilities.

Random Start Monte Carlo Solid Energy Results

When the number of particles in the MC cell is relatively small, say N =54 or 128, one may see the system may freeze into a lattice aligned with the coordinates of the cell even for Γ near Γ_m. For larger values of N the numerical simulation of the freezing process is more difficult near the thermodynamic freezing temperature. Thus with N = 686 our system remained as a supercooled fluid for Γ = 200, and was first observed to freeze only whenwhen Γ = 250. We then did long runs for Γ values at 300, 400, 500, 600, and 700 and observed freezing into usually imperfect crystalline solids. At Γ = 800 the system system settled into a probable supercooled fluid state with the energy constant, and showed no indication of freezing even after half a billion configurations, a very long run. Our energy results are shown in the following table:

Table 1. Monte Carlo Energy Results for N = 686

G	u_{fluid}	u_{bcc}	$u_{MCsolid}$	Du	Nc (10^8)
180	-158.900	-159.675			
200	-176.7774	-177.619			
250	-221.528	-222.425	-222.3481	.077	1.4
300		-267.246	-266.9821	.264	.83
400		-356.842	-356.6764	.166	.59
500		-446.4402	-446.4404	.000	2.05
600		-536.040	-535.0007	1.039	1.8
700		-625.635	-625.0051	.630	1.35
800	-713.1082	-715.230			

The first column gives the MC fluid energies, u_{fluid}, with $\Gamma = 180$ the end of the stable fluid and beginning of the metastable OCP fluid. The value for $\Gamma = 250$ is an estimate since the system froze at 1.4 million configurations. No entries are listed for $\Gamma = 300, --, 700$ because the metastable fluid energy was decreasing; the value for $\Gamma = 800$ is the final state. The values for u_{bcc} are lattice energies obtained from Monte Carlo runs from a bcc start, and checked with Eq. 2. $u_{MCsolid}$ are the final energies for values of Γ that froze. $\Delta u = u_{MCsolid} - u_{bcc}$ is the deviation from the bcc lattice energy of the final MC solid energy due to defects in the solid. For $\Gamma = 500$ there is no difference which indicates that the final state is in a perfect bcc lattice. Note that the values of Δu jump around considerably for different Γ values. This is a reflection of the randomness in the MC process. The final column N_C gives the number of configurations at which the system finished freezing into its final solid state. Again there is a great variation in the number of configurations at which freezing occurs. These values are not expected to be reproducible with different random number generators for the Monte Carlo simulation process. After the freezing occured, we ran the system for several tens of millions more configurations in order to anneal out simple lattice defects and obtain a stable final state. For $\Gamma = 800$ the system gave no indication of freezing, and the internal energy remained flat from 150 to 512 million configurations. The pair distribution function, g(r), was appropriate for a typical liquid, and showed no indication of lattice or glass structure. This seemingly paradoxical result can be understood by considering the system to be first abruptly cooled to nearly T = 0. In this event the system would stay indefinitely as a nearly frozen liquid. In the MC simulation process this situation — too cold to freeze — is evidently reached at $\Gamma = 800$.

Figure 1 shows the evolution of the MC Coulombic energy for the $\Gamma = 500$ from a random start. Each + symbol represents 10^5 configurations. The energy of the supercooled OCP fluid is surprisingly flat with one small drop at 100 million configurations. As seen from Table 1 the duration of the supercooled state is

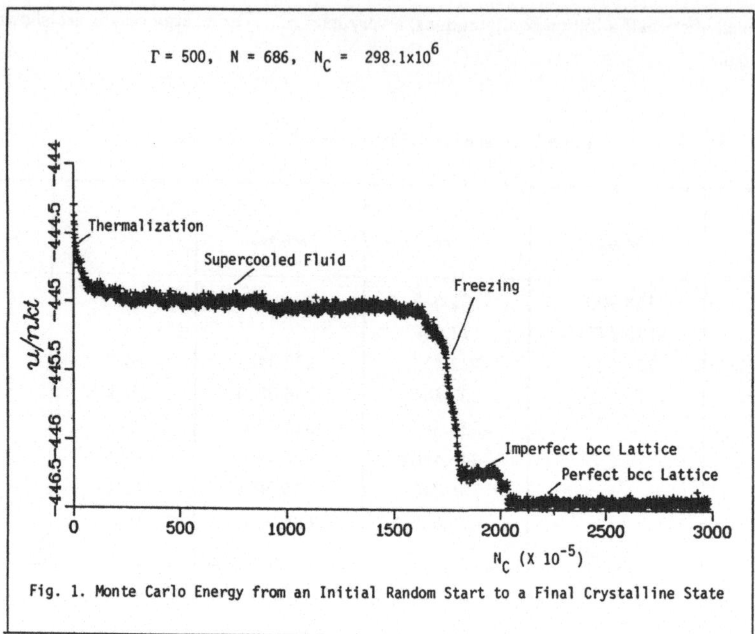

Fig. 1. Monte Carlo Energy from an Initial Random Start to a Final Crystalline State

the longest of any of the six values of Γ that froze. The freezing process once begun at 170 million configurations is largely completed within a few million configurations. This case is interesting in showing a metastable solid which persisted for about 20 million configurations before the MC process annealed out the imperfections. At a approximately 200 million configurations the energy dropped again to the energy of the perfect bcc lattice, as indicated in Table 1.

Figure 2 shows the g(r) obtained from the MC particle position data for both the final state (the perfect lattice) and the short intermediate solid state (the imperfect lattice) as compared with g(r) obtained from a lattice start at $\Gamma = 500$. The + signs for the final state fall on top of the g(r) for the lattice start. The g(r) for the imperfect lattice differs from the final state only in that the peaks are slightly smaller and the valleys less deep. A small number of lattice defects can account for this difference. Note that the structure of the bcc lattice is sharply indicated in the peaks. Thus the first peak corresponds to the 8 nearest neighbors, the hump on the rightr side of the first peak is from the very close 6 next nearest neighbors, the next peak is due to the 12 third nearest neighbors, and the following high peak is due to the 24 4rth nearest neighbors.

In order to study the lattice structure and the defects in the common occurrence of formation of imperfect lattices we developed software to display the MC simulation cube, rotate around any one of the three cartesian axes in order to find the lattice planes. Having one example of a perfect lattice formed at odd angles from the simulation cube axes provided a very helpful example of this process. Figure 3 shows the simulation cube for the $\Gamma = 500$ rotated 6 about the y axis. Figure 4 shows an example of one of these planes in Fig. 3 rotated 90° so that it is seen face on. This plane is an example of a 110 plane of the bcc lattice. which is obtained by cutting a basic cubic cell through its diagonal. We also could find

14

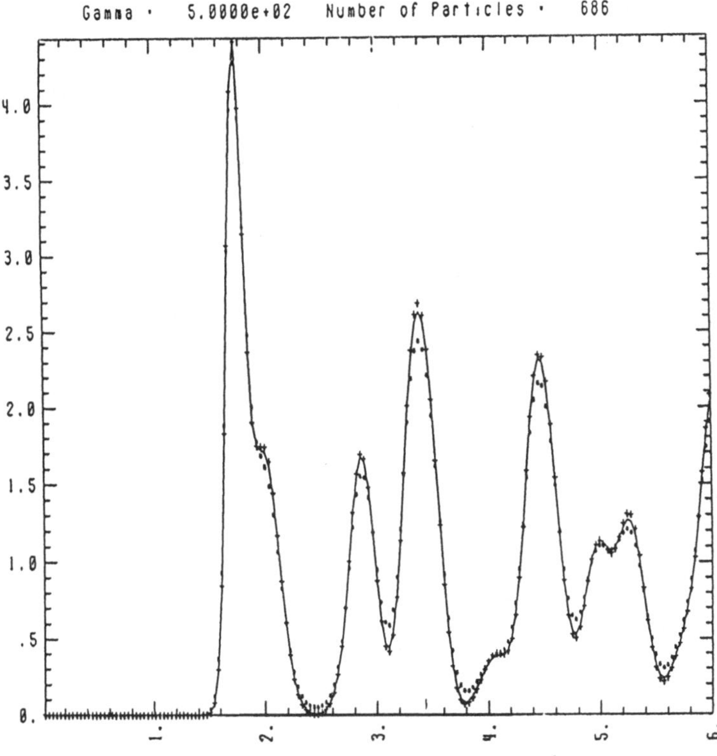

Fig. 2. bcc lattice g(r); the solid line is from a lattice start,
+ for the final state from a random start, and ••• is the
g(r) for the imperfect g(r) at 190 million configurations.

examples of the 100 and 111 planes. When seen edge on the lattice particles are clearly scattered around a straightline with their deviation from their equilibrium positions determined by the thermal fluctuations at $\Gamma = 500$.

The capability to look at any and all lattice planes make it possible to determine directly the extent that the final frozen state was either bcc or fcc or a mixture of the two lattices. Usually we also saw a number of simple lattice defects either interstitial particles or vacancies. Sometimes there would be more involved defect structures. An example is shown in Fig. 5 for $.\Gamma = 250$ in which the lower part of the plane is 110 for bcc, but the upper three lines are distorted and shifted from the 110 pattern. Defects of this sort can account for the energy of the final state being slightly higher than the bcc lattice energy as indicated in Table 1.

We expected that with N =500 (an fcc number) and $\Gamma = 500$ that the system might freeze into a perfect fcc lattice. Instead it froze into a very distorted bcc lattice at 85 million configurations. The final energy was U/NkT = -446.062 at 144 million configurations which is .378 above the fcc lattice energy. In order to

15

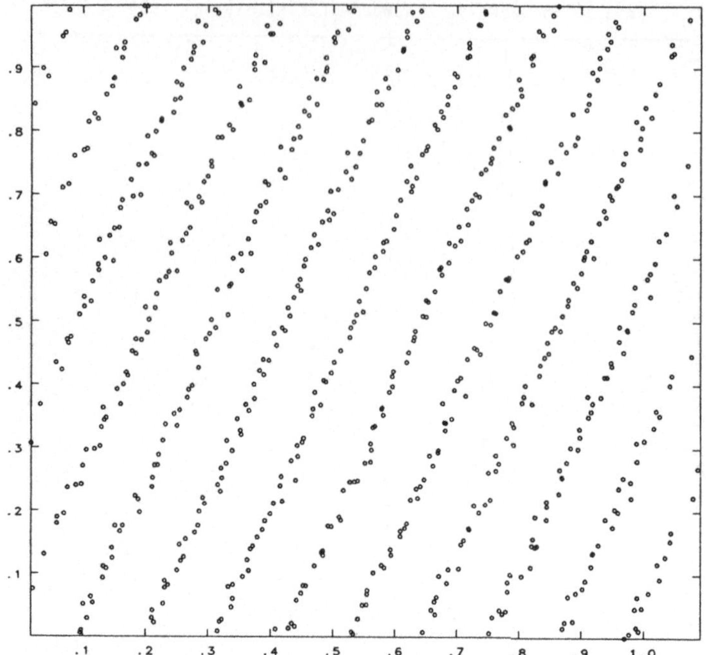

Fig. 3 $\Gamma = 500$ final state. The simulation cube was rotated
to show the lattice planes edge on.

further study the N dependence of our freezing results we went to N = 1458 (a bcc number) and $\Gamma = 500$. This system began to freeze at 120 million configurations and finished at 150 million with a final energy of -446.003. An examination of the lattice planes indicated that instead of one primary nucleation crystal, that there were at least two crystals that froze into contact. Figure 6 shows one view of the simulation cube in which planes from both crystals appear and are joined at a grain boundary. An examination of the lattice planes for the two crystals showed that one was bcc and the other fcc.

In order to further distinguish the lattice structures of the final frozen states and to follow the freezing process we developed a cluster program which examines the the 12 to 14 nearest neighbors of each of the N particles to determine whether they were in bcc, fcc, hcp, or fluid configurations. The program would locate the best hexagon in a plane with the given particle at the center. The bcc configuration has four particles above and four below; the fcc and hcp configuration have three above and three below. Also the program would identify the interior particles of a microcrystal and the surface particles. All particles not in one of the three lattice configurations were deemed to be fluid particles. In the earliest stages of the supercooled fluid phase most of the particles would be fluid, but several clusters with 13 to 15 particles would form. Some of these would grow slightly then contract. As bigger clusters formed the fluid energy would decrease as illustrated in Fig. 1. When a cluster reached some critical size of roughly 60 particles for N = 686, this cluster would grow very rapidly and the system would freeze with the bulk of the particles in that single cluster. In all cases the number of fluid particles decreased to zero when the freezing

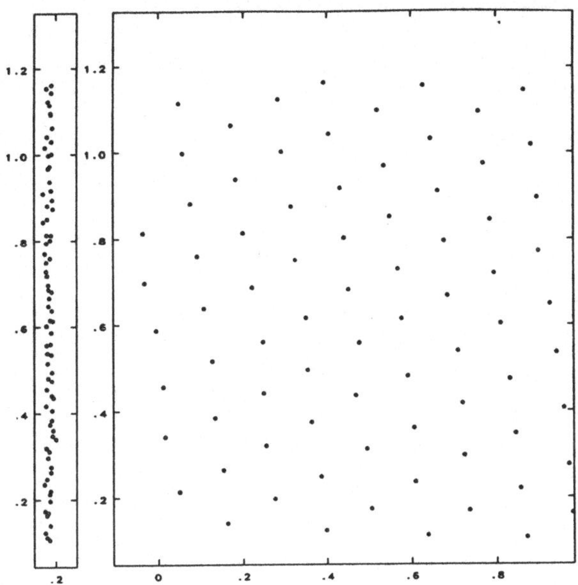

Fig. 4.   One of the lattice planes from Fig. 3 shown face on.
This is a 110 plane of the bcc lattice.

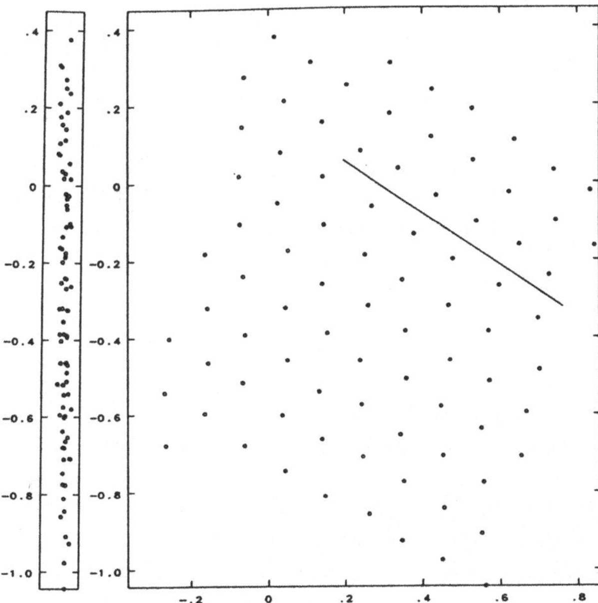

Fig. 5 $\Gamma = 250$, a lattice plane with defects. Note the different structure
above the straight line.

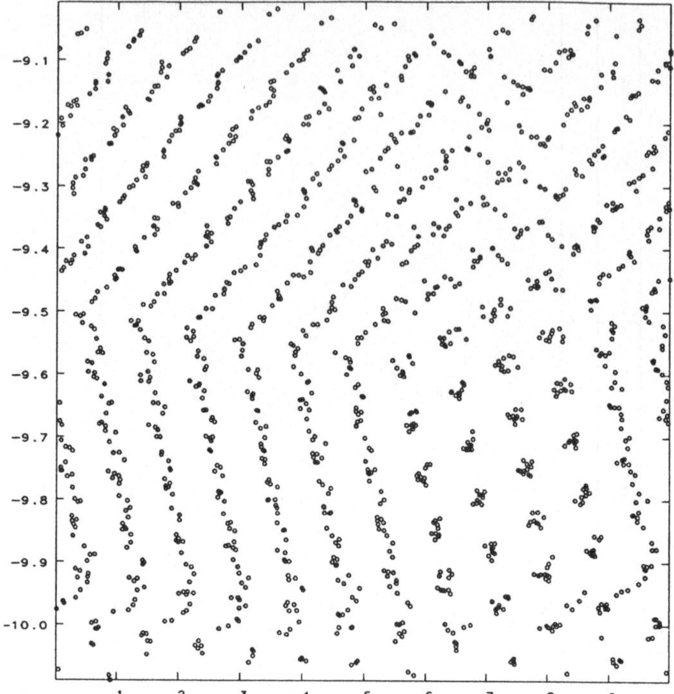

Fig. 6 A view of the simulation cube for N = 1458 and
        Γ = 500 in which two crystals, one fcc and one bcc
        join at a grain boundary.

was nearly complete. For Γ = 250 the final biggest cluster consisted of 488 interior bcc particles and 194 surface particles. For N=1458, Γ = 500 the cluster code indicated an fcc cluster with 678 interior particles, a bcc cluster with 335 interiors particles, and several smaller clusters each with less than 20 particles.

Conclusions

Freezing of the OCP by MC simulation from a random start is possible in a wide range of Γ values. For several hundred particles in the simulation the most likely outcome is a single imperfect crystal, usually bcc for smaller values of Γ but can also be fcc for larger values of Γ. The perfect bcc crystal that we obtained for Γ = 500 with N = 686 is probably a rare event. With several hundred particles the axes of the simlulatin cube do not markedly influence the orientation of the final crystal. The process appears to be highly number dependent since using a larger number of particles, N = 1458, results in two crystals. In the one example we have the larger of the two was fcc., while the smaller was bcc. This result suggests that going to several thousand particles would lead to multiple nucleation sites and thus a final frozen state with several microcrystals. This corresponds to what is normally seen in nature.

The crystalline cluster program enabled us to follow the freezing process in some detail. What we see is in general agreement with the classical nucleation theory [7]. In the supercooled fluid that persists for a considerable period in the simulations it is difficult to identify a single cluster that is going to ultimately grow into a final crystal. Freezing seems to take place when some one cluster gets sufficiently big to overcome a potential barrier and then to grow rapidly without limit. We could see some partial planes of a freezing cluster comprising a sizeable fraction of N, but this was always during the steep drop in energy when the system is actually freezing. Once in the frozen state the final lattice or lattices of the one or more nucleating clusters usually have defects that cannot be annealed out. We know that a final state has been reached, at least for computer simulation, when U/NkT remains constant for a few tens of millions of configuration regardless of the number of lattice defects.

Acknowledgements

Work performed under the auspices of the U.S. Department of Energy by the Lawrence Livermore National Laboratory under contract number W-7405-ENG-48.

References

[1] H. M. Van Horn. Science **252**, 384 (19 April, 1991)

[2] S. Ichimaru, H. Iyetomi, S. Tanaka, Physics Reports **149**, 92, (May 1987)

[3] G. S. Stringfellow, H. E. DeWitt, W. L. Slattery, Phys. Rev. A **41**, 1105 (15 Jan.,1990)

[4] D. H. Dubin, Phys. Rev. A **42**, 4972 (15 Oct., 1990)

[5] V. I. Yukalov, Physics Reports **208**, 396 (Nov. 1991)

[6] S. Ogata and S. Ichimaru, Phys. Rev. A**39**,, 1333 (1989)

[7] E. M. Lifshitz and L. P. Pitaevsky, Physical Kinetics, Vol. XI of Landau & Lifshitz, Chapt. XII

MONTE CARLO SIMULATION STUDY OF
DENSE PLASMAS

Shuji Ogata and Setsuo Ichimaru

Department of Physics, University of Tokyo,

7-3-1 Bunkyo-ku, Tokyo 113, Japan

Development of solidification processes in rapidly supercooled one-component plasmas is studied by a Monte Carlo (MC) simulation method with a number of particles 1458. Evolutions of interparticle correlations and of local and extended bond-orientational orders are investigated microscopically. Layered structures emerge at pre-nucleation stages in the system and expedite a subsequent evolution into metastable states. The metastable states are *bcc* monocrystalline states with an admixture of a few defects in the form of intralayer interstitials, which differ significantly from the glassy states obtained in the previous simulations with a smaller number of particles. The role of MC periodic boundary conditions in the freezing transitions is discussed in conjunction with the number of MC particles.

Introduction

One of the most interesting in the thermal evolution of a dense plasma is a possibility of solidification such as crystallization and glass transition. In the interiors of white dwarfs and neutron stars [1], such a transition may drastically alter the nuclear reaction rates [2] and the physical properties including conductivities [3] and viscoelasticity [4] and therefore have a significant influence on the evolution of the stars [5].

Classical one-component plasma [6,7] (OCP) is a statistical system where N particles of single species with charge Ze interact via Coulomb potentials in a volume V with uniform neutralizing charges. Thermodynamic state of the OCP with number density $n = N/V$ is specified by the Coulomb coupling parameter

$$\Gamma = \frac{(Ze)^2}{ak_\mathrm{B}T} \tag{1}$$

where $a = (3/4\pi n)^{1/3}$ is the ion-sphere radius. The OCP has been treated not only as a basic model in the statistical mechanics but as a realistic model for the dense matter in the outer crust of a neutron star [1,5]. Equation of state for the OCP has been investigated accurately in the fluid [8-11] as well as in the *bcc* [8-10] and *fcc* [12] crystalline phases mainly by the MC simulation method. Comparing the Helmholtz free energies, it has been found [10,11] that the fluid OCP freezes into the *bcc* crystals at $\Gamma_\mathrm{m} = 178$–180.

It is not clear, however, what the final state of a OCP is when a rapid quench is applied to $\Gamma > \Gamma_\mathrm{m}$. It is instructive in these connections to compare the Madelung energies [8] of the OCP between the crystalline structures: $E_\mathrm{M}/Nk_\mathrm{B}T = -0.895929\Gamma$ (bcc), -0.895874Γ (fcc), -0.895838Γ (hcp). We thus find $(E_\mathrm{fcc} - E_\mathrm{bcc})/N = 0.010k_\mathrm{B}T$ and $(E_\mathrm{hcp} - E_\mathrm{bcc})/N = 0.016k_\mathrm{B}T$ at $\Gamma = 180$; the differences are only 1–2% of the thermal energy.

The MC simulations for rapidly quenched OCPs with $N = 432$ were performed and reported earlier in Ref. 13 (referred to as Paper I). The resultant final states corresponded to glasses characterized by random polycrystalline mixtures of *fcc*, *hcp*, and *bcc* crystalline structures. In those glasses, we found development of layered structures [14] over the MC cell. Emerged out of those simulations are the problems: Do the layered structures have any relation with the periodic boundary conditions? How and in what stage are the layered structures formed? Is it possible to obtain a monocrystalline state, rather than a polycrystalline state, by the MC simulation method?

To answer these problems we have performed new MC simulations [15-17] for rapidly supercooled OCPs with a significantly increased value of N. The periodic boundary conditions depending on N may have two kinds of effects on the solidification. The boundary conditions for a smaller N may hinder the motion of particles and the resultant ordering of particles since any particle must move collectively with all its images. However, if N takes on a value specific to the lattice structures, such as a *bcc* number $2I^3$ or a *fcc* number $4I^3$ (I is an integer), the boundary conditions may assist in transforming the system into the respective lattice structure; such an effect may be more efficient for a smaller N. We remark that $N = 432$ in Paper I is one of the *bcc* numbers. To examine those two effects separately, we have chosen $N = 1458$, another *bcc* number.

Freezing transitions

We perform MC simulations with the usual Metropolis algorithm [18]. For the probability density of trial displacements of a randomly selected particle, we have adopted a form similar to that in Paper I. If the concept of "MC elapsed time" applies, we may estimate the elapsed time via the relation [19], $\omega_p t = 0.12\, c/N$, where c denotes the sequential number of configurations in the MC run.

The quenching processes are schematically depicted in Fig. 1. Starting with a fluid equilibrium state at $\Gamma = 160$, we apply stepwise quenches by $\Delta\Gamma = 50$ at every $c/N \times 10^{-4} = 1.25$, excepting for $\Delta\Gamma = 40$ at $c = 0$, until Γ reaches 300 [Fig. 1 (top)] and 400 [Fig. 1 (bottom)]. Phase evolution is monitored until $c/N \times 10^{-4} = 25.0$ in both cases. Stepwise quench to $\Gamma = 800$ is applied subsequently to the final state obtained with the quench to $\Gamma = 400$.

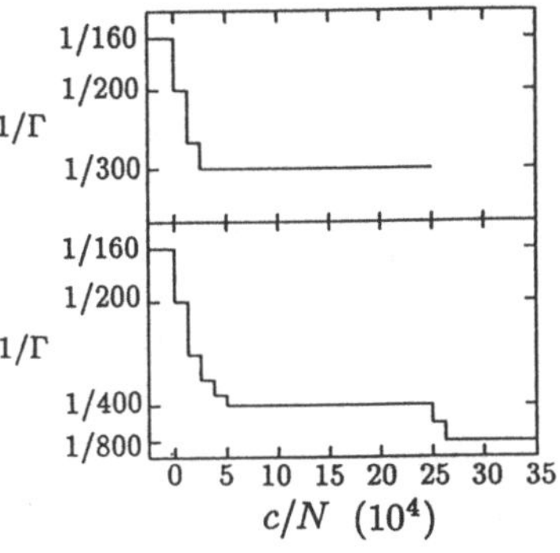

Fig. 1. Variation of $1/\Gamma$ in the MC simulation runs.

(A) Excess internal-energies

Excess internal-energy [6,7] is a primary quantity in characterizing the states of OCP since its volume is fixed. Evolution of the normalized excess internal-energy, averaged over a sequence of $\Delta c/N \times 10^{-4} = 0.007$,

$$u \equiv \frac{U}{Nk_{\mathrm{B}}T} = \left\langle \frac{n}{2k_{\mathrm{B}}T} \int d\vec{r}\, \frac{(Ze)^2}{r} [\hat{g}(r) - 1] \right\rangle , \tag{2}$$

where ($\vec{m}$ is an integer vector)

$$n\hat{g}(\vec{r}) = \frac{1}{N} \sum_{i=1}^{N} \left\{ \sum_{j\neq i}^{N} \delta(\vec{r}_i - \vec{r}_j - \vec{r}) + \sum_{\vec{m}\neq\vec{0}} \sum_{j=1}^{N} \delta(\vec{r}_i - \vec{r}_j + \vec{m}L - \vec{r}) \right\} \tag{3}$$

is shown in Fig. 2 for the quench to $\Gamma = 300$ and in Fig. 3 for the quench to $\Gamma = 400$. In each figure, dashed lines imply the extrapolation values of the fluid internal energy formulas due to Slattery et al. [10] for the upper line and due to Ogata and Ichimaru [11] for the lower; the dot-dashed line means the *bcc* crystalline value [10].

The deviations of u at the metastable states from those in the *bcc* crystalline phase are 0.08 ($\Gamma = 300$) and 0.21 ($\Gamma = 400$). Those are far smaller than the values 0.25 ($\Gamma = 300$) and 0.4 ($\Gamma = 400$) for the glassy states in Paper I.

Five stages for the quench to $\Gamma = 300$ are defined in Fig. 2: (a) at $c/N \times 10^{-4} = 4.6$, (b) 7.0, (c) 8.6, (d) 9.5, (e) 24.9. For the quench to $\Gamma = 400$, four stages are also defined in Fig. 3: (α) 7.3, (β) 12.1, (γ) 15.8, (δ) 19.5. Stages (c) and (γ) correspond to a midst of the abrupt decreases in u for the quenches to $\Gamma = 300$ and 400, respectively.

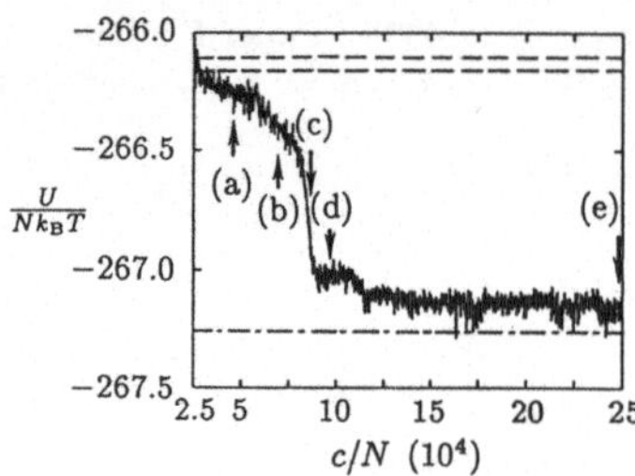

Fig. 2. Evolution of the normalized excess internal-energy for the quench to $\Gamma = 300$.

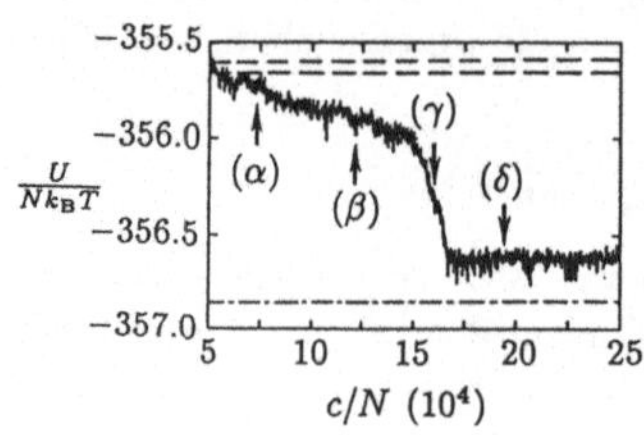

Fig. 3. Same as Fig. 2, but for the quench to $\Gamma = 400$.

Evolution of the radial distribution function, $g(r) = \langle \hat{g}(\vec{r}) \rangle$ averaged over $\Delta c/N \times 10^{-4} = 0.21$, is displayed in Fig. 4 for the quench to $\Gamma = 300$ and in Fig. 5 for the quench to $\Gamma = 400$. For both quenches $g(r)$ exhibits a smooth feature analogous to that in a fluid simulation [cf. Fig. 2 in Ref. 17] at the stages before the abrupt decreases in u ((a), (b), (α), and (β)). The system seems to have acquired a substantial degree of the *bcc* local structures at the stage (c) since the second and the third peaks of $g(r)$ have shoulders at radii corresponding to the *bcc* peaks. For the quench to $\Gamma = 400$, $g(r)$ retains a smooth feature at the stage (γ), though u decreases abruptly at this stage. At the metastable states in both quenches ((e) and (δ)), $g(r)$ is quite similar to the one in the *bcc* crystalline simulation [cf. Fig. 2 in Ref. 17]; finiteness of $g(r)$ at the first bottom, however, indicates a slight deviation from the *bcc* crystalline structures.

Features of $g(r)$ for the metastable states in both quenches are quite different from those for the simulations with $N = 432$. We found no peaks corresponding to the *bcc* structures in all the cases of the quench with $N = 432$.

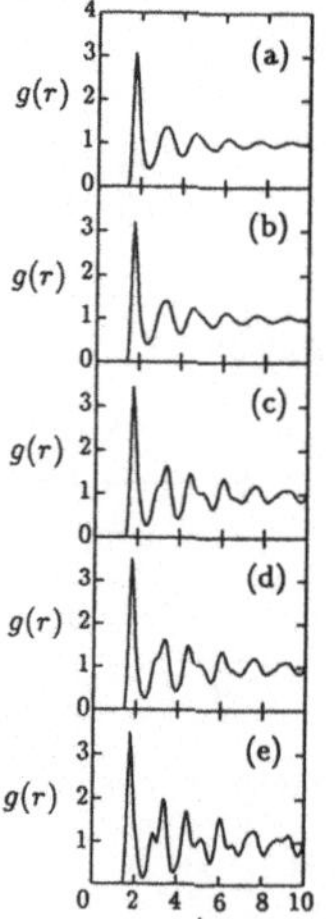

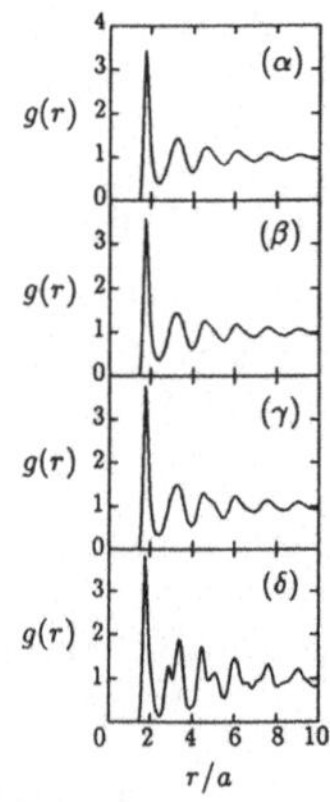

Fig. 4. Evolution of $g(r)$ for the quench to $\Gamma = 300$.

Fig. 5. Same as Fig. 4, but for the quench to $\Gamma = 400$.

(B) Bond-orientational orders

Orders in the particle configurations are described in terms of the orientational correlations between "bonds," which are connecting lines between a particle and its "neighboring particles." The number N_c of neighboring particles, which we shall call "coordination number," is 14 for the *bcc* cluster; 12 for the *fcc*, *hcp*, and *icosahedral* clusters. We associate a set of quantities to each bond in terms of the spherical harmonics. The local bond-orientational order parameters [17,20] W_4 and W_6 for a cluster, which are rotationally invariant combinations in the third order, play a key role in the cluster "shape spectroscopy" in fluids and solids. Comparing the quantity (W_4, W_6) between *bcc*, *fcc*, *hcp*, and *icosahedral* clusters, we find that the local bond-orientational symmetries around a particle can be discerned through its location on the two-dimensional (W_4, W_6) map. Extended bond-orientational symmetries [17,20] are studied in terms of the correlation function $G_6(r)$ where r is the distance between the central positions of bonds.

Evolutions of the two-dimensional (W_4, W_6) maps for clusters with $N_c = 12$ and 14 are plotted in Fig. 6 for the quench to $\Gamma = 300$ and in Fig. 7 for the quench to $\Gamma = 400$. In the figures, the (W_4, W_6) values for the reference clusters are likewise displayed by the diamond markers.

At the stages (a), (b), (α), and (β), (W_4, W_6) values are distributed almost uniformly in the region $|W_4| \leq 0.15$ and $|W_6| \leq 0.16$; this is a typical behavior in a fluid [cf. Fig. 3 in Ref. 17]. We find that substantial proportion of clusters with $N_c = 14$ has local *bcc* symmetry at the stage (c); it is consistent with the emergence of several sub-peaks of $g(r)$ at the positions corresponding to the *bcc* structures at this stage. At the stage (γ), some fractions of clusters with $N_c = 12$ coalesce at the *fcc* marker; at this stage, a concentration to $W_6 \simeq W_6(bcc) = 0.013$ takes place for clusters with $N_c = 14$. The smooth feature of $g(r)$ observed at the stage (γ) may be attributed to such a coexistence of the local *fcc* and *bcc* symmetries. At the metastable states ((e) and (δ)), only a small number of clusters with $N_c = 12$ are remaining, and clusters with $N_c = 14$ have the *bcc* symmetry to a large extent. The metastable states in both quenches have almost perfect local *bcc* structures, which are quite different from the mixture of local *fcc*, *hcp*, and *bcc* structures for the glass states obtained in Paper I.

Evolution of the extended bond-orientational symmetries are displayed in terms of $G_6(r)$ in Fig. 8 for the quench to $\Gamma = 300$ and in Fig. 9 for the quench to $\Gamma = 400$. No extended order exists at the stages (a) and (α). Bond correlation lengths extend themselves to about a half of L at the stages (b) and (β). Final values of $G_6(r)$ in both quenches are about two thirds of the *bcc* value [cf. Fig. 5 in Ref. 17].

(C) Layered structures

One of the main concerns in the present simulations has been a possible formation of the layered structures and their relation to the periodic boundary conditions. For elucidation of these issues, two-dimensional projection maps of particles have been constructed from various directions, to illustrate possible layered structures. We begin with isolating those particles inside a sphere of radius $L/2$, then rotate these as a whole by an angle ζ around the y axis and by η around the z axis where $(\zeta, \eta) = (\frac{\pi}{40}i, \frac{\pi}{40}j)$ with $i, j = 0, 1, \cdots, 20$; the resulting configuration is projected onto the y–z plane. (Note that the particle positions are averaged over a sequence of $\Delta c/N \times 10^{-4} = 0.069$.)

At pre-nucleation stages (a) and (α) in both quenches, we find particles forming a layer-like structure at angles $(\zeta, \eta) = (\frac{\pi}{4}, \frac{\pi}{5})$ and $(\frac{11\pi}{40}, \frac{2\pi}{5})$, respectively.* Evolutions of such particle layers are depicted in Fig. 10 for the quench to $\Gamma = 300$ viewed at the same angles $(\zeta, \eta) = (\frac{\pi}{4}, \frac{\pi}{5})$ and in Fig. 11 at $(\zeta, \eta) = (\frac{11\pi}{40}, \frac{2\pi}{5})$ for the quench to $\Gamma = 400$. Particle layers develop over a half of the sphere at the stages (b) and (β). We find nearly perfect layers already at the stage (c) for

---

* These are illustrative *examples* of layer-forming angles. See Fig. 18 of Ref. 17 for observations at various angles.

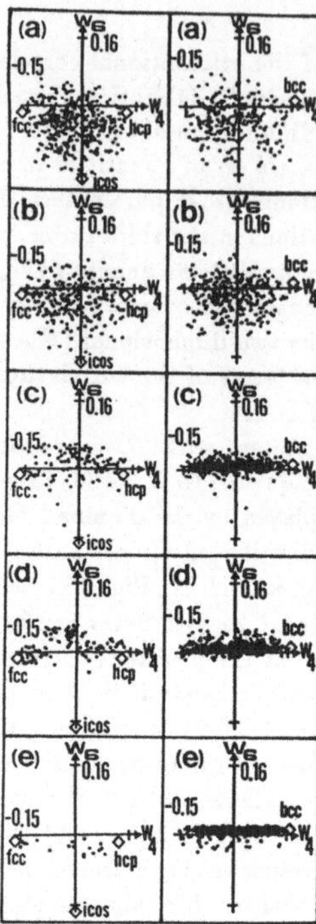

Fig. 6. Evolution of (W_4, W_6) maps for the quench to $\Gamma = 300$: (left), with $N_c = 12$; (right), with $N_c = 14$.

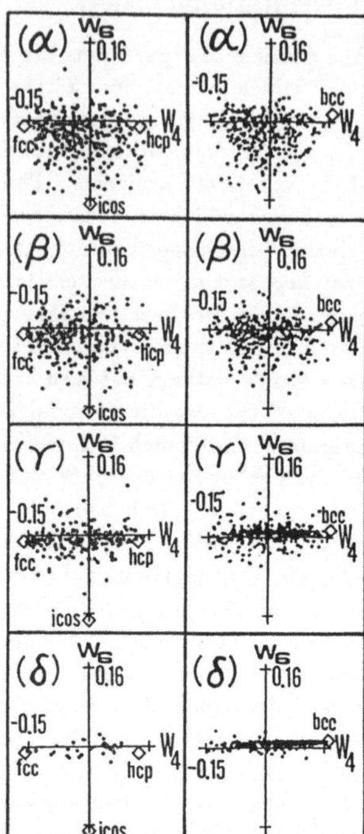

Fig. 7. Same as Fig. 6, but for the quench to $\Gamma = 400$.

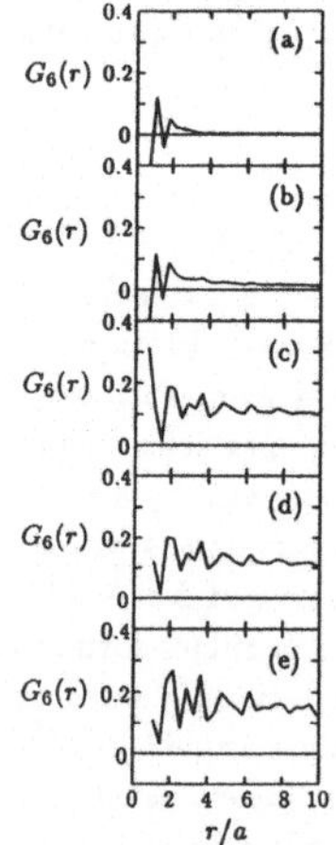

Fig. 8. Evolution of $G_6(r)$ for the quench to $\Gamma = 300$.

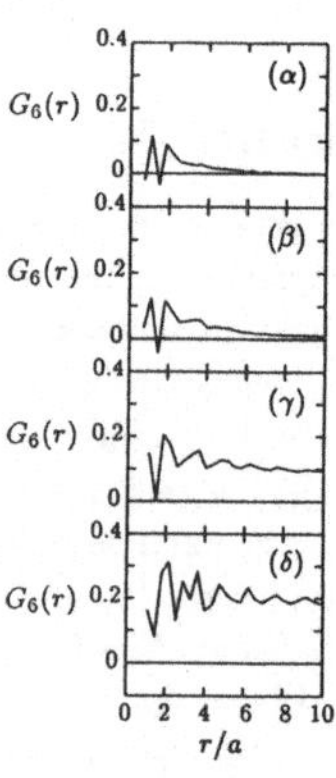

Fig. 9. Same as Fig. 8, but for the quench to $\Gamma = 400$.

the quench to $\Gamma = 300$. For the quench to $\Gamma = 400$, we find two domains at the stage (γ): in one domain, particles are well-ordered forming a nearly perfect layers; in the other domain, particles show a rather disordered feature. At the stage (δ), the domain of well-ordered particles spread over the cell. Such an emergence of two domains at the stage (γ) may be understood as a result of the coexistence of local *fcc* and *bcc* symmetries seen in Fig. 7.

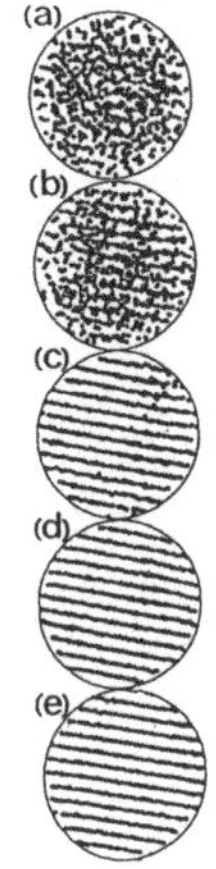

Fig. 10. Evolution of layered structures for the quench to $\Gamma = 300$.

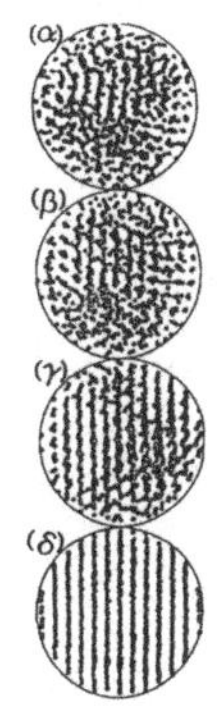

Fig. 11. Same as Fig. 10, but for the quench to $\Gamma = 400$.

Since all the particles reside on layers at the final stages (e) and (δ) as in the case of a lattice structure, two-dimensional particle positions on the layers should contain some imperfections. We investigate the character of the intralayer imperfections and their dynamics during the metastable state by tracing the motion of particles.

Following three periods are defined in terms of $c/N \times 10^{-4}$: (1) 12.6–16.7, (2) 16.7–20.8, (3) 20.8–24.9. In each of the three periods, we first single out particles in a layer (exemplified in Fig. 10 and 11) which extends continuously into the surrounding image cells at the final configuration, and project these (solid circles) on the x–z plane. Then their positions are traced back to the beginning of the period (open circles) with an equal interval of $\Delta c/N \times 10^{-4} = 0.21$. Such maps, extended to the size $2L \times 2L$ to see the linked motion of particles, are depicted in Fig. 12 for the quench to $\Gamma = 300$: (top) corresponds to the period (1); (middle), for the same layer as top, but to the period (2); (bottom), for the same layer as top, but to the period (3). The pairs of vertical noisy lines depict the projections of the particle positions during the corresponding period on a y–z plane: the left line is for $x < 0$ particles; the right, for $x > 0$ particles.

The imperfections found in these analyses can be classified into three types: (A) a particle outside the layer (belonging mostly to one of the adjacent layers) in the open-circle configurations, (B) an interstitial (or an extra particle outside the adjacent layers) in the layer, and (C) a vacancy in the open-circle configurations. For the quench to $\Gamma = 300$, we find real transitions during the period (1) [top of Fig. 12] in the forms of merges between (A) and (C) and between (B) and (C), and of a transformation (or a settling) from (A) to (B), as well as virtual transitions within (B); the virtual transition usually ceases at a return to the original interstitial configurations. We attribute these real transitions as the causes of the transient behaviors of u around $c/N \times 10^{-4} = 14.0$. During the periods (2) and (3) [middle and bottom of Fig. 12], however, no imperfections of the types (A) and (C) appear to remain in the system. The jitters observed in the metastable states shown in Figs. 2 and 3 are attributed to those incidents of the virtual transitions.

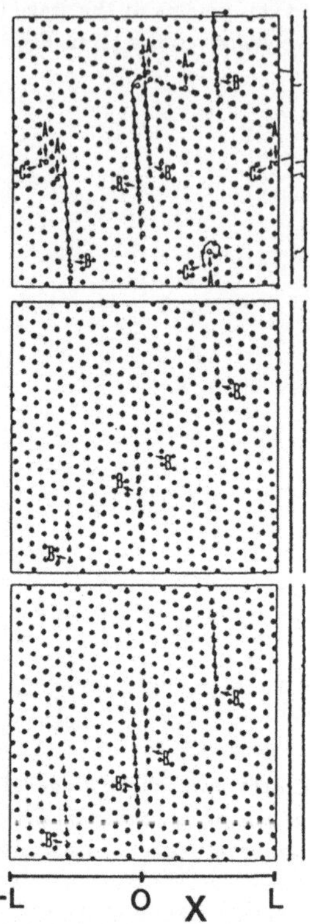

Fig. 12. Motions of particles in and near a layer projected onto a x–z plane for the quench to $\Gamma = 300$ during the three periods specified in the text. The pairs of vertical noisy lines depict the motion of particles (left for $x < 0$, right for $x > 0$) perpendicular to the x–z plane.

We thus characterize the imperfections at the final states in both quenches as intralayer interstitials. An annihilation of such an isolated interstitial by the MC sampling processes would call for a slight but homogeneous compression of a configuration of particles that surround the interstitial in the layer, concurrent with appropriate compressions of the particle configurations in the neighboring layers, in such a way as to preserve the overall *bcc*-crystalline structures. We speculate the probability of such a sampling, if not zero, would be extremely small.

One may expect that a further increase in Γ or decrease in T might introduce some additional changes to the final state. To examine such a possibility, we have applied an additional quench to the final state for the quench to $\Gamma = 400$ as shown in Fig. 1 (bottom). The result has indicated that particles remain locked around their original positions and no structural changes have been observed.

Supercooled fluid state

The OCP has the lowest free energies for the *bcc* phase at $\Gamma > \Gamma_m = 180$. MC simulations with $N \leq 1024$, however, indicates that the supercooled fluid OCP at $\Gamma = 200$ appears stable. The stability of the OCP in the supercooled fluid state would be attributed to a limited span of simulations and to the periodic boundary conditions. We study the stability of the supercooled fluid OCP with $N = 1458$ by performing an additional long run. Starting with the fluid state at $\Gamma = 160$, we apply a sudden quench to $\Gamma = 200$ at $c = 0$. The evolution is monitored until $c/N \times 10^{-4} = 7.5$.

We find that u stays at the fluid extrapolation level from immediately after the quench to the final configuration. Particles behave diffusively; no indication of solidification has been observed during the run.

Discussion

In the present simulations, rapidly quenched OCPs have solidified into *bcc* monocrystalline states with defects. Microscopic structures in these final states are significantly different from those in the polycrystalline glasses obtained in the former simulations with $N = 432$. Since the quenching procedures are almost the same in both simulations, the difference in the final states should be attributed to the difference in N. The choice of N at a *bcc* number appears to bear no essential consequences to the solidification processes since substantial numbers of imperfections from the perfect *bcc* crystalline states have been observed in both metastable states. By considering the effects of the periodic boundary conditions as to hinder the motion of particles for smaller N, we may explain the mechanism of the formation of the metastable states in both simulations: When the OCP is rapidly quenched to $\Gamma > \Gamma_m$, particle layers emerge first in an arbitrary direction, which would favor a *fcc-hcp* local structures. Subsequently the system may transform itself into a *bcc* crystalline state if N is large enough, since the Helmholtz free energies assume the lowest values in the *bcc* phase. In case that N is not large enough, however, the motion of particles would be hindered by the boundary conditions resulting in a mixture of *fcc*, *hcp*, and *bcc* structures; this would be the case for the simulations with $N = 432$.

References

[1] S. L. Shapiro and S. A. Teukolsky, *Black Holes, White Dwarfs, and Neutron Stars* (John Wiley & Sons, New York, 1983).

[2] S. Ogata, H. Iyetomi, and S. Ichimaru, Ap. J. **372**, 259 (1991).

[3] S. Ogata and S. Ichimaru, Ap. J. **361**, 511 (1990).

[4] S. Ichimaru and S. Tanaka, Phys. Rev. Lett. **56**, 2815 (1986); S. Ogata and S. Ichimaru, Phy. Rev. A **42**, 4867 (1990).

[5] H. M. Van Horn, in *Strongly Coupled Plasma Physics*, ed. by S. Ichimaru (North-Holland, Amsterdam, 1990), p. 3.

[6] M. Baus and J.-P. Hansen, Phys. Rep. **59**, 1 (1980).

[7] S. Ichimaru, H. Iyetomi, and S. Tanaka, Phys. Rep. **149**, 91 (1987).

[8] S. G. Brush, H. L. Sahlin, and E. Teller, J. Chem. Phys. **45**, 2102 (1966).

[9] J.-P. Hansen, Phys. Rev. A **8**, 3096 (1973).

[10] W. L. Slattery, G. D. Doolen, and H. E. DeWitt, Phys. Rev. A **21**, 2087 (1980); A **26**, 2255 (1982)

[11] S. Ogata and S. Ichimaru, Phys. Rev. A **36**, 5451 (1987).

[12] H. L. Helfer, R. L. McCrory, and H. M. Van Horn, J. Stat. Phys. **48**, 397 (1981).

[13] S. Ogata and S. Ichimaru, Phys. Rev. A **39**, 1333 (1989) [referred to as Paper I].

[14] S. Ogata and S. Ichimaru, Phys. Rev. Lett. **62**, 2293 (1989).

[15] S. Ogata and S. Ichimaru, J. Phys. Soc. Jpn. **58**, 356 (1989).

[16] S. Ogata and S. Ichimaru, J. Phys. Soc. Jpn. **58**, 3049 (1989).

[17] S. Ogata, Phys. Rev. A **44** (in press, 1992 January 15, issue).

[18] N. Metropolis, A. W. Rosenbluth, M. N. Rosenbluth, A. H. Teller, and E. Teller, J. Chem. Phys. **21**, 1087 (1953).

[19] S. Ogata and S. Ichimaru, Phys. Rev. A **38**, 1457 (1988).

[20] P. J. Steinhardt, D. R. Nelson, and M. Ronchetti, Phys. Rev. B **28**, 784 (1983).

Large Coulomb and Lennard-Jones Crystals and Quasicrystals [1]

Rainer W. Hasse [2]

ESR Division, Gesellschaft für Schwerionenforschung GSI,
D-6100 Darmstadt, Germany

With the help of molecular dynamics computer simulations we study the equilibrium configurations at low temperature of large systems of strongly correlated particles either under the influence of their mutual long-range Coulomb forces and a radial harmonic external confining force or of the short-range Lennard-Jones potential. The former is a model for charged particles in ion traps and the latter for clusters.

For the Coulomb plus harmonic force, the particles arrange in concentric spherical shells with hexagonal structures on the surfaces. The closed shell particle numbers agree well with those of multilayer icosahedra (mli). A Madelung (excess) energy of -0.8926 is extracted which is larger than the bcc value. The results are compared with those obtained by a simple onion shell model.

For the Lennard-Jones force we employ various initial configurations like multilayer icosahedra or hexagonal closed packed (hcp) spheres. Cohesive (volume) and surface energies per particle are extracted and compared to the energies of scaled mli quasicrystals and of spherical scaled crystals with N up to $36\,000$. It is shown that relaxed mli are the dominant structures for $N < 5\,000$ and hcp spheres for larger particle numbers. For $N < 22\,000$, hcp crystals have about the same closed shell numbers as mli quasicrystals but smaller ones for $N > 22\,000$. The same magic numbers obtain with other short range Mie potentials.

1. Coulomb plus harmonic forces

The strongly correlated infinite one-component plasma (OCP) crystallises in an bcc lattice if the plasma parameter, i.e. the ratio of Coulomb to thermal energy, reaches the value [1] $\Gamma \approx 171$. Finite systems, on the other hand, exhibit hexagonal structures on the surfaces [2, 3], however imperfect due to the incompatibility between a perfect lattice and a curved surface and due to the incommensurability of two adjacent shells. A change of structure from the plane hexagonal into the bcc one occurs only if the dimensions of the system become as large as about 100 interparticle distances [4].

In order to study large but finite Coulomb systems we extend the calculations on small systems by Rafac et al. [5] and employ the molecular dynamics (MD) technique to solve the classical equations of motion under the external harmonic confining force

$$\mathbf{F}^i_{\text{conf}} = -K\mathbf{r}_i,$$

and the Coulomb force,

$$\mathbf{F}^i_{\text{Coul}} = -q^2 \sum_{j \neq i} \frac{\mathbf{r}_i - \mathbf{r}_j}{|\mathbf{r}_i - \mathbf{r}_j|^3}.$$

The initial momenta are chosen at random as to give an initial kinetic energy corresponding to $\Gamma \approx 1$ and the initial coordinates of N ions usually are chosen as those of the $(N{-}1)$-ion system with one ion added at random. The system then is followed in time thereby cooling by reducing the momenta until thermal equilibrium has been reached; for details see ref. [3]. In this chapter distances are measured in units of the Wigner-Seitz radius $a_{\text{WS}} = (q^2/K)^{1/3}$ and energies in units of q^2/a_{WS}. The total energy, i.e. the sum of confining and Coulomb energies per particle, then reads

$$\epsilon = \frac{1}{2N} \sum_{i=1}^{N} \mathbf{r}_i^2 + \frac{1}{N} \sum_i \sum_{j<i} |\mathbf{r}_i - \mathbf{r}_j|^{-1}$$

---

[1]Talk given at the 6th Int. Workshop on Physics of Nonideal Plasmas – Theory and Experiment, Berlin-Gosen, Germany, 18-21 Nov. 1991

[2]Email: UL29@DDAGSI3.earn

and the excess energy is defined by subtracting the homogeneous value of $\bar{\epsilon} = \frac{9}{10} N^{2/3}$.

The resulting magic numbers are listed in Table 1 where the excess energy is minimal as compared to the neighbouring particle numbers and the excess energy is shown in Fig. 1. For systems with $N < 64$, there are strong shell effects in the excess energy with peaks and minima if a new shell opens, in particular the transitions at $N = 12, 60, 146$. One notes that the smaller magic numbers are even as compared to those of the mli because the center particle is always missing. In particular, the second shell starts at $N = 13$, the third one at $N = 61$, and the next ones at or around $N = 147$, 309, 565, 900, 1400, 2100, respectively.

These closed shell particle numbers are compatible with the magic numbers of Mackay's multi-layer icosahedra (mli) [6],

$$N = (2M + 1) \left[\tfrac{5}{3} M(M + 1) + 1\right] \ , \tag{1}$$

notably $N = 13, 55, 147, 309, 561, 923,...$ which have been verified experimentally as closed shell particle numbers in metal clusters, see e.g. refs. [7, 8]. The subshell numbers are given by $10 N^2 + 2 = 12, 42, 92, 162, 242,...$ They are thus more general in the sense that they not only are identical to those of stacked cuboctahedra but also show up in Coulomb systems. Due to the long range nature of the Coulomb force, Coulomb quasicrystals tend to be as spherical as possible. This is not in contradiction to the edged nature of the mli because here the plane surfaces of the mli become curved.

Table 1:
Structures, rms radii and excess energies of closed shell N-particle systems

N	Structure	R_{rms}	$\epsilon_{\mathrm{excess}}$
12	12	1.6002	-0.87663
60	48+12	2.9335	-0.88498
146	93+41+12	4.0054	-0.88745
308	163+93+42+10	5.7404	-0.88919
561	255+161+93+42+10	6.3418	-0.88994
899	356+247+154+92+38+12	7.4361	-0.89059
1414	491+365+244+163+94+44+13	8.6600	-0.89124
2057	641+491+363+257+158+93+44+10	9.8209	-0.89132
2837	805+634+480+356+246+173+143*	10.9384	-0.89172
3871	992+801+639+1439*	12.1379	-0.8915†

* The remaining shells cannot be resolved
† At small temperature

An infinite Coulomb system can take advantage of all long range interactions in order to arrange in an bcc lattice and to minimize the Madelung energy by summing up all long range contributions. A finite system, on the other hand, can explore only the short and intermediate range parts of the Coulomb interaction. It is well known that systems with short range interactions arrange in fcc or hcp lattices with coordinations (the number of almost equal nearest neighbour distances) of 12 but with higher Madelung energies. A finite Coulomb system, hence, will arrange in such a way as to maximize the coordination i.e. to achieve the maximum possible number of equilateral triangles. In Fig. 2 is shown the front hemisphere of the outer shell of the 5000-particle system. Here the overall hexagonal structure is well pronounced, however with dislocations and pentagonal point defects. The particles in the next inner shell most often sit below the line connecting two particles rather than below the center of the triangle.

A fit for $N > 300$ gives

$$\epsilon_{\mathrm{MD}} = -0.8926 \pm 0.0001 + 0.0088 N^{-1/3} + 0.096 N^{-2/3}$$

with an asymptotic Madelung energy of -0.8926 which is higher than the one of the infinite bcc OCP of -0.895929, thus indicating that even at very large particle numbers the hexagonal surface structure is still dominating over the bcc structure. This is due to the small surface energy of the hexagonal lattice on the curved surfaces. The surface energy extracted from the MD data is the average over the whole surface and also includes relaxation and reconstruction of the surface that

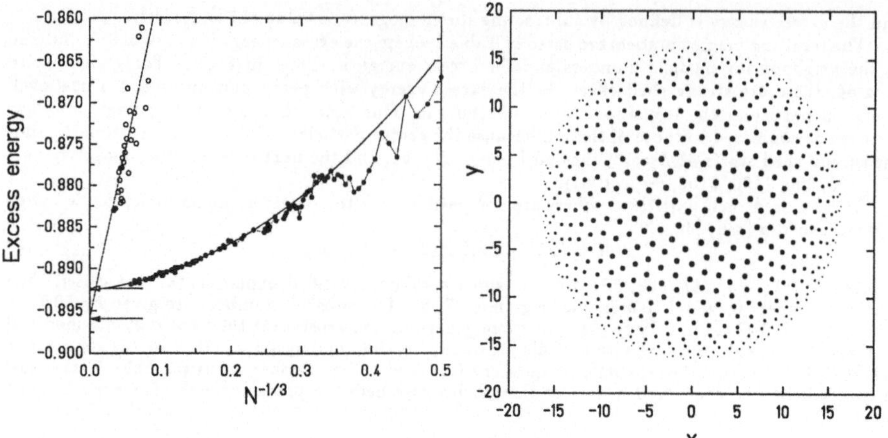

Figure 1:
Excess (Madelung) energy of Coulomb quasicrystals. The dots are MD data and the full line is the best fit. The open arrow points to the asymptotic value and the full arrow to the Madelung energy of the infinite bcc OCP. Open circles are results of spherical bcc matter together with a fit (dashed line).

Figure 2:
The front hemisphere of the outer shell of the 5000 ion system. Note the haxagonal structure with approximately equilateral triangles and the point defect below the center.

minimize the total energy. To evaluate the order of magnitude of surface relaxation we calculated the excess energies of unrelaxed spherical fragments containing up to 25 000 particles, sliced out of infinite ideal bcc matter with the same density, see the open circles in Fig. 1. Extrapolation of these data to the bcc Madelung energy (dashed line) gives a surface energy coefficient of ≈ 0.4. It crosses the fit of the MD data at $N \approx 4 \times 10^6$. This number of particles at which the infinite bcc lattice takes over energetically is compatible with the estimate of Dubin [4].

In order to explain the results of the computer simulations we compare them with those of an onion shell model, where the quasicrystals are supposed to consist of M homogeneously charged shells of radius R_ν and number of particles N_ν with the constraint $\sum_\nu N_\nu = N$. The Coulomb and confining energies then become

$$\epsilon_{\text{Coul}} = \frac{1}{N}\left[\sum_{\nu=1}^{M}\frac{N_\nu}{R_\nu}\left(\frac{1}{2}N_\nu + \sum_{\nu<\mu}N_\mu\right)\right],$$

$$\epsilon_{\text{conf}} = \frac{1}{2N}\sum_\nu N_\nu R_\nu^2.$$

This is corrected for the energy per particle of the plane hexagonal lattice[9, 10],

$$\epsilon_{\text{Mad}}^{\text{p-hex}} = -\frac{\alpha}{\sqrt{f}},$$

where $\alpha = 1.960515789$ and f is the unit area of an ion. With the geometrical value of $f = a^2\sqrt{3}/2$, where $a = (16\pi/9)^{1/3}$ is the minimum distance, it takes on the value $\epsilon_{\text{Mad}}^{\text{p-hex}} = -1.187397$. Minimization of the total energy with respect to the radii yields

$$R_\nu^3 = \frac{1}{2}N_\nu + \sum_{\mu<\nu}N_\mu, \tag{2}$$

which, in turn gives the total energy per particle

$$\epsilon_{\text{total}} = \frac{3}{2N} \sum_\nu N_\nu R_\nu^2 + \epsilon_{\text{Mad}}^{\text{p-hex}} .$$

Demanding that the area of a given shell is occupied by N_ν areas of equilateral triangles of sides a, thus also allowing for noninteger particle numbers N, N_ν,

$$4\pi R_\nu^2 = \frac{4\pi}{3d} N_\nu + \frac{5}{24} a^2 \qquad (\nu > 0), \tag{3}$$

eqs. (2,3) are solved iteratively with initial values $N_0 = R_0 = 0$. The excess energies without $\epsilon_{\text{Mad}}^{\text{p-hex}}$-corrections, of course, here would be positive, i.e. higher than the true minimum of the homogeneous system. The agreement of the shell model results with the MD results is rather good even for the first shell. The opening of new shells is accurately predicted to parts of a particle number and the radii and energies are well reproduced. This yields

$$\epsilon_{\text{Mad}}^{\text{3-hex}} = -0.894383, \tag{4}$$

above the bcc (-0.895929), fcc (-0.895874) and hcp (-0.895838) values but below the MD result (-0.89280) and of the sc lattice (-0.880059).

2. Lennard-Jones force

The largest magic number identified in cluster experiments is around 21 300 [8]. They are well in agreement with those obtained from purely classical geometrical packing of N particles for instance into mli or cuboctahedra (mlc). However, the energetical stability of large crystals and quasicrystals under short-range forces has only been studied extensively by Raoult et al. [11]. Here icosahedral, octahedral and mono-twinned fcc, decahedral, tetrakaidecahedral, hexakaiicosahedral (hki), dodecahedral pentakaitetrakontahedral (dpk) and other truncated structures are considered.

We extend these calculations to very large particle numbers and, as an approximation to the effective two-body force in clusters, we employ the short range LJ potential,

$$V_{\text{LJ}}(r) = \epsilon_0 \left[\left(\frac{\sigma}{r}\right)^{12} - 2 \left(\frac{\sigma}{r}\right)^6 \right] .$$

In this chapter we will use $\epsilon_0 = \sigma = 1$. The energy per particle, hence, is given by

$$E_{\text{LJ}} = \frac{1}{N} \sum_i^N \sum_{j<i} V_{\text{LJ}}(|\mathbf{r}_i - \mathbf{r}_j|) \quad . \tag{5}$$

Due to the very short range nature of the LJ force a start with initial random coordinates never yields a stable configuration. Therefore we employ mli, mlc, spheres of hexagonal closed packed (hcp), face centered (fcc) or body centered cubic (bcc) matter, or even stacked rhombic dodecahedra (srd) proposed by Kepler, see ref. [12]. Systems with particle numbers up to $N = 6\,525$ are completely relaxed by MD and those with $2 < N < 36\,000$ are studied by uniform scaling and energy minimization. Systems with nonmagic numbers of particles were started with a magic core and the extra particles at random in the next shell; for details see ref. [13].

With inverse power-law potentials scaling effects can be calculated by scaling the dimensions in eq. (5) and minimizing with respect to the scaling parameter, $s_0 = (e_{12}/e_6)^{1/6}$ to yield the minimum energy

$$E_{\text{LJ}}^{\text{min}} = -e_6^2/e_{12}.$$

For infinite bcc, fcc and hcp matter, the inverse-power sums e_n are known, see [14], to give the scaling parameters and cohesive energies of Table 2.

Table 2:
Cohesive and surface energies and scaling factors of different Lennard-Jones crystals and of infinite mli and mlc quasicrystals.

Structure	V_{LJ}	S_{LJ}	s_0
hcp	-8.611065[1]	15.5[2]	0.971228[1]
fcc	-8.610201[1]	15.4[2]	0.971234[1]
mlc	-8.59[3]	15.6[3]	0.974[3]
hki	-8.545[4]	14.23[4]	
mli	-8.54[3]	14.18[3]	0.953[3]
dpk	-8.538[4]	14.20[4]	
bcc	-8.237292[1]	15.1[2]	0.979204[1]

[1] Analytical values
[2] From minimizing scaled large spheres
[3] Estimated from MD relaxation
[4] Lower limits extrapolated from ref. [11]

Under the LJ force, hcp matter is lowest in energy, followed by fcc and bcc matter. We repeated those scaling calculations for spheres of such matter with up to 36 000 particles in order to also obtain the surface energy coefficient in the expansion

$$E_{LJ} = V_{LJ} + S_{LJ} N^{-1/3} ...$$

which are also listed in Table 2. Relaxation of finite systems is achieved by solving the coupled classical equations of motion with standard MD as described above.

The resulting energies are listed in Table 3 and shown in Fig. 3. For small systems it can be seen that the energy of mli with the magic numbers 13, 55, 147 and 309 attains minimal values as compared to surrounding nonmagic clusters.

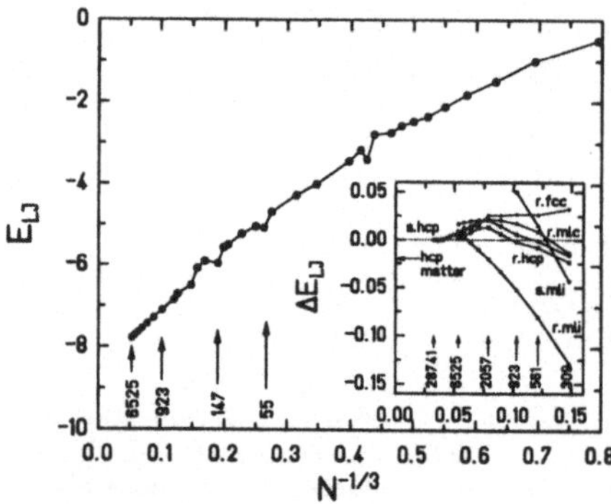

Figure 3:
Energies per particle of LJ crystals and quasicrystals. Full (open) circles are relaxed MD results with mli initial configurations at magic (nonmagic) particle numbers. In the insert are shown various energies (r: relaxed by MD, s: scaled and minimized) with the scaled-hcp average of $-8.591 + 15.035 N^{-1/3}$ subtracted. The open arrow points to the hcp matter value.

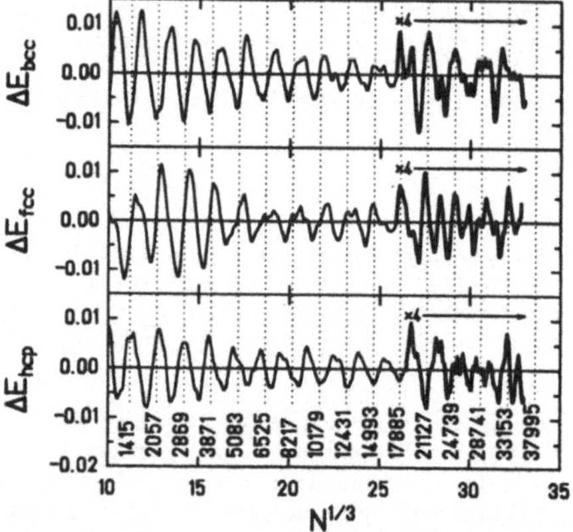

Figure 4:
Shell energies of scaled spherical LJ crystals, smoothed and the mean subtracted. The heavy parts are magnified.

Table 3:
Closed shell particle numbers N of shell M and their energies of relaxed mli, mlc, hcp and fcc structures and of scaled mli, hcp and fcc.

M	N	$E_{\mathrm{mli}}^{\mathrm{MD}}$	$E_{\mathrm{mlc}}^{\mathrm{MD}}$	$E_{\mathrm{hcp}}^{\mathrm{MD}}$	$E_{\mathrm{fcc}}^{\mathrm{MD}}$	$E_{\mathrm{mli}}^{\mathrm{scaled}}$	$E_{\mathrm{hcp}}^{\mathrm{scaled}}$	$E_{\mathrm{fcc}}^{\mathrm{scaled}}$
1	13	-3.4098					-3.4098	-3.1466
2	55	-5.0772	-4.8778	-4.9114	-4.7660	-5.0373	-4.9052	-2.4041
3	147	-5.9623	-5.8121	-5.8636	-5.7888	-5.8938	-5.8576	-4.0776
4	309	-6.4959	-6.3805	-6.3907	-6.3347	-6.4102	-6.3834	-5.3022
5	561	-6.8492	-6.7595	-6.7707	-6.7522	-6.7530	-6.7621	-6.1041
6	923	-7.0994	-7.0295	-7.0490	-7.0202	-6.9965	-7.0409	-6.6062
7	1 415	-7.2854	-7.2312	-7.2261	-7.2454	-7.1781	-7.2184	-6.9529
8	2 057	-7.4291	-7.3875	-7.3956	-7.3831	-7.3185	-7.3880	-7.1823
9	2 869	-7.5432	-7.5121	-7.5204	-7.5168	-7.4304	-7.5137	-7.3676
10	3 871	-7.6361	-7.6138	-7.6257	-7.6209	-7.5216	-7.6193	-7.5085
11	5 083	-7.7128	-7.6982	-7.7133	-7.7037	-7.5973	-7.7071	-7.6219
12	6 525	-7.7775	-7.7696	-7.7832	-7.7808	-7.6611	-7.7774	-7.7121
13	8 217					-7.7156	-7.8404	-7.7849
14	10 179					-7.7628	-7.8933	-7.8515
15	12 431					-7.8041	-7.9420	-7.9000
16	14 993					-7.8404	-7.9807	-7.9441
17	17 885					-7.8726	-8.0158	-7.9865
18	21 127					-7.9013	-8.0477	-8.0238
19	24 739					-7.9272	-8.0748	-8.0554
20	28 741					-7.9505	-8.1003	-8.0851
21	33 153						-8.1223	-8.1086

Relaxed icosahedrons are lowest in energy up to $N = 3\,871$ with either multilayer or stacked cuboctahedrons being much higher. In addition, since simple spherical hcp crystals always have lower energy than mlc and since even fcc crystals are lower in most of this interval of particle numbers, we exclude the posssibility of the existence of large cuboctahedral clusters. Due to their bcc structure the same holds for rhombic dodecahedra.

Despite the fact that mli have a very low surface energy, the volume energy and, hence, the total energy becomes too large for large particle numbers. Around particle numbers of $5\,083$, hence, there is a transition to hcp crystals which, however, have larger surface energy but lower volume energy. Similar reasoning holds for the hki and dpk. The corresponding values for the cohesive and surface energies of Table 2 have been obtained by extrapolating the results of ref. [11]. They have to be taken as lower limits. However, from Table 2 it can also be seen that although hki have a slightly larger surface energy than dpk, their cohesive energy is slightly less. Hki, hence becomes the more stable configuration for $N > 40\,000$.

In order to follow the magic numbers beyond $N = 6\,525$ the shell energies of scaled (but not relaxed) spherical simple cubic (sc), bcc, fcc and hcp crystals have been calculated. By subtracting the mean and smoothing we obtain the shell energies of Fig. 4. All of them exhibit characteristic shell oscillations due to the fact that there exist spherical crystalline configurations with minimum (maximum) number of surface particles at the same radius, hence with small (large) surface energies. However, for large systems sc, bcc and fcc do not show the characteristic shell spacing of eq. (1) whereas hcp does for $N = 10\,179, 12\,431, 14\,993, 17\,885$ and $21\,127$ at the correct positions (also, at $8\,217$ there is a local minimum). The next two minima appear at $N \simeq 23\,600$ and $27\,500$ rather than at $24\,793$ and $28\,741$. Here the magic numbers were not yet identified experimentally. For larger particle numbers the shell energy becomes rather small and one expects that magic numbers cease to exist.

In order to assure that the effect of magic numbers in hcp spheres is not an artifact of the LJ force we repeated the scaling calculations with the Mie potential $r^{-4n} - 2r^{-2n}$ ($n = 3$ is the LJ potential) with $n = 1...4$. For the short range potentials $n = 2, 3, 4$ we found the minima of the energy essentially at the same magic numbers as discussed above. In the long range case $n = 1$, on the other hand, the shell oscillations are very irregular and one cannot associate magic numbers. However, we cannot exclude that macroscopic structures formed of hcp matter other than the sphere might have even less surface energy.

References

[1] W. Slattery, G. Doolen and H. DeWitt, Phys. Rev. **A 21** (1980) 2087
[2] R.W. Hasse and J.P. Schiffer, Ann. Phys. **203** (1990) 419
[3] R.W. Hasse and V.V. Avilov, Phys. Rev. **A 65** (1991) 4506
[4] D.H.E. Dubin, Phys. Rev. **A 40** (1989) 1140
[5] R. Rafac, J.P. Schiffer, J.S. Hangst, D.H.E. Dubin and D.J. Wales, Proc. Natl. Acad. Sci. USA **88** (1991) 483
[6] A.L. Mackay, Acta Crystallogr. 15 (1962) 916
[7] H. Göhlich, T. Lange, T. Bergmann and T.P. Martin, Phys. Rev. Lett. **65** (1990) 748
[8] T.P. Martin, T. Bergmann, H.Göhlich and T. Lange, *in* Proc. 5th Int. meeting on small particles and inorganic clusters, Konstanz 10-14 Sept. 1990, Z. Phys. **D 19** (1991) 25
[9] H. Totsuji and J.-L. Barrat, Phys. Rev. Lett. **60** (1988) 2484
[10] D.H.E. Dubin and T.M. O'Neil, Phys. Rev. Lett. **60** (1988) 511
[11] B. Raoult, J. Farges, M.F. De Feraudy and G. Torchet, Phil. **Mag. B 60** (1989) 881
[12] I. Stewart, New Scientist **13 July** (1991) 29
[13] R.W. Hasse, Phys. Lett. **A** in print
[14] C. Kittel, Introduction to solid state physics (Wiley, New York, 5th ed. 1976) p. 80.

LOCALIZATION OF QUANTUM ELECTRONS IN ONE, TWO AND THREE DIMENSIONAL DISORDERED SYSTEMS OF SCATTERERS. MONTE CARLO INVESTIGATIONS.

V. S. Filinov

Institute for High Temperatures, USSR Academy of Sciences,
Izhorskaya 13/19, Moscow 127412, USSR

L. I. Podlubnyi

Moscow Power Institute, Krasnokazarmenay 14,
Moscow 111250, USSR

Abstract

Path integrals and complex Monte Carlo method are formulated for the system of noninteracting electrons in a random potential of scatterers and external electrical field. The ensemble-averaged probability for electron with energy E to travel a distance $|r-r'|$ have been calculated. The numerical results for 1D, 2D and 3D disordered systems have been obtained and compared with analytical results. The unwaited sharp peaks, which interrupt the exponential decay of the mean squared value of the Green function modulus and delocalization of electrons in external electrical field have been also obtained.

Introduction.

The most important and actual problem of quantum theory of disordered systems and dense plasma is the consistent treatment of multiple scattering. In the process of studying of this problem the concept of localization was first proposed by Anderson [1] and Mott [2,3] in the context of electron diffusion in random potentials, its validity as a generic phenomenon for particles and waves propagation in random media is by now well recognized.

It is known for three-dimensional systems that particles moving in a random potential will be localized if either the potential fluctuation exceed a certain threshold value or if the electron energy is low enough. For one-dimensional systems the quantum particles will be always localized irrespective of how small the disorder may be [1-5].

Substantial progress in understanding of the critical properties of the localization length near the Anderson transition was made, when the problem was formulated in terms of the renormalization group and one-parameter scaling hypothesis was introduced.

However ten years after the development of the scaling theory of localization most of researchers working in the field agree that in contrast to the initial euphorism, the problem of Anderson transition is not yet solved.

This situation motivated us to develop a new numerical approach for solving the problem. In this approach the path integral technique and complex Monte Carlo method have been combined to calculate the moments of the Green function of the stationary Schrödinger equation for electron in potential field of randomly distributed scatterers.

A new representation of the Green function in the form of complex path integral, where the integrand is Hunkel function of the first kind has been developed [6]. The limiting transition of this representation to Feynman path integrals [7] has been also considered. The well known in literature Monte Carlo method based on the Metropolis algorithm is employed extensively in lattice gauge theory and quantum mechanics only for euclidean version of the Feynman path integrals. This method is applicable only for evaluating the integrals of real functions. In the present approach the Monte Carlo method is extended to evaluation of the integrals of complex-valued functions and complex path integrals [8-11].

Path Integrals and Laplace transformation.

Our analysis employs the model of independent electrons moving in a potential provided by randomly distributed static impurity scattering center. The interaction of electron with the impurities is described by the following dimensionless Hamiltonian:

$$\hat{H} = -\frac{1}{2}\Delta + \sum_{j=1}^{N} \zeta\, V\, (\,|r-R_j|\, /\, \sigma'\,) \tag{1}$$

In (1), the wave number k is defined by $k^2 = 2mE/\hbar^2$, m and E are the electron mass and energy, $\hbar$ is Planks constant, σ' is a characteristical length of the scattering potential. We assume that $\sigma = 2\sigma'$ is the diameter of the scatterer, R_j denotes its random position, all lengths are expressed in units of k^{-1}. The short range interaction potential $V(|r-R|/\sigma')$ vanishes at the large distance from the scatterer $(>\sigma')$ and acts jump like at the value unity on a scatterer.

The Green function of equation (1) can be expressed through the Laplace transform

$$G(r,r',z) = \int_0^\infty \exp(izl)\, G(r,\,r',\,l)\, dl \qquad (2)$$

of the Green function $G(r,r',l)$ of the nonstationary Schrödinger equation. The Green function $G(r,r',l)$ can be presented to be the Feynman integral, whose finite-dimensional approximation is

$$G(r,r',l) = i\theta(l)(2\pi i\tau)^{-M\nu/2}\int..\int \exp(i\tau^{-1}S[r_0,r_1,..,r_M])\, d\{r_m\} \qquad (3)$$

In (3), $\theta(l)$ is the Heaviside step-function; the paths are approximated by broken lines with the path apexes at the νD points r_m at the time moments $l_m = \tau m$, $\tau=1/M$, $m=0,1,...,M$, $r_0 = r'$, $r_M = r$, $d\{r_m\} = dr_1...dr_{M-1}$, $V(r_m,R_j) = \sum_j \xi V(|r_m-R_j|/\sigma')$, $\langle r|K|r\rangle = \sum_m |r_{m+1}-r_m|^2$ -quadratic form, $|r\rangle - \nu*(M+1)$ - dimensional vector $\{r_0,...,r_m\}$. $S = \langle r|K|r\rangle - \tau^2\sum_m \xi V(r_m,R_j)$ is finite difference approximation of the "action integral" along the path.

After performing the procedure of averaging over scatterer position, as was proposed in [12], and Laplace transformation we can obtain the final approximative expression for the mean squared value of the Green function modulus in the following form [6]:

$$\overline{G(r,r',1/2)\, G^*(r,r',1/2)} = \qquad (4)$$

$$= M \int...\int d\{r'_m\} \int...\int d\{r''_m\} (2\pi i\zeta')^{-(M\nu-1)/2}(-2\pi i\zeta'')^{-(M\nu-1)/2}$$

$$\exp(i\langle r'|K|r'\rangle/\zeta'-i\langle r''|K|r''\rangle/\zeta''+n\int dR\, F(R)+\ln(\hat S(z')\hat S^*(z'')))$$

where $F = \exp(-i\zeta'_+ \sum_{m=0}^{M-1} \xi\, V(|r'_m-R|/\sigma)+i\zeta''_+\sum_{m=0}^{M-1} \xi\, V(|r''_m-R|/\sigma))-1$, $z^2 = M\langle r|K|r\rangle$, $\zeta = z/M$, $\hat S(z) = \sum_{a=0}^{A-1} (-2iz)^a (A+a-1)! / (a!(A-a-1)!)$,

$A = (M\nu-1)/2-1$, $\zeta_+ = \sqrt{M\langle r|K|r\rangle-(M\nu/2)^2}-iM\nu/2 \approx \zeta$. This expression is valid for large distance from the electron source compared to the wave length and small density of scatterers $(n\sigma^3 << 1)$.

The criteria for electron localization is chosen to be the absence of diffusion in disordered systems. The ensemble-averaged probability for electron with energy E to travel a distance $|r-r'|$ is $\overline{G(r,r',1/2)\, G^*(r,r',1/2)} = p(|r-r'|,E)G_0 G_0^*$ (G_0 for $n\sigma^3 = 0$). The exponential decay of $p(|r-r'|,E)$ with the distance $|r-r'|$ has been assumed to be the criteria for localization [1-5].

Numerical results.

Let us consider the numerical results for one-dimensional case. Fig.1 presents the results for 10 * log(p(x-x',E)) versus the modulus of ξ at the fixed values of |x-x'| = 120, nσ = 0.1π/6 and σ = 4π/12. The points obtained by the Monte Carlo method lie on the straight line with the small spread within the statistical error bar. The results are also insensitive to ξ's sign changes.

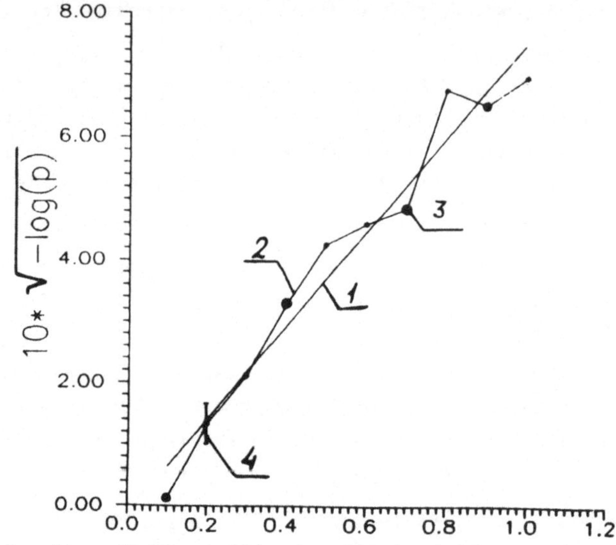

Fig.1. p(|x-x'|,E) vs |ξ|: 1 - the best linear fit of Monte Carlo results; 2,3 - Monte Carlo results for ξ > 0 and ξ < 0 respectively; 4 - statistical error bar.

Fig. 2 presents the Monte Carlo results for 10 * log(p(x-x',E)) at |ξ|=0.03 as the function of |x-x'| at the fixed above values of others parameters. Note that for small distance |x-x'| the calculated points lie on the weekly oscillating curve in the vicinity of zero (|x-x'| < 100), so we can conclude that in this region the effect of the first cluster integral n$\int$F(R)dR is negligible.

At the larger distance Monte Carlo points oscillate near the straight line with negative slope. Note that the exponential decay is formed only when the distance |x-x'| is more than five or six average distance between scatterers ($(n\sigma)^{-1}$ $\approx$ 20).

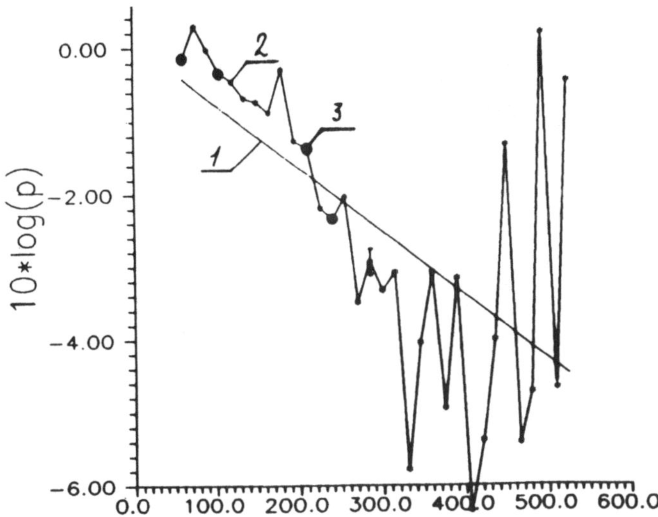

Fig.2. p(|x-x'|,E) vs |x-x'|. All symbols are the same as on Fig.1.

As p(|x-x'|,E) is not the self-averaged value the very seldom resonance configuration may give the dominant contribution compared to contribution of the typical configurations [13] and that may be the reason of sharp peaks of p(|x-x'|,E) at the large distance $280 < |x-x'| < 600$.

Fig. 3 presents the Monte Carlo results for 3D disordered system for the following values of basic parameters: $n\sigma^3 = 0.1$, $\sigma = 4\pi/12$, $\xi = 0.01$. Now the above mentioned peaks are smaller. Note that for 3D space the amplitudes of scattered waves decrease as inverse distance from the scattering center and the interference phenomenon is not so important as in 1D case, when the scattered wave amplitudes do not vanish (are constant).

Let us consider the dependence of p(|x-x'|,E) on the diameter of the scatterer at the fixed above values of basic parameters: $n\sigma = 0.1\pi/6$, $|x-x'| = 120$, $\xi = 0.03$. On Fig. 4 we show the Monte Carlo results as a function of σ. It is known from the literature [14], that for small potential fluctuation and large distance |x-x'| the value p(|x-x'|,E) is proportional to the following expression: $\exp(-D|x-x'|/4)$, where the diffusion coefficient D is defined by the relation [14]:

$$D = \sigma_v^2 * \sigma/2(1+4\sigma^2) \qquad (5)$$

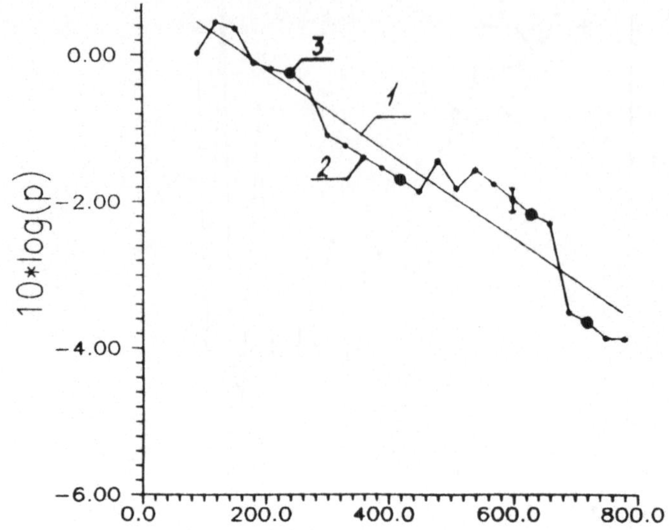

Fig.3. $p(|r-r'|,E)$ vs $|r-r'|$ for three dimensional case ($n\sigma^3$= = 0.1, $\sigma = 4\pi/12$, $\xi = 0.01$). All symbols are the same as on Fig.1.

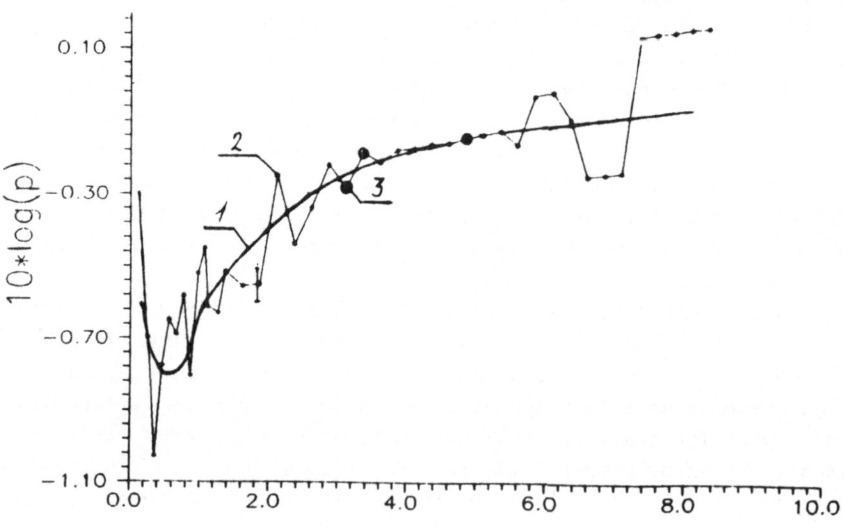

Fig.4. $p(|x-x'|,E)$ vs σ. All symbols are the same as on Fig.1.

Now let us try to verify the possibility to approximate the Monte Carlo results by the simple relation (5). To calculate the unknown factor σ_v^2 in formula (5) we have used only one of the Monte Carlo points which belongs to the smooth part of the Monte Carlo curve ($\sigma_v^2 \approx 0.05$). On Fig. 4 line 1 denotes this functional dependence. Practically all the other Monte Carlo points oscillate near the line 1.

For the large values of σ ($\sigma > 5.5$) at the fixed values of parameters $|x-x'| = 120$ and $n\sigma = 0.1\pi/6$ the average distance between scatterers $\sigma/n\sigma$ becomes larger than $| x-x' | = 120$. So the above mentioned formula (5) for this region is inconsistent. It is important that Monte Carlo results in this region display a sharp change of functional behavior (the unwaited increase of the oscillation).

On the other hand, when $\sigma \ll 1$ the average distance between scatterers is very small compared the fixed distance $| x-x' | = 120$. As we mentioned above in this region there exist the sharp peaks of $p(|r-r'|,E)$ due to the contribution of seldom resonance configuration.

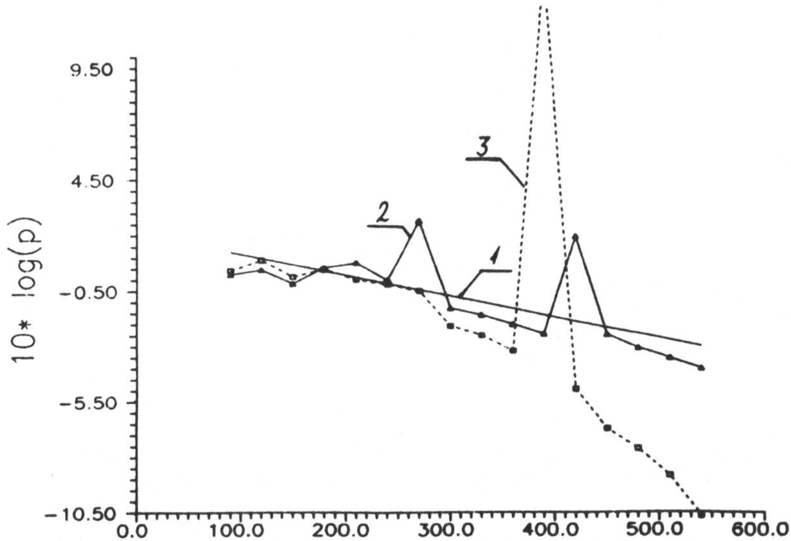

Fig.5. $p(|\rho-\rho'|,E)$ vs $|\rho-\rho'|$: 1-the best linear fit of Monte Carlo results; 2,3-Monte Carlo results for $\delta\sigma / 2E = 0$ and $\delta\sigma / 2E = 0.0005$ respectively

The effect of an electrical field $\mathcal{E}$ for 2D disordered system
is shown on the Fig. 5 for the following values of basic
parameters:$n\sigma^2 = 0.2$, $\sigma = 4\pi/12$, $\xi = 0.015$, $\mathcal{E}\sigma / 2E = 0.0005$. Lines
2, 3 display the Monte Carlo results for bariers without and in
presence electrical field respectively. For small distance the
effect of electrical field is negligible, as the perturbation
effect is defined by the average potential of electrical field on
the paths from ρ' to ρ ($|\rho-\rho'| < 200$). On the middle distance
the electrical field implies the increase of above mentioned
peaks. For very large distance ($|\rho-\rho'| > 550$) formally calculated
$p(|\rho-\rho'|,E)$ is equal to unity, that may be caused by
delocalization of electrons, but Monte Carlo calculation in this
region not reliable. In [15,16] for 1D disordered systems
delocalization of electrons was predicted for conditions, when
energy supported to electron by electrical field is sufficiently
large for collisionless character of its moving.

References

[1] P.W. Anderson, Phys. Rev. 109, 1492 (1958)

[2] N.F.Mott, W.D.Twose, Adv.Phys. 10, 107 (1961)

[3] N.F.Mott, Adv.Phys. 16, 49 (1967)

[4] S. A. Gredeskul, V. D. Freilicher, Usp. Fiz.Nauk 160, 239
(1990)

[5] I. M. Lifshits, S. A. Gredeskul, L. A. Paster, Introduction to
the Theory of Disordered Systems (Nauka, M., 1982) p. 112

[6] V.S.Filinov, Waves in Random Media 2, 141 (1991)

[7] R. P. Feynman, A. R. Hibbs, Quantum Mechanics and Path
Integrals (McGraw-Hill, N.Y., 1965)

[8] V.S. Filinov, Nucl.Phys. B271, 717 (1986)

[9] J.D. Doll, T.L. Beek, J.Chem.Phys. 89, 5753 (1988)

[10] J.D. Doll, D.L. Freeman, M.J. Gillian, Chem.Phys.Lett. 143,
277 (1988)

[11] N. Makri, W.H. Miller J.Chem.Phys. 89, 2170 (1988)

[12] S.F. Edwards, J.Non-cryst. 4, 417 (1970)

[13] I. M. Lifshits, V. Ya. Kirpichenkov, Zh. Eksp. Teor. Fiz. 77,
989 (1979)

[14] V. I. Klyatskin, Invariant imbedding method in the theory of
wave propagation (Nauka, M., 1986) p. 156

[15] V.N. Prigodin, Zh. Eksp. Teor. Fiz. 79, 2338 (1980)

[16] T.R. Kirkpatrick, Phys.Rev. B33, 780 (1986)

N log N Code for Dense Plasma Simulations

S. Pfalzner

Gesellschaft für Schwerionenforschung,

Planckstr.1, 6100 Darmstadt, Germany

P. Gibbon

IBM Deutschland GmbH, Wissenschaftliches Zentrum,

Tiergartenstr.15, 6900 Heidelberg, Germany

Abstract

In recent experiments lasers have produced plasmas around solid density and above. Collision processes play an important role in such dense plasmas ($\Gamma \leq 1$). For calculating collective effects and transport processes the usual N-body ('Molecular Dynamics') codes would be impractical, due to the N^2-dependence of the computation time. A hierarchical algorithm is suggested, previously used in astrophysics, which has a N log N- scaling.

1   Introduction

There are three types of simulation codes that are used in plasma physics: fluid, particle-in-cell(PIC) and particle kinetics.

The macroscopic behaviour neccessary for the analysis of experiments can be modelled by fluid codes. Collisions are treated with a simple damping term, which assumes that the thermal distribution remains Maxwellian. This is of course insufficient if one wants a more detailed description like wave breaking etc.

PIC codes [1] describe plasmas on a more microscopic scale. For dense plasmas their disadvantage is that although electron-ion collisions can be included [2], the electron-electron collisions can not easily be considered and the plasma becomes unphysically non-Maxwellian. For this reason PIC is mainly used to study collisionless plasmas for short time scales.

Particle-kinetic codes partly fill the gap between these two approaches. Solving the Fokker-Planck equation directly for the electron velocity distribution, these codes include electron-ion and electron-electron collisions. Particle-kinetic codes have proved successful in long-scale length plasmas with extended underdense coronas, but the treatment of the collisions becomes inadequate for dense plasmas where the correlations between particles become significant. In this case it is invalid to assume that the collision rate is determined by the sum of many small-angle scattering events: possibility of very large deflections must be taken into account.

"

Numerical simulations of the collision frequency by 'measuring' the heating rate when an electric field is applied employ a direct force summation to solve the motion of each particle [4],[5]. Unfortunately the required computation time is proportional N^2 and is therefore not practicable for more than a few hundred particles. Here a 3D N-body code is presented which reduces the computation time to a N log N scale [6]. Similiar codes have been used by a few authors to study the development of galaxies [7], but so far this method has not been used to study collisional plasmas.

2 N log N Algorithm

For simulating N-body problems the force calculation requires most of the computation time. For N-body systems with a $1/r^2$- dependence of the force like in gravitational or electrostatic problems the nearest neighbours contribute most to the total force on a individual particle, whereas more distant particles contribute much less. In plasmas this effect is even stronger because of the Debye screening.

The hierarchical algorithm is based on representing the force seen by a specific particle as the sum of the forces due to the individual particles in its immediate neighbourhood plus coarser groupings of more distant particles. The calculation is performed without a grid: instead the physical space occupied by the particles of the plasma is recursively divided up until there is only one particle per box. The resulting structure is called a 'tree'. Fig.1 shows an example of how this tree structure is built.

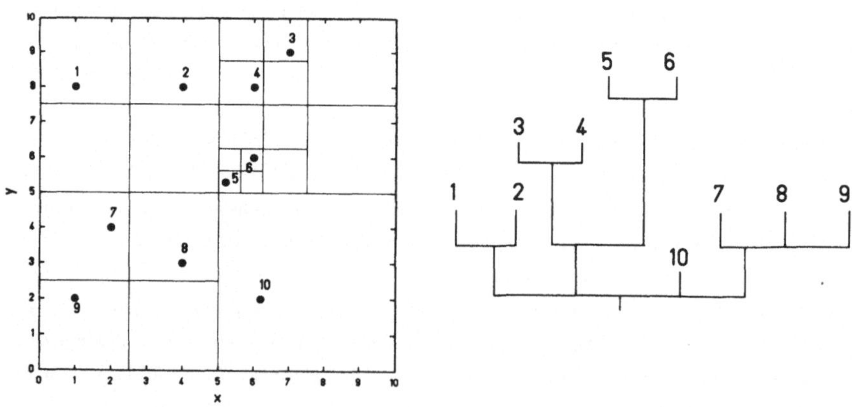

Fig.1: Example of the division of the physical space and the resulting tree structure.

In the force calculation, more distant particles are grouped together to form pseudoparticles. Ergo, it is neccessary to evaluate these centres of charge. The centres of charge at each level of the tree are evaluated starting from the 'leaves' (single particles) down to the root (whole system). The force is then calculated for each individual particle.

For each particle the force calculation starts at the root of the tree. Comparing the box size of the current pseudoparticle s with the distance the distance d to the particle the relation

$$s/d \leq \theta$$

where θ is a tolerance parameter, decides whether to subdivide the box or to calculate the force contribution of the pseudoparticle. The force on each particle is then computed the following way:

```
start at root
do
      if s/d < θ then
            evaluate force
      else
            resolve cell into daughter cells
      endif
until whole tree searched
```

The computation time and the accuracy of the calculation depend on the choice of θ. Independently of θ, the accuracy of the force calculation of the pseudoparticle can be improved using a multipole expansion. Unlike in gravitational problems, this multipole expansion is essential for the force calculation in plasmas, because it could happen that the number of electrons and ions is equal, so that the nett charge would be zero and the pseudoparticle thus ignored. Our calculations have shown that taking into account up to quadrupole terms and $\theta \sim 1$, a sufficient accuracy of the result is obtained with a N log N computation time.

Having calculated the force, the positions and velocities can be up dated, and for the next time step the new tree can be built again. The overall algorithm can be summarized as follows:

1. Construct tree

   - divide into cells

   - evaluate centres of charge

2. For each particle:

   - evaluate force

   - update velocity and position

3. Do diagnostics

4. Goto 1

The technique has an additional advantage over PIC (besides treating collisions accurately)
that it is gridless, which enables complicated density distributions to be modelled. Fig.2 shows
the tree structure for a given particle distribution in 3 dimensions. Resolution is applied where
it is needed.

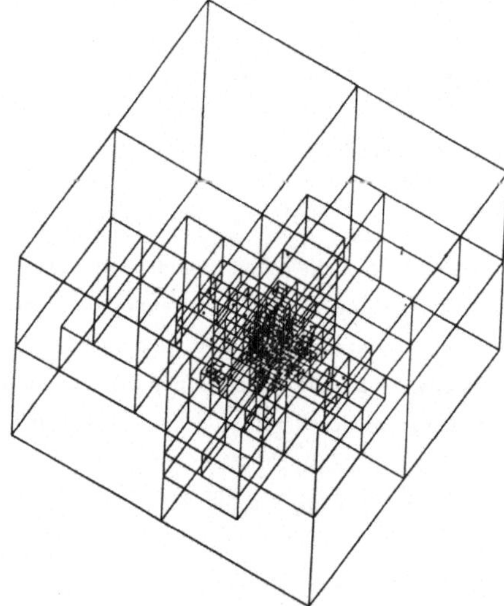

Fig.2: Example of a 3 dimensional tree structure.

3 Application

The application for this kind of code in plasma physics will lie in the (n_i, T_e) parameter range, where collisional effects become important. This means when collisions are not only determined by small-angle scattering but by large-angle deflections too. This is the case for dense plasmas ($\sim 10^{23}$ cm^{-3}), which have recently been produced by laser beams. For these plasmas, where $\Gamma \sim 1$, the above algorithm can be used to calculate collision rates which are difficult to describe otherwise.

As one example that this code produces plasma behaviour in a realistic way, we show the relaxation of a plasma with a non-maxwellian velocity distribution:

$$f_x(v_x) = f_o exp(-(v_x + v_o)^2/2v_{te})$$

The calculation has been performed with periodic boundary conditions. Fig.3 shows the initial condition and the relaxation into a maxwellian plasma.

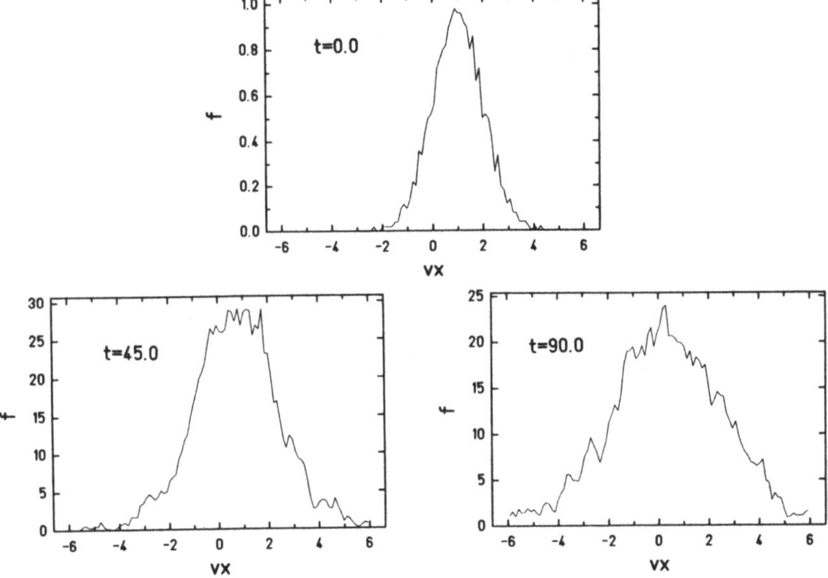

Fig.3: Relaxation of a initially non-maxwellian distribution of a fully ionized plasma with $\Gamma = 0.12$ and Z=1.

4 Conclusion

A numerical algorithm has been presented which has a N log N dependence of the computation time. In addition the code has the advantage of a gridless treatment, which enables to evaluate problems with a complex geometry.

This code is currently used to study dense plasmas. A future extension could be to include the degeneracy of the electron gas. On the computational side the code could be improved by individual time steps.

The macroscopic effects which we want to investigate in future using the above algorithm are collisional processes such as inverse bremsstrahlung absorption for lasers and stopping power for heavy ion beams. Here the high - as well as low-temperature case can be treated, the latter is very difficult by most other approaches.

References

[1] C.K.Birdsall and A.B.Langdon, *Plasma Physics via Computer Simulation*, McGraw-Hill, New York,(1985).

[2] J.M.Wallace, D.W.Forslund, J.M.Kindel, G.L.Olson and J.C.Comly, Phys. Fluids B **3**, 2337, (1991)

[3] R.Cauble, W.Rozmus, Phys.Fluids **28**, 3387, (1985).

[4] H.Furukawa, K.Nishihara, Phys.Rev. A **42**, 3532, (1990).

[5] T.Katsouleas, C.Decker, W.B.Mori, 21th Anomalous Absorption Conference, 2P12 (1991).

[6] J.Barnes, P.Hut, Nature, **324**, 446, (1986).

[7] L.Greengard, Computers in Physics, Mar/Apr, 142, (1990).

STATISTICAL THERMODYNAMICS OF BOUND STATES
AND PHASE TRANSITIONS IN NONIDEAL PLASMAS

Werner Ebeling

Humboldt University, Institute of Theoretical Physics
Invalidenstr. 42, D-01040 Berlin, Germany

Abstract

The two principal methods: Grand canonical and canonical
description are introduced and compared with respect to bound states.
It is shown that a fugacity expansion (grand ensemble) is well suited
to describe bound states of a few particles. Therefore the grand
ensemble is the natural representation for bound state systems.
However on the other hand, fugacity expansions are not appropriate to
describe the effects of short range forces. But already a few terms of
a density expansion (canonical ensemble) which is not appropriate for
bound states give a good representation of short range interactions.
The chemical picture, with is to be interpreted as a mixed
representation, is combining the advantages of both descriptions. The
stability analysis of the thermodynamic functions shows the existence
of additional phase transition at high temperatures and high pressures
due to nonideality effects. It is shown that phase transitions are
much easier to detect in mixed representations (chemical picture) than
in the grand canonical ensemble (fugacity expansions).

1. Introduction

Investigations of bound states and ionization processes in dense
plasmas are of interest for the study of stars and planets as well as
for technical applications. We will show in this work that Greens
function methods or direct fugacity expansions which are based on the
grand ensemble yield the natural representation for bound state
systems [1]. However on the other hand, fugacity expansions are not
appropriate to describe the effects of short range forces. Already a
few terms of a density expansion (canonical ensemble) give a good
representation of short range interactions but they do not reflect
bound state effects. We want to show that the chemical picture, which

is to be interpreted as a mixed representation based on the summation
of classes of contributions, is combining the advantages of both
descriptions. In the last part the thermodynamic functions are
analyzed. We consider hydrogen and helium plasmas and pay special
attention to the existence of phase transitions. It is shown that in
systems with long-range Coulombic interactions, besides the classical
first-order transition typical for neutral gases, a second first-order
transition may appear. Along the coexistence line a system undergoing
a plasma transition is divided into two phases of different density
and different degree of ionization. The experimental and theoretical
efforts to check the existence of this transition are discussed.

2. Canonical, Grand Canonical and Chemical Descriptions

Let us first look at low-density hydrogen plasmas from a very
elementary point of view, taking into account the formation of
hydrogen atoms as bound states of electrons and protons

$$p^+ + e^- = H \tag{2.1}$$

We assume that the densities of free electrons n_e^*, free protons n_p^*
and free atoms n_H^* are connected by a mass action law

$$n_H^* / (n_e^* n_p^*) = K(T) = \{ h^3 / (2\pi mkT)^{3/2} \} \exp \{ - E_{10}/kT \} \tag{2.2}$$

where m is the reduced mass and E_{10} the ground state energy of the
bound electron. Then the pressure is in the ideal approximation given
by

$$p = kT [n_e^* + n_p^* + n_H^*] \tag{2.3}$$

Formally this chemical picture of the formation of bound states is
fully equivalent to a fugacity expansion in the physical picture were
only electrons and protons are considered as the constituents. In
order to prove this let us write eqs. (2.1-3) in the form

$$p = kT [z_e + z_p + z_e z_p K(T)] \tag{2.4}$$

$$n_e = z_e + z_e z_p K(T); \quad n_p = z_p + z_e z_p K(T) \tag{2.5}$$

where n_e and n_p are the total numbers of electrons and protons
respectively. This can be interpreted as a fugacity expansion in the
physical picture which includes the second virial coefficient. The

understanding of the relation between fugacity expansions and chemical picture we owe to the work of HILL [2] and other workers [3,4]. Fugacity expansions arise in a natural way from the grand canonical ensemble, which is based on chemical potentials instead of densities. Summarizing we come to the conclusion, that the appropriate way to include bound states are fugacity expansions in the physical picture which include the virial coefficients up to certain order. If we want to take into account two particle bound states (atoms), we must go up to the second virial coefficients. However if we want to include even molecules or higher complexes, we have to consider virial coefficients which include the corresponding elementary constituents (as e.g. two protons and two electrons in the case of hydrogen molecules). Based on these considerations we have a clear prescription how to proceed: We have to find a fugacity expansion (grand canonical representation) that includes all relevant bound state configurations and calculate the pressure. However this is only half of the story. For some reason (to get more physical insight and to improve the convergence with respect to short range interactions) we shall go back from the formally simple fugacity expansions (corresponding to eqs. (2.4-5)) to a chemical picture (corresponding to eqs. (2.1-3)). It is just this transition which contains the most difficult (and in part still unsolved) problems of the theory of bound states.

A stronger formulation of the ideas explained above requires the quantum statistical theory, which we can survey only in brief following [1] and [4]. The great advantage of the technique of Green function is that it covers the entire "corner of correlation" and further that it operates from the very beginning in the grand canonical ensemble. Let us define the n-order Green function (GF) by

$$G_n(1 \ldots n, 1' \ldots n') = (1/i\hbar)^n \langle T\{a(1)\ldots a^+(n')\}\rangle \qquad (2.6)$$

The single particle GF and the two particle GF can be used to derive the thermodynamic functions. To evaluate the GF we may use the theorem that the n-particle GF is given as the sum of the contributions corresponding to all topologically nonequivalent connected Feynman diagrams starting from the open ends $1,2,\ldots,n$ and leading to the open ends $1',2',\ldots,n'$. For practical purposes one has to restrict to certain subclasses of the diagrams, which correspond to virial coefficients in fugacity expansions. In order to get realistic descriptions of bound states of $s > 1$ elementary particles, we have to consider GF with at least s ends. Further we must take into account screening which is essential for all Coulombic systems. Technically that means, the basic diagrams with at least s ends have to be considered and further, infinite classes of internal parts of these basic diagrams have to be summed up which correspond to the effects:

(a) screening of the interaction (chain summation),

(b) formation of bound states (summation of ladder diagrams), and

(c) chemical equilibria (quasi particle summations).

Strong correlations between the particles and especially the formation of bound states require the summation of all ladder diagrams connecting s outer lines. For s=2 this corresponds to the equation

$$G_2^L = G_2^O + G_2^O \, V \, G_2^L \qquad\qquad (2.7)$$

We call G_2^L the ladder approximation to G_2. Equation (2.7) which is also called the Bethe-Salpeter equation, determines the two particle bound states in the plasma. In order to select out diagrams which are of relevance. we use (based on our elemntary considerations at the beginning), the principle of equivalence between bound states and composite particles, which may be formulated as: Bound states are equivalent to new species (composites). Elementary particles and composites should be treated on the same footing [3,4]. This means that a diagram containing free particle propagators G_1 should be completed by the corresponding diagram containing a propagator of the composites [5]. The influence of the surrounding plasma on the two particle states is discussed by many authors [5-7]. In order to calculate the thermodynamic functions we may use the formulae

$$p(\beta, \mu) = \int_{-\infty}^{\mu} d\mu' \, n(\beta, \mu') \; ; \; \beta = 1/kT \; ; \; \mu = kT \ln z + const \qquad (2.8)$$

$$n(\beta, \mu)) = \frac{1}{V} \sum_{p} \int \frac{d\omega}{2\pi} \, f_1(\omega) \, \text{Im} \, G_1(p\omega) \qquad (2.9)$$

With the approximations described above, we arrive at expressions with s-particle contributions as [5]

$$n(\beta, \mu) = \sum_{p} f_1(\varepsilon_1) + \sum_{nP} g_2(E_{nP}) + \sum \int d\omega \, g_2(\omega) \, \frac{d}{d\omega} \sin \delta(\omega) \qquad (2.10)$$

where $\delta(\omega)$ is a generalized scattering phase. The energy levels include shifts

$$\varepsilon_1 = \frac{p^2}{2\,m} + \Delta_p^{free} + \Delta_P^{bound}$$

$$E_{nP} = E_{nP}^{\circ} + \Delta_{nP}^{free} + \Delta_{nP}^{bound} \qquad (2.11)$$

By integration of eq. (2.10) one obtains

$$p = p_{id}^{(1)}(\beta,\mu) + p_{id}^{(2)}(\beta,\mu) + \ldots + p_{int}(\beta,\mu) \qquad (2.12)$$

Here the first term corresponds to free one-particle states, the second one to two-particle bound states (second virial coefficient), and the last one corresponds to the interaction of quasi particles. So far the theory has been worked out in the grand canonical ensemble. This representation is appropriate for the description of chemical bonds, however additive forces are not described well. Furthermore, phase transitions of first order are difficult to recognize in the grand canonical technique.

3. Transition to the Chemical Picture

The analysis given above suggests that the transition to a mixed representation (between grand canonical and canonical ensemble), the chemical picture may be of advantage. Following the ideas formulated in the principle of equivalence (bound states are to be treated on the same footing as free particles) we introduce chemical potentials of quasi particles by

$$\mu_1^{(\nu)} = \mu + \Delta_\nu(0)$$

$$\mu_2^{(\nu)} = \mu(1) + \mu(1') + E_n^{\circ} + \Delta_n(0) \qquad (3.1)$$

where $\Delta_\nu(0)$ and $\Delta_n(0)$ are shifts taken at the bottom of the energy bands. The band index ν has been introduced for the case that G_1 shows an energy band structure as, e.g., in crystalline states. In the new representation we may easily go to the canonical ensemble

$$f(\beta,n_1^{(\nu)},n_2^{(\nu)},\ldots) = \sum_\nu f_{id}^{(\nu)}(\beta,\mu_{1,id}^{(\nu)}) + \sum_n f_{id}^{(n)}(\beta,\mu_{2,id}^{(n)} -$$

$$- p_{int}(\beta,\mu_{1,id}^{(\nu)}). \qquad (3.2)$$

$$n_1^{(\nu)} = \sum_p f_1(\varepsilon_1^{\circ},\beta,\mu_{1,id}^{(\nu)})$$

$$n_2^{(n)} = \sum_P g_2(E_{nP}^{\circ},\beta,\mu_{2,id}^{(n)}) \qquad (3.3)$$

Finally we have to minimize the free energy with respect to the division into the different chemical species

$$f(\beta, n_1^{(\nu)}, n_2^{(\nu)}, \ldots) = \min,$$

$$n = \sum_{\nu} n_1^{(\nu)} + \sum_{n} n_2^{(n)} + \ldots \tag{3.4}$$

The relations (3.1 - 3.4) represent the "chemical quasi particle picture" which describes effects as ionization, formation of atoms and molecules, band structure of electronic states etc.

Let us still mention that there exists an alternative approach for the transition to a chemical description [1,8]. This approach is based on a "chemical" reinterpretation of the density (eq. (2.10)) instead of the pressure (eq. (2.12)). Both approaches lead to different expressions for the thermodynamic functions and especially for the mass action laws. For practical calculations these differences are negligible and the experimentator may not care about it. A deeper analysis shows, that the approach based on the reinterpretation of the density [1,8] leads to certain physical difficulties. The reason for the problems arising in reinterpretations of the density is the fact, that a pressure representation may well be splitted into ideal and interaction terms, a total density however can have only ideal contributions (partial densities). In other words a density can consist only of a sum of other densities, the physics requires strict additivity. A reinterpretation based on eq. (2.10) therefore leads to the (incorrect) conclusion that the composites are ideal particles. The reinterpretation of eq. (2.12) is more elastic since interaction terms are admitted. This leads to a picture including also interacting composites.

4. Thermodynamical Effects of Nonideality

Explicite calculations of the fugacity expansions for real plasmas were first given by BARTSCH et al. [9] and were reproduced and discussed in detail in the monograph [4]. It was shown that fugacity expansions give a correct representation of the formation of atoms and molecules in hydrogen plasmas if the fourth virial coefficient is taken into account. Further several exact results for the first orders in the fugacity expansions were given. Another result given in [4] is, that in the fugacity expansion picture the thermodynamical functions are stable. In other words, the thermodynamic stability conditions are never violated. For example the pressure is monotonously increasing with the density. This is a correct result, however, if one is interested in a search for phase transitions one would better like to see VAN DER WAALS wiggles. As well known, such wiggles may appear only in canonical ensemble representations. In fugacity representations one can only guess a phase transition if the pressure becomes a flat function of the pressure in certain density range. The method of fugacity expansions was worked out then by ROGERS who did also a large amount of numerical calculations [10]. However, since ROGERS was interested in moderate densities, he did not consider the problem of phase transitions. Being concentrated on that problem, we shifted in the seventies to mixed representations, i.e. to the chemical picture [11-13]. In this framework we calculated the thermodynamic functions in the chemical picture including nonideality effects. For practical calculations we have developed Pade approximations for the thermodynamic potentials. Instead of going into details we refer only to some references [11-13]. Let us discuss here only the most dramatic nonideality effect, namely the appearance of additional phase transitions due to Coulomb interactions, which was predicted first by Landau, Zeldovich, Norman and Starostin and was studied then in several theoretical and experimental papers [12-21]. This transition is a discrete form of the density ionization observed principally in all matter beyond some density.

We start with the consideration of pure hydrogen plasmas. In the above mentioned theoretical work the thermodynamical stability criteria as the monotonicity of the pressure were checked. It was shown that besides the classical first-order transition typical for neutral gases a second first-order transition may appear [4,11,12]. Along the coexistence line a system undergoing a plasma transition is divided into two phases of different density and different degree of ionization. Our theory yields the estimate for the second critical point of hydrogen

$$C_2 : \quad T_c = 16500 \text{ K} , \quad p_c = 22.8 \text{ GPa} , \quad \rho_c = 0.13 \text{ g cm}^{-3} ,$$

The coexistence line as well as several other characteristic lines were determined. It was shown that around the critical point C_2 a very quick change in the composition occurs. We also note that near the critical point the coexisting gases and liquids are only partially ionized (about 30%). At lower temperatures a molecular gas may coexist with a metallic liquid. One of the most important results obtained so far is the drastic lowering of the pressures for the transition to metallic hydrogen in comparison with the corresponding transition pressure in the solid state, which is typically in the range 200 - 800 GPa. The expected pressures for the transition from liquid molecular hydrogen to liquid metallic hydrogen are ten times smaller than those for the transition from solid molecular hydrogen to solid metallic hydrogen. This may be of interest for laboratory comparisons with adiabatic compression of hydrogen as well as for astrophysical applications. A first estimate for the critical point of hydrogen mixtures as e.g. (H-He systems) is obtained by considering the other components as a neutral (not ionized) solvent with the pressure p_s, the mass density ρ_s and the dielectric constant ε_s. Then the critical data of the mixture are estimated as:

$$T_c^{mix} = T_c \; / \; \varepsilon_s^2 \; ; \quad p_c^{mix} = p_c + p_s \; ; \quad \rho_c^{mix} = \rho_c + \rho_s$$

Following our estimates the metallization of the interior of Jupiter occurs at much lower pressures, i.e. nearer to the surface of the planet, than assumed so far. For laboratory studies of the plasma phase transition adiabatic compression experiments with initial conditions near to the jovian adiabate are required. GRIGORIEV et al. [14] reported about a density step from 1.1 g / cm^3 to 1.3 g / cm^3 at the pressure of about 280 GPa and temperatures of about 10^4 K. Further NELLIS et al. [15] and ROSS et al. [16] reported about two pressure data at the highest densities (about 0.5 g/cm^3) which are decreasing with the density. The experimental findings could well be a first hint to the existence of a plasma phase transition in hydrogen. So far however there are no reliable data which could be considered as an experimental proof. Further theoretical and experimental work is necessary to decide the open question of the existence of a plasma phase transition in hydrogen. More recent theoretical investigations of the phase transition in hydrogen plasmas were performed by SCHLANGES and KREMP [17], SAUMON and CHABRIER [18].

Let us discuss now the case of helium plasmas [13,19-21]. The most clear picture of the ionization is obtained by looking at the density dependence of the mean charge of the ions. At low densities where the nonideality is still negligible we observe the standard behaviour known from solutions of the ideal Saha equation, i.e. one observes first full ionization of both electrons $\bar{z} = 2$, then with increasing density ionic bound states $\bar{z} = 2$. Finally in the region $n \sim 10^{19} \ldots 10^{22}$ cm^{-3} atomic bound states $\bar{z} = 2$ are observed. Any further increase of the density leads to a sudden decrease of the binding energies. Due to this nonideality effect, first the atomic bound states and then the ionic bound states break down leading to full ionization again.

The first calculations for Helium plasmas showed only one plasma phase transition [19]. More accurate calculations showed the splitting of C_2 into two critical points, connected with single and double ionization respectively as shown in recent work by Förster et al. [20]. Our estimate gives the critical data [20,21]

$$C_2': \quad T_c'' \cong 35000 \text{ K} , \quad p_c \cong .7 \text{ TPa}; \qquad C_2'': \quad T_c \cong 120000 \text{ K} , \quad p_c \cong 10 \text{ TPa}$$

Since the calculation of critical points requires very high accuracy of the thermodynamic functions, we consider these data as first estimates only. A more comprehensive discussion of phase transitions is given in [21]

5. References

[1] W.D. Kraeft et al.: Quantum Statistics of Charged Particle Systems. Akademie-Verlag Berlin & Plenum Press New York 1986

[2] T.L. Hill, Statistical Mechanics, Mc Graw Hill, New York 1956

[3] W. Ebeling, Z. phys. Chem. (Leipzig) 240(1969)265; Physica A 73(1974)573

[4] W. Ebeling, W.D. Kraeft, D. Kremp, Bound States and Ionization Equilibrium in Plasmas and in the Solid State,Akademie-Verl. 1976

[5] W.D. Kraeft et al., Ann Physik 45 (1988) 429

[6] L. Hitschke, G. Röpke, Phys. Rev A 37(1988) 4991

[7] K. Kilimann, W. Ebeling, Z. Naturforschung 45a(1990)613

[8] R. Zimmermann, Many Particle Theory of Highly Excited Semi-conductors. Teubner-Verlag Leipzig 1987

[9] G.P. Bartsch, W. Ebeling, Beitr. Plasmaphysik 11(1971)393

[10] F.A. Rogers, Phys. Rev. A 10 (1974) 2441, A 38(1988)5007

[11] W. Ebeling et al., Beitr. Plasmaphysik 10(1970)507
Ann. Physik 28(1973)289, 45(1988)529

[12] W. Ebeling, W. Richert, Phys. Lett. 108A(1985)80

[13] W. Ebeling, Contr. Plasma Physics 29(1989)165; 30(1990)553;

[14] F.V. Grigoriev et al., Zh. Eksp. Teor. Fiz. (USSR) 69(1975)743

[15] W.J. Nellis et al., Phys. Rev. Lett. 48(1982)816

[16] M. Ross et al., J. Chem Phys. 79(1983)1487

[17] D. Saumon et al., J. Chem. Phys. 90(1989)7395

[18] D. Saumon, G. Chabrier, Phys. Rev. A 15(1991) in press

[19] W. Ebeling, In: Inside the Sun, Kluwer Dordrecht 1989

[20] A. Förster et al., ICPIG XX, Contributed Papers Vol. 2(1991)385

[21] W. Ebeling et al., Thermophysical properties of Hot Dense Plasmas Teubner-Verlag Stuttgart-Leipzig 1991

A SIMPLE STATISTICAL MECHANICAL MODEL FOR PRESSURE INDUCED IONIZATION

Ph. A. Martin

Institut de Physique Théorique

Ecole Polytechnique Fédérale de Lausanne. PHB-Ecublens, CH- 1015 Lausanne

Switzerland

Abstract

A purely static characterisation of "atomic" versus "ionized" states is given in terms of the spectral properties of the pair reduced density matrix. These spectral properties can be analyzed in a simple model (a "short range electron" with classical "hard core protons"). In the model the Mott effect occurs at high density as a consequence of the reduction of particle fluctuations.

I. Introduction

In these proceedings, I report on the motivations and content of the work developed in ref. [1].

Usually, a pressure induced ionization phenomenon is signalled in a many-particle system by the occurence of a Mott transition. A Mott transition is observed when some effective two-body (density and temperature dependent) hamiltonian $H_{eff}(\rho, \beta)$ looses its boundstates as the density ρ increases ($\beta = (k_B T)^{-1}$ is the inverse temperature). The effective hamiltonian can be obtained in various approximation schemes. The simplest example is given by an electron in a Debye potential with screening length λ_D : when λ_D becomes of the order of the Bohr radius, all eigenvalues merge into the continuum. More refined effective two-body hamiltonians exhibiting the Mott effect have been derived by means of a decoupling of the hierarchy equations for the Green's functions of the many-particle system (see [2] and references quoted therein). In all these approaches, the Mott effect can only be analyzed after the use of some approximation procedure to produce a tractable $H_{eff}(\rho, \beta)$. Although the results obtained in this way are physically sensible, and often very good for practical purposes, it is legitimate to investigate, as a matter of principle, if the concept of Mott transition can be defined intrinsically in the many-body system without the a priori recourse to such approximation procedures. This is the question addressed in ref. [1].

Consider an electron-proton gas in thermal equilibrium. Then the pair reduced density matrix $\rho_2(x_e, X_p \mid y_e, Y_p)$, where x_e, y_e and X_p, Y_p denote the electron and proton coordinates respectively, embodies all informations on electron-proton pairs, as far as pure equilibrium properties are concerned. Following a proposal of Girardeau [3], we identify "atomic" states in terms of eigenvalues and eigenfunctions of $\rho_2(x_e, X_p \mid y_e, Y_p)$. More precisely, we write $\rho_2(x_e, X_p \mid y_e, Y_p) = \rho_2(x, y \mid r-s)$ as a function of the relative $x = x_e - X_p$, $y = y_e - X_p$ and center of mass coordinates $r = \dfrac{m x_e + M X_p}{m + M}$, $s = \dfrac{m y_e + M Y_p}{m + M}$ (m = electron's mass, M = proton's mass). Because of translation invariance, ρ_2 depends only on the difference $r-s$. For a given wave number q, we define

$$\tilde{\rho}_2(x, y \mid q) = \int dr \, e^{iqr} \, \rho_2(x, y \mid r) \tag{1}$$

This quantity is the electron-proton pair reduced density matrix in terms of relative coordinates and with a fixed momentum $\hbar q$ of its center of mass. Clearly, $(\tilde{\rho}_2(x, y \mid q))^* = \tilde{\rho}_2(y, x \mid q)$ is a symmetric positive definite kernel, and for each q, we may ask for solutions of the eigenvalue equation

$$\int dy \, \tilde{\rho}_2(x, y \mid q) \, \chi_{vq}(y) = \lambda_{vq} \, \chi_{vq}(x) \tag{2}$$

with localized (square integrable) wave function $\chi_{vq}(x)$. The positive numbers λ_{qv} are interpreted as occupation numbers for "atomic" states $\chi_{vq}(x)$; $\tilde{\rho}_2(x, y \mid q)$ can also have a continuous part in its spectrum corresponding to extended or "ionized" states. We will say that a Mott transition occurs at density ρ_c and inverse temperature β^{-1} if (2) has no more localized solutions, i.e. if all eigenvalues merge into the continuum. To motivate this definition, we observe that in the low density limit

$$\tilde{\rho}_2(x, y \mid q) \approx \rho^2 \exp\left[-\beta \frac{\hbar^2 |q|^2}{2(m + M)}\right] (x \mid \exp(-\beta H_{e\text{-}p} \mid y) \tag{3}$$

where $H_{e\text{-}p}$ is the hydrogen atom hamiltonian. Then the solutions of (2) are the hydrogen atom bound states with energies E_v and

$$\lambda_{vq} \approx \rho^2 \exp\left(-\beta \left[\frac{\hbar^2 |q|^2}{2(m + M)}\right] + E_v\right) \tag{4}$$

are the corresponding Boltzmann weights. Thus, in this limit, the spectrum of $\tilde{\rho}_2(x, y \mid q)$ will be proportional to the exponential of that of the hydrogen atom, i.e. it has a full interval $[0, \Sigma]$ corresponding to ionized states plus a number of eigenvalues above it, like in the left part of Fig. 1. The definition (2) extrapolates this picture at higher densities, and the problem is to study the spectrum of $\tilde{\rho}_2(x, y \mid q)$ as the density varies and decide for the possible disappearance of eigenvalues. Notice that since the pair reduced density matrix ρ_2 is positive it would be quite natural to formally define a two-body effective hamiltonian by writing $\rho_2 = \rho^2 \exp(-\beta H_{eff}(\rho, \beta))$ where $H_{eff}(\rho, \beta)$ reduces to the hydrogen hamiltonian as $\rho \to 0$. The point is that we prefer a direct study of the spectral properties of ρ_2 without resorting to an approximate calculation of H_{eff} (ρ, β) defined in this way. This will also enable us to establish a link between the occurence of a Mott transition (as it is defined here) and the fluctuation properties in the system.

II. Models

In the present situation, we do not have a sufficient mathematical control on the electron-proton pair reduced density matrix, and we have to test the above ideas in very simplified models, hoping that they capture some of the main features. We consider a single quantum particle (the "electron") in thermal equilibrium with a gas of classical particles (the "protons"). Each "proton" is the source of a short range attractive potential V which can bind the "electron". So the hamiltonian of the "electron" in a given configuration $r_1, ..., r_n$ of protons is

$$H\,[r_1, ..., r_n] = \frac{|p|^2}{2m} + \sum_{j=1}^{n} V\,(x\text{-}r_j) \quad , \quad V(x) \leq 0 \tag{5}$$

where x and p are the quantum mechanical position and momentum of the "electron". In particular, the one particle hamiltonian

$$H[o] = \frac{|p|^2}{2m} + V(x) \tag{6}$$

is supposed to have bound states with negative eigenvalues E_v. The sole interaction between the "protons" themselves is a hard core repulsion of diameter d. If d = 0, the "protons" form a perfect classical gas. In the following, we speak of the hard core model when d > 0 and of the perfect gas model if we set d = 0.

For a given density ρ and temperature β^{-1} of the "proton" gas we introduce

- the "electron" density matrix $\rho_1(x, y)$ (with "electron" coordinates x, y)

- the pair density matrix $\rho_2(x, y \,|\, 0)$ of the "electron" and one "proton" at the origin
  r = 0.

These quantities are defined as usual by integrating out in the Gibbs distribution all "proton" coordinates for $\rho_1(x, y)$ and all but one for $\rho_2(x, y \,|\, 0)$. Because of translation invariance, $\rho_1(x, y) = \rho_1(x\text{-}y, 0)$ and its Fourier transform $\tilde{\rho}_1(k) = \int dx e^{ikx} \rho_1(x, 0)$ is the kinetic energy distribution of the "electron" in the gas. Thus the spectrum of $\rho_1(x, y)$ is continuous and spans the interval [0, σ], $\sigma = \sup_k \tilde{\rho}_1(k)$. For a free electron we would have

$$\tilde{\rho}_1(k) = \exp\left(-\beta \frac{\hbar^2 |k|^2}{2m}\right)$$ and $\sigma = 1$. We can now state our results on the pair density matrix $\rho_2(x, y \mid 0)$.

III. Results

We prove the following exact results.

A. General properties (hard core and perfect gas models)

For any ρ and β the spectrum of $\rho_2(x, y \mid 0)$ consists of

- a continuous part spanning the interval $[0, \Sigma]$ where $\Sigma = \rho \, \sigma$ is the threshold of the continuous spectrum and $\sigma = \tilde{\rho}_1(0)$.

- a finite number of eigenvalues λ_v (if any) all greater than $\Sigma = \rho \sigma$.

B. Low density (hard cores and perfect gas model)

For fixed β and ρ small, the discrete spectrum is not empty. The eigenvalues λ_v satisfy $\lambda_v \approx \rho \sigma e^{-\beta E_v}$, $\rho \to 0$ where E_v are the eigenvalues of the one particle hamiltonian (6). The results A and B establish qualitatively the picture in the low density part of the Fig. 1 : as ρ increases we observe a smooth deformation of the spectrum of an isolated atom. In the high density regime, the situation is different for the two models.

C. High density

Hard core model : for fixed β and ρ large enough the discrete spectrum is empty.

Perfect gas model : for fixed β and ρ large enough there exists always at least one eigenvalue above Σ.

We see that the hard core model exhibits the Mott effect : there exists a critical density ρ_c above which all eigenvalues have merged into the continuum, as indicated in the high density regime of Fig. 1. On the contrary, if the gas of "protons" is perfect, eigenvalues persist at high density, and it can be shown that this will be the case even for an arbitrarily weak potential V.

These two different behaviors at high density can be interpreted as follows. In a system of hard cores the variance of the number of particles in a given region tends to zero as the density increases, so the "proton" configurations become uniform, and the "electron" sees an essentially constant potential : hence it becomes delocalized and bound states disapear. On the other hand, in a perfect gas, the variance of the particle number increases linearly with the density. Typical configurations are rather inhomogeneous, which corresponds on a microscopic scale to large variations in the potential seen by the "electron". Thus the "electron" can always bind with a "proton" or a cluster of "protons" and eigenvalues corresponding to localized states persist at high density. It would be of interest to substanciate this interpretation by a dynamical study of the "electron" wave function interacting with heavy (quasi classical) "protons".

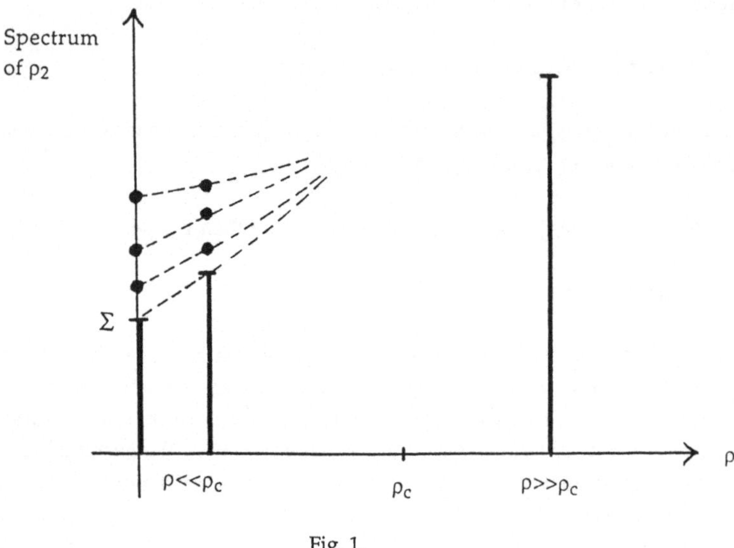

Fig. 1.

IV. Methods of proofs

In [1], we mimic a classical system of hard cores by a "cell model of a fluid" [4]. In the cell model one divides the space into a lattice of cells of linear dimension d, and allows each cell to be occupied by at most one particle with a uniform probability distribution. The main idea is to study ρ_2 as a perturbation of ρ_1 by writing

$$\rho_2 = \rho\, \rho_1 + \rho_T \tag{7}$$

where $\rho_T \equiv \rho_2 - \rho\, \rho_1$ is the truncated pair density matrix. Here, we have suppressed the arguments x, y in ρ_1, ρ_2 and ρ_T, considering these density matrices as operators acting on the electronic wave functions. Then, one can proceed in close analogy with the usual Schrödinger problem where the kinetic energy $H_0 = |p|^2/2m$ is perturbed by a potential V(x). We recall that ρ_1 gives the kinetic energy distribution of the "electron" in the fluid, so it plays the role of the kinetic term in (7) as does H_0 in the Schrödinger equation. The truncated pair density matrix ρ_T embodies the effects of the interactions of the "electron" in the fluid, thus it plays the role of a perturbation in (7). In the Schrödinger problem, it is well known that a short range potential does not modify the location of the continuous spectrum. In the same way, we show that ρ_2 and $\rho\,\rho_1$ have the same continuous spectrum. This follows from the fact that ρ_T is a trace-class operator and from the stability theorem for the absolutely continuous spectrum under trace-class perturbations [5], thus proving the result A. For B, we consider the first term in the low density expansion of ρ_2 (as in (3))

$$\rho_2 = \rho\left(e^{-\beta H[o]} + O(\rho)\right) \tag{8}$$

By the regular perturbation theory of eigenvalues, we conclude that the eigenvalues $e^{-\beta E_v}$ of $e^{-\beta H[o]}$ are only slightly displaced at low density.

To obtain the high density result (C) for the hard core model, we observe that as a consequence of the small particle fluctuations, ρ_T is also a small perturbation of $\rho\,\rho_1$. It is known that in the three dimensional Schrödinger problem, a very weak potential cannot create bound states. The same is true here (one uses the Birman Schwinger estimate on the number of eigenvalues and show that it becomes vanishingly small at high density). In the perfect gas model, we can exhibit a variational wave function φ such that $(\varphi, \rho_2\,\varphi) > \Sigma$ (Σ = threshold of continuum), so that the discret spectrum cannot be empty at high density. I refer to [1] for detailled proofs of these assertions.

V. The binary collision approximation

In order to make plausible the different high density behaviors (C) of the hard core and perfect gas models, let us apply to them the binary collision approximation in the standard formalism of time dependent Green's functions. It is expedient here to consider an assembly of independent "electrons" (in second quantized formalism) interacting with classical "protons" as defined in the models of section II. Let $\psi(x, t)$ be the Fermi electron field at time t evolving in a fixed configuration of "protons" located at $r_1, r_2, \ldots, r_n$. According to (5) we have

$$i\hbar \frac{\partial}{\partial t}\psi(x, t) = -\frac{\hbar^2}{2m}\nabla^2\,\psi(x, t) + \sum_{j=1}^{n} V(x-r_j)\,\psi(x, t) \tag{9}$$

The "protons", being static, are entirely characterized by the classical number density

$$N(R) = \sum_{j=1}^{n} \delta(R-r_j) \tag{10}$$

Then the Green's function for one "electron" and k "protons" is defined by

$$i\, G_{k+1}(xt, x't', R_1, \ldots, R_k) = \langle T(\psi(x, t)\, \psi^*(x', t'))\, N(R_1) \ldots N(R_k)\rangle \tag{11}$$

where $\langle \cdots \rangle$ denotes the equilibrium average with respect to the hamiltonian (5), and T is the time ordering operation. We obtain from (9) and (11) the following equation which couples the Green's functions for $k = 1$ and $k = 2$

$$i\hbar\, \frac{\partial}{\partial t}\, G_2(xt, x't', R_1) + \frac{\hbar^2}{2m}\, \nabla^2\, G_2(xt, x't', R_1) + \int dR_2\, V(x{-}R_2)\, G_3(xt, x't', R_1, R_2)$$

$$= \rho\, \delta(x{-}x')\, \delta(t{-}t') \tag{12}$$

with $\rho = \langle N(R)\rangle$ the density of the classical "protons".

We can single out in G_3 the singular contribution at $R_1 = R_2$ (coming from the self part in $N(R_1)\, N(R_2) = \delta(R_1{-}R_2) \sum_i \delta(R_1{-}r_i) + \sum_{i \neq j} \delta(R_1{-}r_i)\, \delta(R_2{-}r_j))$

$$G_3(xt, x't', R_1, R_2) = \delta(R_1{-}R_2)\, G_2(xt, x't', R_1) + G_3^{(r)}(xt, x't', R_1, R_2) \tag{13}$$

where the regular part $G_3^{(r)}$ involves only pairs of distinct "protons". Since the latter cannot be closer than d in the hard core model, we have

$$G_3^{(r)}(xt, x't', R_1, R_2) = 0\ , \qquad |R_1{-}R_2| \leq d \tag{14}$$

For R_1 fixed and $|R_1{-}R_2| > d$, we approximate $G_3^{(r)}$ by the factorized form

$$G_3^{(r)}(xt, x't', R_1, R_2) = \rho\, G_2(xt, x't', R_1)\ , \qquad |R_1{-}R_2| > d \tag{15}$$

Introducing (13), (14) and (15) in (12), we find that the Green's function $G_2(xt, x't', R_1)$ for the "electron-proton" pair obeys a closed equation of motion corresponding to the effective hamiltonian (setting $R_1 = 0$)

$$H_{eff} = -\frac{\hbar^2}{2m}\, \nabla^2 + V(x) + \rho \int_{|R| \geq d} dR\, V(x{-}R)$$

$$= -\frac{\hbar^2}{2m}\, \nabla^2 + V_{eff}(x) + V_0 \tag{16}$$

with

$$V_{eff}(x) = V(x) - \rho \int_{|R| \leq d} dR\, V(x{-}R)\ , \qquad V_0 = \rho \int dx\, V(x) \tag{17}$$

The negative constant V_0 produces only a global shift of the spectrum. When ρ is small $V_{eff}(x)$ is close to $V(x)$ and H_{eff} has, up to small displacements, the same bound states as those due to a single "proton". On the other hand, writing $\rho = \eta\,\rho_c$ where η is the fraction of the close-packing density ρ_c ($\rho_c \geq d^{-3}$), one has (we recall that $V(x) \leq 0$)

$$V_{eff}(x) \geq V(x) - \eta\,\frac{1}{d^3} \int\limits_{|R| \leq d} dR\; V(x-R)$$

$$\approx \left(1 - \frac{4\pi}{3}\eta\right) V(x), \qquad d \to 0 \tag{18}$$

This shows that $V_{eff}(x)$ can be made arbitrarily small, or even positive, by taking d sufficiently small and η close to 1. Hence the bound states disapear at high density.

If there is no hard core (perfect gas of "protons"), we set $d = 0$ in (17) and obtain $V_{eff}(x) = V(x)$. Thus, up to the shift V_0, H_{eff} has the same spectrum as in the single "proton" case and bound states remain at all densities. This qualitatively accounts for the exact results stated in section III C.

References

[1] J. L. Lebowitz, N. Macris, Ph. A. Martin : "Atomic versus Ionized States in Many Particles Systems and the Spectra of Reduced Density Matrices : a Model Study", Rutgers University, preprint (1991), to appear in J. Stat. Phys.

[2] W. D. Kraeft, D. Kremp, K. Kilimann, H. E. DeWitt, Phys. Rev. A **42**, 2340 (1990)

[3] M. D. Girardeau, Phys. Rev. A **41**, 6935 (1990).

[4] J. M. H. Levet, E. G. D. Cohen, "Studies in Statistical Mechanics", Vol. II, Ed. de Boer and G. E. Uhlenbeck, chap. 4, North Holland (1964).

[5] T. Kato, "Perturbation Theory for Linear Operators", Springer Verlag (1984).

Single and Two Particle Properties in Dense Plasmas

B.Strege and W.D. Kraeft
Fachbereich Physik der Universitaet Greifswald
Domstr. 10 a, O-2200 Greifswald, Germany

Abstract

It is shown how single and two particle properties are used for the description of the behaviour of dense plasmas. Numerical values for the self energy are given.

1   Introduction

The macroscopic behaviour of (dense) plasmas may most effectively be described using single and two particle properties. For instance ,the SAHA equation contains effective two particle bound state energies which are density dependent and thus quite different from the bound state energies of an isolated atom. Also the continuum edge which is given by the sum of two single particle energies for zero momentum is strongly density dependent. Thus the effective ionization , i.e., the difference between the bound state energy and the continuum ,can tend to zero; this fact leads to the phenomenon of vanishing bound states which is referred to as the Mott effect .See, i.e., [1]. So the ionization equilibrium is essentially gouverned by the few particle properties mentioned. Another example in which single and two particle properties are of high interest is the kinetic theory. Here it is quite usual to consider the behaviour of interacting quasiparticles, i.e. single particles the properties of which are determined by the surrounding medium. Again, the two particle energies play a role if ionization and recombination kinetics is dealt with. See [2], this volume. Also the equation of state is successfully described in terms of single and two particle properties; here bound and scattering states of pairs are of interest. For the reasons given, it is rather essential to have an exact knowledge of certain few particle properties. In section **2** we will outline briefly the methods and will give analytic expressions, and section 3 presents certain results for single particle properties.

2   Basic equations and approximations

The application of the Green's function theory turns out to be very appropriate for the description of the single and two particle behaviour. For the introduction into the theory the reader is referred to the literature, eg.[1] and [3]. In the usual way, the single and two particle Green's functions, which give, of course the properties we are looking for, are determined as the grand canonical ensemble averages of time ordered products of annihilation and creation operators,

$$G_1(11^*) = \frac{1}{i\hbar} < T\{\psi(1)\psi^+(1^*)\} >, \tag{1}$$

$$G_2(121^*2^*) = \frac{1}{(i\hbar)^2} < T\{\psi(1)\psi(2)\psi^+(2^*)\psi^+(1^*)\} > . \tag{2}$$

The equation for the determination of the single particle Green's function is given by the Dyson equation

$$G_1(1\bar{1}) = G_1^0(1\bar{1}) + \int_0^{-i\hbar\beta} d1^* d2 G_1^0(11^*)\Sigma(1^*2)G_1(2\bar{1}). \tag{3}$$

The self energy Σ replaces the coupling to the next higher order Green's function and is defined by

$$\int d\bar{1}\,\Sigma(1\bar{1})G_1(\bar{1}1^*) = i\hbar \int d2\, V(12)G_2(121^*2^+), \tag{4}$$

where

$$2^+ = (r_2, t_2 + i\epsilon), \quad \epsilon \to +0.$$

From (4) we see that approximations of G_2 are intimately connected with those of σ. A simple but physically extremely meaningful approximation to Σ is the screened-potential approximation, i.e.,

$$\overline{\Sigma}_a(1\bar{1}) = i\hbar V_{aa}^s(1\bar{1}G_a(1\bar{1}). \tag{5}$$

where V_{aa}^s is the dynamically screened potential which reads in Matsubara Fourier representation

$$V_{aa}^s(k,\Omega_\nu) = \frac{V_{aa}(k)}{\epsilon(k,\Omega_\nu)}, \tag{6}$$

where $\epsilon(k,\Omega_\nu)$ are the corresponding Fourier coefficients of the dielectric function which will be used later. If we want to include the influence of bound states into the self energy Σ we should use a G_2 in (4) which includes all ladder type diagrams (see, e.g., [1]).
If we want to discuss the two particle properties, we have to consider in any case the ladder equation for G_2 , which we write in a representation using momenta and Matsubara frequencies $z_\nu^a = \frac{\pi\nu}{-i\beta} + \mu_a$, $\Omega_\lambda^{ab} = \frac{\pi\lambda}{-i\beta} + \mu_a + \mu_b$, μ_a - chemical potential, ν - odd , λ - even [1]

$$\begin{aligned}
G_2(121^*2^*; z_\nu, \Omega_\lambda - z_\nu) &= G_1(1, z_\nu)G_1(2, \Omega_\lambda - z_\nu) * \\
&\quad * \left(\delta_{11^*}\delta_{22^*} + \int d1^{**} d2^{**} K(121^{**}2^{**}, \Omega_\lambda)G_2(1^{**}2^{**}1^*2^*\Omega_\lambda)\right).
\end{aligned} \tag{7}$$

The generalized two particle potential K is approximately replaced by the dynamically screened potential V^s according to (6) with ϵ being, e.g., the random phase dielectric function to be used later. A more general dielectric function was derived by Roepke and Der, see [1]. The V^s-approximation for Σ reads explicitly (omitting the Hartree-Fock contribution)

$$\begin{aligned}
Re\Sigma_a^{corr}(k,\omega) &= -\hbar \int \frac{dp^*}{(2\pi)^3} P \int_{-\infty}^{+\infty} \frac{d\omega^*}{2\pi} 2V_{aa}(p^*) * \\
&\quad * \frac{1 - f(p^* + k + n_B(\omega^*))Imc^{-1}(p^*\omega^* + i0)}{\hbar(\omega - \omega^*) - \frac{\hbar^2}{2m_a}(\vec{p^*} + \vec{k})^2}.
\end{aligned} \tag{8}$$

In the nondegenerate case we neglect Fermi functions as compared to unity, and we have for the simplest solution of the dispersion relation, $\omega = \frac{\hbar k^2}{2m_a}$,

$$Re\,\Sigma_a^{corr}\left(k,\frac{\hbar k^2}{2m_a}\right) = -\int\frac{d\vec{p}}{(2\pi)^3}P\int_{-\infty}^{+\infty}\frac{d\hbar\omega}{2\pi}\;\frac{2ImV_{aa}^s(p\omega)n_B(\omega)}{\hbar\omega+\frac{\hbar^2 k^2}{2m_a}-\frac{(\vec{k}+\vec{p})^2}{2m_a}}. \tag{9}$$

This equation will be the starting point for our numerical discussion given in the next section. If we consider the nondegenerate case for the two particle problem, too, we get after a lengthy calculation among which the SHINDO-approximation leads to an essential simplification [1] for the two particle Green's function

$$\left(\frac{p_1^2}{2m_a}+\frac{p_2^2}{2m_b}+\Delta_{ab}-z_\lambda^{ab}\right)G_{ab}(p_1...p_4 z_\lambda^{ab})\;+ \tag{10}$$

$$\int\frac{dq}{(2\pi)^3}V_{ab}^{eff}(qp_1p_2z_\lambda^{ab})G_{ab}(p_1-q,p_2+q,p_3p_4z_\lambda^{ab})\;=$$

$$=\;-\left(\delta(p_1p_3)\delta(p_2p_4)-\delta_{ab}\delta(p_1p_4)\delta(p_2p_3)\right)\frac{1}{-i\hbar\beta}\sum_\nu\;\left(G_a(p_1z_\nu^a)+G_b(p_2z\lambda^{ab}-z_\nu^a)\right).$$

In this equation, many particle effects are included in the self energy Δ_{ab}, in the effective potential which is more general than the dynamically screened one (eq.(6)), and in the phase space occupation factor (last bracket of (10)).
The quantities Δ_{ab} and V_{ab}^{eff} are intimately connected with self energies [1] to be discussed in section 3.
New results for the eigenvalues of eq.(10) where presented in the poster sessions of the Gosen meeting to which this volume is devoted, and will be published elsewhere [4]. The case of the static screening was dealt with, again, in [5]. In [6] it was shown that it is possible to use a pertubation procedure to solve an equation of the type (10) if one has a static potential as a reference model which gives results near to the desired ones.

3 Numerical evaluation

The interest in a more detailed knowledge of the self energy for small momenta is connected with the fact that one needs in applications expressions simple enough for simple handling. For this reason, one often uses the rigid shift approximation, [7] and [8] and references quoted therein. However, we want to point out that the application of a rigid shift which corresponds to certain mean value with respect to the momenta does not represent the self energy for zero momentum. In contrast, for an electron ion system, the zero momentum self energy of electrons has very large value at small momenta deviating strongly from mean values. In [1], [8], the expansion for small momenta of the self energy is given, namely

$$Re\,\Sigma_a^{corr}\left(k,\frac{\hbar k^2}{2m_a}\right) =$$

$$= -\frac{e_a^2}{4\pi\epsilon r_D}\left(\frac{1}{2}+\frac{1}{2}\sum_c\frac{m_c}{2m_a}-\frac{\pi}{32}\left(\sum_c\frac{1}{2}\left(\frac{m_c}{m_a}\right)^{\frac{1}{2}}\right)^2\right)$$

$$+\frac{\hbar^2 k^2}{2m_a}\frac{e_a^2}{4\pi\epsilon_0 k_B T r_D}\left(\frac{1}{12}\sum_c\frac{m_c^2}{2m_a^2}-\frac{1}{16}\sum_c\left(\frac{m_c}{2m_a}\right)^2-\frac{\pi}{48}\left(\sum_c\frac{1}{2}\left(\frac{m_c}{m_a}\right)^{\frac{1}{2}}\right)^2\right.$$

$$+\frac{1}{6}\sum_c\frac{m_c}{2m_a}-\frac{\pi}{64}\left(\sum_c\frac{1}{2}\left(\frac{m_c}{m_a}\right)^{\frac{3}{2}}\right)\left(\sum_c\frac{1}{2}\left(\frac{m_c}{m_a}\right)^{\frac{1}{2}}\right)$$

$$\left.-\frac{\pi^2}{64}\frac{5}{16}\left(\sum_c\frac{1}{2}\left(\frac{m_c}{m_a}\right)^{\frac{1}{2}}\right)^4\right)+\cdots. \tag{11}$$

In [8], an evaluation of eq.(9) was carried out for an electron system only. The limiting results are in agreement with expression (11) for equal masses. The dielectric function involved in (9) was taken in the analytic shape derived in [9], see also [1],

$$Re\epsilon(X,Y) = 1 + \frac{K^2}{4X^2} \sum_a \sum_{j=0}^{1} (-1)^j \left(\delta_a \frac{Y}{X^2} + (-1)^j \right) *$$

$$_1F_1 \left(1, \frac{3}{2}; -\frac{1}{4\delta_a} \left(\delta_a \frac{Y}{X^2} + (-1)^j \right)^2 K^2 X^2 \right); \tag{12}$$

$$Im\epsilon(X,Y) = (\frac{\pi}{16})^{\frac{1}{2}} \frac{K}{X^3} \sum_{j=0}^{1} \delta_a^{\frac{1}{2}} (-1)^j exp \left(-\frac{1}{4\delta_a} \left(\delta_a \frac{Y}{X^2} + (-1)^j \right) K^2 X^2 \right); \tag{13}$$

$$X^2 = \frac{\hbar^2 k^2}{2m_e} \frac{1}{\hbar\omega_{pl}^e},$$

$$Y = \frac{\hbar\omega}{\hbar\omega_{pl}^e},$$

$$K = \left(\frac{\hbar\omega_{pl}^e}{k_B T} \right)^{\frac{1}{2}},$$

$$\delta_a = \frac{m_a}{m_e}.$$

In [8], the plasmon poles occuring in the integrand of (9) via $\epsilon^{-1}(k,\omega)$ were carefully dealt with using sum rules. In the discussion of the electron-ion system we used another set of dimensionless variables, namely

$$X = \frac{p}{\kappa_D}; \quad Y = \frac{\omega}{\omega_{pl}^e}; \quad Y_0 = \left(\frac{\hbar p^2}{2m_e} + \frac{\hbar kp}{m_e} \cos\vartheta \right) \frac{1}{\omega_{pl}^e}.$$

Here, p is the wave number of integration, and k is the wave number occuring in the self energy. In contrast to [8], in eq.(14) first the frequency integration is carried out. For this purpose we write

$$\Sigma_e^{corr} \left(k, \frac{\hbar k^2}{2m_a} \right) = -\frac{m_e e^2}{\pi\hbar k} \int_0^{+\infty} \frac{dp}{p} \int_{\frac{\hbar p^2}{2m_e} - \frac{\hbar kp}{m_e}}^{\frac{\hbar p^2}{2m_e} + \frac{\hbar kp}{m_e}} d\omega_0 f(\omega_0); \tag{14}$$

here

$$\hbar\omega_0 = \frac{\hbar^2 p^2}{2m_e} + \frac{\hbar^2 \vec{k}\vec{p}}{m_e}$$

and

$$f(\omega_0) = \frac{1}{\pi} P \int_{-\infty}^{+\infty} \frac{d\omega n_B(\omega) Im\epsilon^{-1}(p,\omega)}{\omega - \omega_0}. \tag{15}$$

Taking into account the analytic properties of $n_B(\omega)$ and $\epsilon^{-1}(p,\omega)$ as well, one may use residuum rules in order to bring (15) into a shape which is appropriate for numerical purposes.

Numerical results were derived for the integrand of (14) after carrying out the integration over angles. In fig.1 we see, for fixed modulus of momentum of integration and fixed external momentum, the occurrance of two plasmon peaks for elektrons only and the singular behaviour of single particle peaks of the self energy.

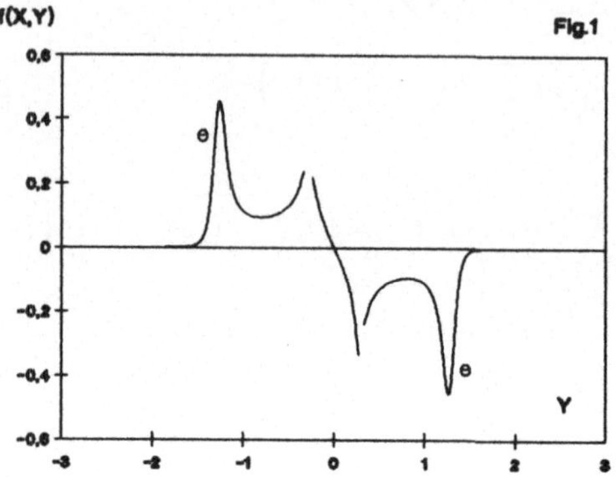

Figure 1: Integrand of (15) for an electron plasma; $T = 10^4 K, n = 10^{16} cm^{-3}$; X, Y see text, $X = 0.4$

In contrast to this behaviour, also proton-plasmon peaks occur in a two component plasma, of course at lower frequency (fig.2).

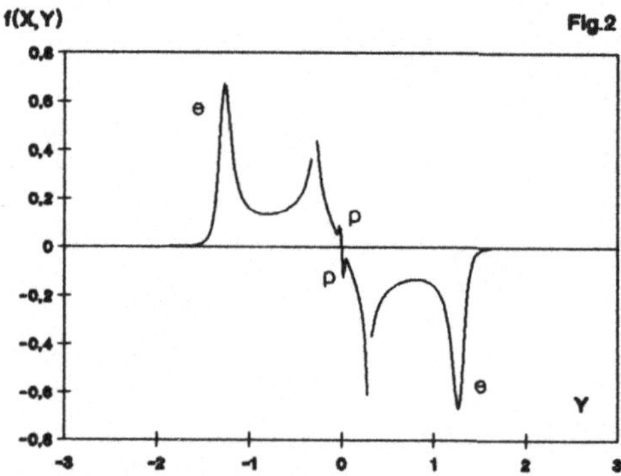

Figure 2: Integrand of (15) for an electron-proton plasma; $T = 10^4 K, n = 10^{16} cm^{-3}$; X, Y see text, $X = 0.4$

The integration over momenta and frequencies shows for electrons only the behaviour known from [8](fig.3). Additionally the Maxwell-Boltzmann distribution is shown to indicate the difference between a mean value (which is $\frac{\kappa e^2}{2}$,see [1], [8]) and the zero momentum value.

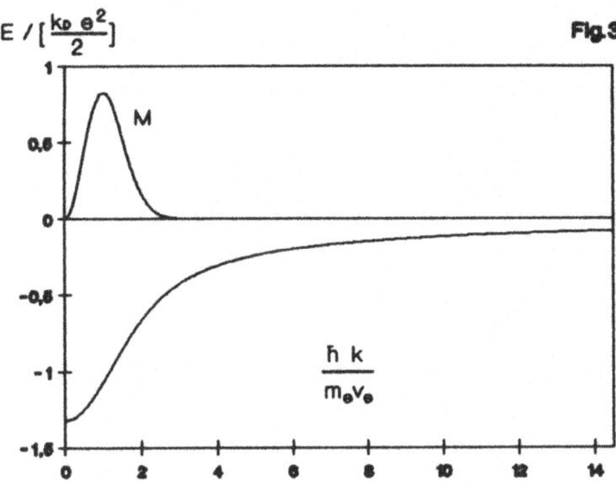

Figure 3: Self energy of an electron in an electron plasma; $T = 10^4 K, n = 10^{16} cm^{-3}$

As pointed out in [8], it could be convenient, to take, instead of a mean value, a negative constant for zero momentum and positive parabolic behaviour using an effective mass. For this reason one could start from the analytical expression (11) in order to get approximations for E_e^0 and m_e^{eff} occuring in the expression for the single particle energy:

$$E_e(k) = -E_e^0 + \frac{\hbar^2 k^2}{2m_e^{eff}}.\tag{16}$$

For small momenta $\hbar k$ such an expression is in agreement with numerical data.
The situation changes drastically if we consider the electron proton system. In agreement with eq.(12), the zero momentum value is very large especially as compared to the mean value which is again $\frac{\kappa e^2}{2}$ (see [1]), what can be estimated already from the position of the Maxwell-Boltzmann curve with respect to the momentum behaviour of the self energy (see fig.4).
Another feature of the self energy is of interest. In contrast to certain estimations which yield $\lim_{k\to\infty} \Sigma(k) \sim \frac{1}{k^2}$ (see, e.g.Kudrin [10]), we found analytically the asymptotic behaviour to be

$$\lim_{k\to\infty} \Sigma_e^{corr}\left(k, \frac{\hbar k^2}{2m_e}\right) = -\frac{c^2 \kappa_D}{2} \frac{v_e m_e}{\hbar k}\left(1 + \frac{v_e m_e}{2\hbar k}\right),\tag{17}$$

where v_e is the thermal velocity of electrons.

Acknowledgement

The authors are grateful to Chr. Pratsch for valuable discussions.

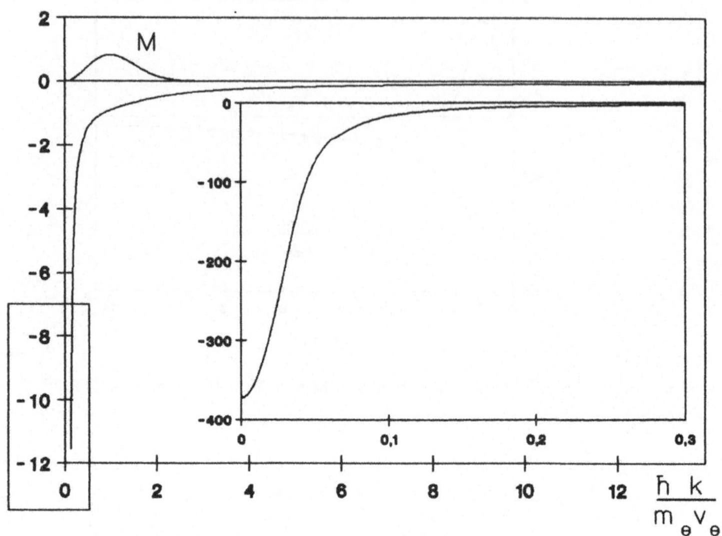

Figure 4: Self energy of an electron in an electron-proton plasma; $T = 10^4 K, n = 10^{16} cm^{-3}$; insert: enlarged scale for small momenta

References

[1] W.D. Kraeft, D. Kremp, W. Ebeling and G. Roepke;
Quantum Statistics of Charged Particle Systems;
Akademie-Verlag, Berlin, 1986; Plenum, London/New York, 1986; Mir, Moscow, 1988.
[2] M. Schlanges, T. Bornath, D. Kremp; this volume
[3] L.P. Kadanoff and G. Baym; Quantum Statistical Mechanics;
Benjamin, New York, 1962
[4] J. Seidel, S. Arndt and W.D. Kraeft; in preparation
[5] W.D.Kraeft, D. Kremp, K. Kilimann, H.E. DeWitt;
Phys. Rev. A 42, 2340 (1990)
[6] S. Arndt, J.Seidel, W.-D. Kraeft, T. Bornath;
Pisa 1991, ICPIG, Contr.Papers
[7] R. Zimmermann;
Many Particle Theory of Highly Excited Semiconductors;
Teubner, Leipzig, 1987
[8] W.D. Kraeft, B. Strege, M.D. Girardeau, Contr.Plasma Phys.30 (1990) 5, 563
[9] Yu.L.Klimontovich, W.D. Kraeft;
High Temp. Physics (in Russian) 12 (1974) 239
[10] L.P.Kudrin;
Statistical Physics of Plasmas (in Russian) Nauka, Moscow 1974

SECOND-MOMENT SUM RULES FOR CORRELATION FUNCTIONS IN A CLASSICAL IONIC MIXTURE

L.G. Suttorp

Institute for Theoretical Physics, Valckenierstraat 65,
1018 XE Amsterdam, The Netherlands

Abstract

The complete set of second-moment sum rules for the correlation functions of arbitrarily high order describing a classical multi-component ionic mixture in equilibrium is derived from the grand-canonical ensemble. The connection of these sum rules with the large-scale behaviour of fluctuations in an ionic mixture is pointed out.

1 Introduction

The equilibrium correlations in classical Coulomb systems are dominated by screening effects, as has first been demonstrated by Debye and Hückel for dilute systems. These screening effects give rise to sum rules that determine the first few moments of the correlation functions. An example of such a sum rule is the second-moment rule obtained by Stillinger and Lovett [1]. For a one-component plasma it gives an exact expression for the second moment of the pair correlation function. If the system is a mixture of several species of charged particles, the Stillinger-Lovett rule contains a weighted sum of partial pair correlation functions for each of the species. A systematic treatment of second-moment sum rules for both two- and three particle correlation functions of ionic mixtures has been given a few years ago [2], [3]. As shown in these papers generalized Stillinger-Lovett sum rules, with different weights entering the summation over the species, can be formulated as well. These generalized sum rules are indispensable in deriving fluctuation formulas for an ionic mixture.

Recently, second-moment sum rules for correlation functions of arbitrarily high order have been discussed for the special case of a one-component plasma [4], [5]. It is the purpose of the present paper to generalize these results and derive the complete set of second-moment sum rules for the higher-order correlation functions of general multi-component ionic mixtures. Moreover, we wish to establish the connection of these sum rules with the large-scale behaviour of the fluctuations in ionic mixtures. An important tool in our derivation will be a set of differentiation formulas that determine the partial derivatives of average physical quantities with respect to the thermodynamic variables characterizing an ionic mixture in equilibrium.

2 Grand-canonical ensemble

A multi-component ionic mixture with s species of charged particles in a neutralizing background is described by the Hamiltonian

$$
\begin{aligned}
H = & \sum_{\sigma\alpha} \frac{p_{\sigma\alpha}^2}{2m_\sigma} + \frac{1}{2} \int^V d\mathbf{r} \int^V d\mathbf{r}' v(|\mathbf{r}-\mathbf{r}'|) \Bigg[\sum_{\sigma_1\alpha_1,\sigma_2\alpha_2}' e_{\sigma_1} e_{\sigma_2} \delta(\mathbf{r}-\mathbf{q}_{\sigma_1\alpha_1}) \delta(\mathbf{r}'-\mathbf{q}_{\sigma_2\alpha_2}) \\
& - q_v \sum_{\sigma_1\alpha_1} e_{\sigma_1} \delta(\mathbf{r}-\mathbf{q}_{\sigma_1\alpha_1}) - q_v \sum_{\sigma_2\alpha_2} e_{\sigma_2} \delta(\mathbf{r}'-\mathbf{q}_{\sigma_2\alpha_2}) + q_v^2 \Bigg] \quad .
\end{aligned}
\tag{2.1}
$$

The phase space of the system is determined by the positions $\mathbf{q}_{\sigma\alpha}$ and the momenta $\mathbf{p}_{\sigma\alpha}$ of the particles, which are labeled by a double index $\sigma\alpha$, with σ indicating the species. The charges and masses of the particles of species σ are e_σ and m_σ, respectively. The interaction between the particles is given by the Coulomb potential $v(r) = 1/(4\pi r)$. Self-interactions are excluded, as indicated by the prime at the first summation sign between the square brackets. The charge density $-q_v$ of the neutralizing background is uniform throughout the volume V of the system.

Equilibrium averages can be evaluated with the use of the grand-canonical ensemble the weight function of which is

$$\rho_{gr} = Z_{gr}^{-1} \left[\prod_\sigma N_\sigma! h^{3N_\sigma} \right]^{-1} \exp(\beta \sum_\sigma \mu_\sigma N_\sigma - \beta H) \quad , \tag{2.2}$$

with β the inverse temperature, μ_σ the chemical potentials, and Z_{gr} the grand-canonical partition function. Since H depends on q_v, the weight function depends on a thermodynamically overcomplete set of $s + 2$ variables, namely β, $\{\beta\mu_\sigma\}$ and q_v. The same is true for the grand-canonical partition function. However, as a consequence of the long-range nature of the Coulomb interactions non-neutral configurations for which $\sum_\sigma e_\sigma N_\sigma$ differs from Vq_v, are strongly suppressed in the thermodynamic limit, so that q_v is in fact determined once β and $\{\beta\mu_\sigma\}$ are given. It can be proved [3] that the grand-canonical partition function is related to the pressure p in the usual way, namely as

$$p = \lim_{V \to \infty} \frac{1}{\beta V} \log Z_{gr} \quad , \tag{2.3}$$

if the background charge density is chosen such that the constraint

$$\left(\frac{\partial}{\partial q_v} Z_{gr} \right)_{\beta, \{\beta\mu_\sigma\}} = 0 \tag{2.4}$$

is satisfied identically. This constraint fixes q_v as a function of the independent variables β and $\{\beta\mu_\sigma\}$. Likewise, the pressure p is a function of these variables, which fulfils the standard differential relation $d(\beta p) = -u_v d\beta + \sum_\sigma n_\sigma d(\beta\mu_\sigma)$, with u_v the internal energy density and n_σ the partial densities.

On account of the homogeneity of the Coulomb potential the pressure obeys a scaling relation which reads

$$p(\beta, \{\beta\mu_\sigma\}) = \lambda^4 p(\lambda\beta, \{\beta\mu_\sigma - \frac{3}{2} \log \lambda\}) \quad , \tag{2.5}$$

for all positive λ. Likewise, the partial densities fulfil the scaling relation

$$n_\sigma(\beta, \{\beta\mu_\sigma\}) = \lambda^3 n_\sigma(\lambda\beta, \{\beta\mu_\sigma - \frac{3}{2} \log \lambda\}) \quad . \tag{2.6}$$

Differentiation of (2.5) and (2.6) with respect to λ yields the well-known equation of state for Coulomb systems

$$p = \frac{1}{3} u_v + \frac{n}{2\beta} \quad , \tag{2.7}$$

(with n the total particle density) and an identity for the partial densities of the form

$$\left[\beta \frac{\partial}{\partial \beta} - \frac{3}{2} \sum_{\sigma'} \frac{\partial}{\partial \beta\mu_{\sigma'}} + 3 \right] n_\sigma = 0 \quad . \tag{2.8}$$

The equilibrium average $\langle f \rangle$ of a microscopic quantity f depends on the variables β and $\{\beta\mu_\sigma\}$ via the weight function ρ_{gr}. Formal expressions for the partial derivatives of $\langle f \rangle$ follow by employing (2.2). In taking the derivatives one should be careful in accounting for the dependence through the charge density q_v [3]. If the latter is kept fixed differentiation of $\langle f \rangle$ with respect to $\beta\mu_\sigma$ yields the fluctuation expression $\langle (f - \langle f \rangle)(N_\sigma - \langle N_\sigma \rangle) \rangle$. The complete expression for the partial derivative $\partial \langle f \rangle / \partial \beta \mu_\sigma$ contains an additional term:

$$\left(\frac{\partial \langle f \rangle}{\partial \beta \mu_\sigma} \right)_{\beta, \{\beta \mu_{\sigma'}\}} = \langle (f - \langle f \rangle)(N_\sigma - \langle N_\sigma \rangle) \rangle + \left(\frac{\partial \langle f \rangle}{\partial q_v} \right)_{\beta, \{\beta \mu_{\sigma'}\}} \left(\frac{\partial q_v}{\partial \beta \mu_\sigma} \right)_{\beta, \{\beta \mu_{\sigma'}\}} \quad . \tag{2.9}$$

When this identity is summed over σ, with weights e_σ, the first term at the right-hand side yields an expression involving the fluctuation of the total charge $\sum_\sigma e_\sigma N_\sigma$, which is negligible in the thermodynamic limit. Hence, one may solve for the partial derivative $\partial \langle f \rangle / \partial q_v$. Substitution in (2.9) then gives:

$$\frac{\partial \langle f \rangle}{\partial \beta \mu_\sigma} - \frac{\beta}{S} \frac{\partial q_v}{\partial \beta \mu_\sigma} \sum_{\sigma'} e_{\sigma'} \frac{\partial \langle f \rangle}{\partial \beta \mu_{\sigma'}} = \langle (f - \langle f \rangle)(N_\sigma - \langle N_\sigma \rangle) \rangle \quad , \tag{2.10}$$

with the abbreviation $S = \beta \sum_\sigma e_\sigma \partial q_v / \partial \beta \mu_\sigma$. All partial derivatives are taken with respect to the independent variables β and $\beta \mu_\sigma$. Likewise, one may derive a formal expression for the derivative with respect to β:

$$\frac{\partial \langle f \rangle}{\partial \beta} - \frac{\beta}{S} \frac{\partial q_v}{\partial \beta} \sum_\sigma e_\sigma \frac{\partial \langle f \rangle}{\partial \beta \mu_\sigma} = -\langle (f - \langle f \rangle)(H - \langle H \rangle) \rangle \quad . \tag{2.11}$$

These differentiation formulas will be used to evaluate partial derivatives of the correlation functions.

3 Correlation functions

The equilibrium correlation functions $g^{(k)}_{\sigma_1 \ldots \sigma_k}(\mathbf{r}_1, \ldots, \mathbf{r}_k)$ are defined as

$$n_{\sigma_1} \ldots n_{\sigma_k} g^{(k)}_{\sigma_1 \ldots \sigma_k}(\mathbf{r}_1, \ldots, \mathbf{r}_k) = \left\langle \sum_{\alpha_1, \ldots \alpha_k}{}' \delta(\mathbf{r}_1 - \mathbf{q}_{\sigma_1 \alpha_1}) \ldots \delta(\mathbf{r}_k - \mathbf{q}_{\sigma_k \alpha_k}) \right\rangle \quad . \tag{3.1}$$

Often we shall abbreviate the left-hand side as $G^{(k)}(1, \ldots, k)$.

The correlation functions depend on β and $\{\beta \mu_\sigma\}$. They satisfy the scaling relation:

$$g^{(k)}_{\sigma_1 \ldots \sigma_k}(\mathbf{r}_1, \ldots, \mathbf{r}_k, \beta, \{\beta \mu_\sigma\}) = g^{(k)}_{\sigma_1 \ldots \sigma_k}(\lambda \mathbf{r}_1, \ldots, \lambda \mathbf{r}_k, \lambda \beta, \{\beta \mu_\sigma - \frac{3}{2} \log \lambda\}) \quad . \tag{3.2}$$

As a consequence the derivatives of the correlation function $g^{(k)}$ are linearly dependent. Using (2.8) we may write the relation of linear dependence as:

$$\left[\sum_{i=1}^{k} \mathbf{r}_i \cdot \nabla_i + \beta \frac{\partial}{\partial \beta} - \frac{3}{2} \sum_\sigma \frac{\partial}{\partial \beta \mu_\sigma} + 3k \right] G^{(k)}(1, \ldots, k) = 0 \quad , \tag{3.3}$$

with $\nabla_i \equiv \partial / \partial \mathbf{r}_i$.

The spatial partial derivatives in (3.3) are determined by the hierarchy equations:

$$\left\{ \beta^{-1} \nabla_j + \sum_{i=1(i \neq j)}^{k} [\nabla_j v(i,j)] \right\} G^{(k)}(1, \ldots, k) =$$

$$= -\sum_{\sigma_{k+1}} \int d\mathbf{r}_{k+1} G^{(1|k)}(k+1|1, \ldots, k) \nabla_j v(j, k+1) \quad , \tag{3.4}$$

with $v(i,j) \equiv e_{\sigma_i} e_{\sigma_j} v(|\mathbf{r}_i - \mathbf{r}_j|)$ and with the truncated correlation function defined as

$$G^{(1|k)}(k+1|1, \ldots, k) = n_{\sigma_1} \ldots n_{\sigma_{k+1}} \left[g^{(k+1)}_{\sigma_1 \ldots \sigma_{k+1}}(\mathbf{r}_1, \ldots, \mathbf{r}_{k+1}) - g^{(k)}_{\sigma_1 \ldots \sigma_k}(\mathbf{r}_1, \ldots, \mathbf{r}_k) \right] \quad . \tag{3.5}$$

The derivatives with respect to the thermodynamic variables in (3.3) can be evaluated with the help of the diferentiation formulas (2.10) and (2.11). Let us first consider (2.10). Writing N_σ as the integral of $\sum_\alpha \delta(\mathbf{r} - \mathbf{q}_{\sigma\alpha})$ over the volume one may express the fluctuation expression at the right-hand side in terms of correlation functions as well, with the result:

$$
\left[\frac{\partial}{\partial\beta\mu_{\sigma_{k+1}}} - \frac{\beta}{S}\frac{\partial q_v}{\partial\beta\mu_{\sigma_{k+1}}}\sum_\sigma e_\sigma \frac{\partial}{\partial\beta\mu_\sigma}\right] G^{(k)}(1,\ldots,k) =
$$

$$
= G^{(k)}(1,\ldots,k)\sum_{j=1}^k \delta_{\sigma_j,\sigma_{k+1}} + \int d\mathbf{r}_{k+1} G^{(1|k)}(k+1|1,\ldots,k) \quad . \tag{3.6}
$$

These are $s - 1$ independent relations for the s partial derivatives with respect to the chemical potentials. Upon summing over σ_{k+1}, with a weight factor $e_{\sigma_{k+1}}$, the left-hand side yields 0, so that one finds

$$
\sum_{\sigma_{k+1}} e_{\sigma_{k+1}} \int d\mathbf{r}_{k+1} G^{(1|k)}(k+1|1,\ldots,k) = -\left(\sum_{j=1}^k e_{\sigma_j}\right) G^{(k)}(1,\ldots,k) \quad . \tag{3.7}
$$

This is the well-known charge sum rule, which may be derived from the hierarchy equations as well.

We now turn to the derivative of $G^{(k)}$ with respect to β as given by (2.11). Substituting (2.1) at the right-hand side and rewriting the average in terms of correlation functions we find

$$
\left[\frac{\partial}{\partial\beta} - \frac{\beta}{S}\frac{\partial q_v}{\partial\beta}\sum_\sigma e_\sigma\frac{\partial}{\partial\beta\mu_\sigma}\right] G^{(k)}(1,\ldots,k) =
$$

$$
= -\frac{3}{2\beta}\left[kG^{(k)}(1,\ldots,k) + \sum_{\sigma_{k+1}}\int d\mathbf{r}_{k+1} G^{(1|k)}(k+1|1,\ldots,k)\right]
$$

$$
- \frac{1}{2}\sum_{\sigma_{k+1},\sigma_{k+2}}\int d\mathbf{r}_{k+1}d\mathbf{r}_{k+2} G^{(2|k)}(k+1,k+2|1,\ldots,k)v(k+1,k+2)
$$

$$
- \sum_{\sigma_{k+1}}\sum_{j=1}^k\int d\mathbf{r}_{k+1} G^{(1|k)}(k+1|1,\ldots,k)v(j,k+1)
$$

$$
- \frac{1}{2}G^{(k)}(1,\ldots,k)\sum_{i,j=1(i\neq j)}^k v(i,j) \quad . \tag{3.8}
$$

Here we introduced the generalized truncated correlation function

$$
G^{(2|k)}(k+1,k+2|1,\ldots,k) = n_{\sigma_1}\ldots n_{\sigma_{k+2}}\left[g^{(k+2)}_{\sigma_1\ldots\sigma_{k+2}}(\mathbf{r}_1,\ldots,\mathbf{r}_{k+2})\right.
$$

$$
- g^{(k+1)}_{\sigma_1\ldots\sigma_{k+1}}(\mathbf{r}_1,\ldots,\mathbf{r}_{k+1}) - g^{(k+1)}_{\sigma_1\ldots\sigma_k\sigma_{k+2}}(\mathbf{r}_1,\ldots,\mathbf{r}_k,\mathbf{r}_{k+2})
$$

$$
- g^{(2)}_{\sigma_{k+1}\sigma_{k+2}}(\mathbf{r}_{k+1},\mathbf{r}_{k+2})g^{(k)}_{\sigma_1\ldots\sigma_k}(\mathbf{r}_1,\ldots,\mathbf{r}_k) + 2g^{(k)}_{\sigma_1\ldots\sigma_k}(\mathbf{r}_1,\ldots,\mathbf{r}_k)\right] \quad . \tag{3.9}
$$

The truncated correlation function $G^{(2|k)}$ can be eliminated from (3.8) by using the hierarchy equation (3.4). In fact, writing (3.4) for $k + 1$ instead of k, and choosing $j = k + 1$ one derives an expression for the derivative of $G^{(1|k)}$ (defined in (3.5)) with respect to $\mathbf{r}_{k+1}$. It contains an integral with a function $G^{(1|k+1)}(k + 2|1,\ldots,k + 1)$ that can be written as a linear combination of $G^{(2|k)}(k + 1, k + 2|1,\ldots,k)$, $G^{(1|k)}(k + 2|1,\ldots,k)$ and $G^{(k)}(1,\ldots,k)$. In this way one finds the auxiliary relation:

$$
\sum_{\sigma_{k+2}}\int d\mathbf{r}_{k+2} G^{(2|k)}(k+1,k+2|1,\ldots,k)\nabla_{k+1}v(k+1,k+2) =
$$

$$
= -\beta^{-1}\nabla_{k+1}G^{(1|k)}(k+1|1,\ldots,k)
$$

$$
- G^{(1|k)}(k+1|1,\ldots,k)\sum_{j=1}^k \nabla_{k+1}v(j,k+1)
$$

$$- \; n_{\sigma_{k+1}} \nabla_{k+1} \left[G^{(k)}(1,\ldots,k) \sum_{j=1}^{k} v(j,k+1) \right.$$

$$\left. + \sum_{\sigma_{k+2}} \int d\mathbf{r}_{k+2} G^{(1|k)}(k+2|1,\ldots,k) v(k+1,k+2) \right] \quad . \tag{3.10}$$

We now take the inner product of this vector identity with $\mathbf{r}_{k+1}$, integrate over $\mathbf{r}_{k+1}$ and sum over σ_{k+1}. At the left-hand side we use the symmetry with respect to the interchange of $k+1$ and $k+2$, and the homogeneity of the potential function. As a result we find an expression that is equal to the term with $G^{(2|k)}$ in (3.8). In the first, third and fourth terms at the right-hand a partial integration can be performed, while the second term at the right-hand side may be transformed with the use of the identity

$$\mathbf{r}_{k+1} \cdot \nabla_{k+1} v(j,k+1) = -v(j,k+1) - \mathbf{r}_j \cdot \nabla_j v(j,k+1) \quad . \tag{3.11}$$

The contribution with $\nabla_j v(j,k+1)$ that arises in this way can be rewritten once more with the help of (3.4). The identity that is finally obtained from (3.10) may be used to eliminate $G^{(2|k)}$ from (3.8). As a result the latter gets the form

$$\left[\frac{\partial}{\partial \beta} - \frac{\beta}{S} \frac{\partial q_v}{\partial \beta} \sum_\sigma e_\sigma \frac{\partial}{\partial \beta \mu_\sigma} \right] G^{(k)}(1,\ldots,k) =$$

$$= \; -\frac{3}{2\beta} \left[k G^{(k)}(1,\ldots,k) - \sum_{\sigma_{k+1}} \int d\mathbf{r}_{k+1} G^{(1|k)}(k+1|1,\ldots,k) \right]$$

$$- \; \beta^{-1} \sum_{j=1}^{k} \mathbf{r}_j \cdot \nabla_j G^{(k)}(1,\ldots,k)$$

$$+ \; 3 \sum_{\sigma_{k+1}} n_{\sigma_{k+1}} \int d\mathbf{r}_{k+1} \left[G^{(k)}(1,\ldots,k) \sum_{j=1}^{k} v(j,k+1) \right.$$

$$\left. + \sum_{\sigma_{k+2}} \int d\mathbf{r}_{k+2} G^{(1|k)}(k+2|1,\ldots,k) v(k+1,k+2) \right] \quad . \tag{3.12}$$

The term with the spatial derivatives in (3.12) can be eliminated with the help of the scaling relation (3.3). The derivative with respect to β then drops out as well. Furthermore, the last two terms in (3.12) may be transformed with the help of the Legendre expansion for the electrostatic potential. In doing so it is convenient to choose as the integration domain for the integral over $\mathbf{r}_{k+1}$ a large sphere S around the origin (with a radius tending to ∞). Using moreover the charge sum rule (3.7) we write the last two terms of (3.12) in such a way that the integral over $\mathbf{r}_{k+1}$ can be carried out, with the result:

$$-\frac{1}{2} q_v \left[G^{(k)}(1,\ldots,k) \sum_{j=1}^{k} e_{\sigma_j} r_j^2 + \sum_{\sigma_{k+2}} e_{\sigma_{k+2}} \int d\mathbf{r}_{k+2} G^{(1|k)}(k+2|1,\ldots,k) r_{k+2}^2 \right] \quad . \tag{3.13}$$

Insertion of this expression in (3.12) finally leads to the identity

$$\sum_\sigma \left[\frac{3}{2} - \frac{\beta^2}{S} e_\sigma \frac{\partial q_v}{\partial \beta} \right] \frac{\partial}{\partial \beta \mu_\sigma} G^{(k)}(1,\ldots,k) =$$

$$= \; \frac{3}{2} \left[k G^{(k)}(1,\ldots,k) + \sum_{\sigma_{k+1}} \int d\mathbf{r}_{k+1} G^{(1|k)}(k+1|1,\ldots,k) \right]$$

$$- \; \frac{1}{2} \beta q_v \left[G^{(k)}(1,\ldots,k) \sum_{j=1}^{k} e_{\sigma_j} r_j^2 + \sum_{\sigma_{k+1}} e_{\sigma_{k+1}} \int d\mathbf{r}_{k+1} G^{(1|k)}(k+1|1,\ldots,k) r_{k+1}^2 \right] \quad . \tag{3.14}$$

This identity comes in addition to the $s - 1$ relations (3.6) found before. Together they constitute s relations from which the partial derivatives of $G^{(k)}$ with respect to the chemical potentials can be found.

Solution of the s linear equations (3.6) and (3.14) for the partial derivatives $\partial G^{(k)}/\partial \beta \mu_\sigma$ yields the following result

$$\frac{\partial}{\partial \beta \mu_{\sigma_{k+1}}} G^{(k)}(1, \ldots, k) = G^{(k)}(1, \ldots, k) \sum_{j=1}^{k} \delta_{\sigma_j, \sigma_{k+1}} + \int d\mathbf{r}_{k+1} G^{(1|k)}(k+1|1, \ldots, k)$$

$$- \frac{\beta}{6} \frac{\partial q_v}{\partial \beta \mu_{\sigma_{k+1}}} \left[G^{(k)}(1, \ldots, k) \sum_{j=1}^{k} e_{\sigma_j} r_j^2 + \sum_{\sigma_{k+2}} e_{\sigma_{k+2}} \int d\mathbf{r}_{k+2} G^{(1|k)}(k+2|1, \ldots, k) r_{k+2}^2 \right] \quad , \quad (3.15)$$

where we used the equality

$$\beta \frac{\partial q_v}{\partial \beta} = \frac{3}{2} \sum_{\sigma} \frac{\partial q_v}{\partial \beta \mu_\sigma} - 3 q_v \quad , \tag{3.16}$$

which follows from (2.8).

The complete set of partial derivatives of the correlation functions $G^{(k)}$ with respect to the chemical potentials has now been found. The expressions for these derivatives contain integrals of the truncated correlation functions $G^{(1|k)}$.

4 Second-moment sum rules

A different way to present the results of the previous section consists in writing them as sum rules for the zeroth and second moments of the truncated correlation functions. In fact, the zeroth moment of $G^{(1|k)}$ is found directly from (3.6). Returning to the notation of (3.1) we get

$$n_{\sigma_1} \ldots n_{\sigma_{k+1}} \int d\mathbf{r}_{k+1} \left[g_{\sigma_1 \ldots \sigma_{k+1}}^{(k+1)}(\mathbf{r}_1, \ldots, \mathbf{r}_{k+1}) - g_{\sigma_1 \ldots \sigma_k}^{(k)}(\mathbf{r}_1, \ldots, \mathbf{r}_k) \right] =$$

$$= -n_{v_1} \ldots n_{v_k} g_{\sigma_1 \ldots \sigma_k}^{(k)}(\mathbf{r}_1, \ldots, \mathbf{r}_k) \sum_{j=1}^{k} \delta_{v_j, v_{k+1}}$$

$$+ \left[\frac{\partial}{\partial \beta \mu_{\sigma_{k+1}}} - \frac{\beta}{S} \frac{\partial q_v}{\partial \beta \mu_{\sigma_{k+1}}} \sum_{\sigma} e_\sigma \frac{\partial}{\partial \beta \mu_\sigma} \right] n_{\sigma_1} \ldots n_{\sigma_k} g_{\sigma_1 \ldots \sigma_k}^{(k)}(\mathbf{r}_1, \ldots, \mathbf{r}_k) \quad . \tag{4.1}$$

As remarked already below (3.6) this sum rule leads to the charge sum rule for the truncated correlation functions if a suitable weighted sum over the species is carried out. The zeroth-order sum rule found here does not contain such a sum: it is valid for each species separately. We will refer to it as a *generalized charge sum rule*. Alternatively, we may write it in terms of the Ursell functions $h_{\sigma_1 \ldots \sigma_k}^{(k)}$ that are related to the correlation functions through the standard cluster decomposition.

The second-moment sum rule for the truncated correlation functions follows from (3.15). The simplest form is obtained by summing that equation over σ_{k+1}, with weights e_{k+1}, since then the first two terms at the right-hand side drop out on account of the charge sum rule (3.7). Adopting again the notation of (3.1) we get

$$n_{\sigma_1} \ldots n_{\sigma_k} \sum_{\sigma_{k+1}} e_{\sigma_{k+1}} n_{\sigma_{k+1}} \int d\mathbf{r}_{k+1} \left[g_{\sigma_1 \ldots \sigma_{k+1}}^{(k+1)}(\mathbf{r}_1, \ldots, \mathbf{r}_{k+1}) - g_{\sigma_1 \ldots \sigma_k}^{(k)}(\mathbf{r}_1, \ldots, \mathbf{r}_k) \right] r_{k+1}^2 =$$

$$= -n_{\sigma_1} \ldots n_{\sigma_k} g_{\sigma_1 \ldots \sigma_k}^{(k)}(\mathbf{r}_1, \ldots, \mathbf{r}_k) \sum_{j=1}^{k} e_{\sigma_j} r_j^2 - \frac{6}{S} \sum_{\sigma} e_\sigma \frac{\partial}{\partial \beta \mu_\sigma} n_{\sigma_1} \ldots n_{\sigma_k} g_{\sigma_1 \ldots \sigma_k}^{(k)}(\mathbf{r}_1, \ldots, \mathbf{r}_k) \quad . (4.2)$$

By using the cluster decomposition one may rewrite this identity in terms of the Ursell functions. The second-moment sum rule found here may be called a *generalized Stillinger-Lovett relation*. The reason for that becomes obvious if the special case $k = 1$ is considered. Using translation invariance and isotropy we may write (4.2) in that case as

$$n_{\sigma_1} \sum_{\sigma_2} e_{\sigma_2} n_{\sigma_2} \int d\mathbf{r}_{12} h^{(2)}_{\sigma_1\sigma_2}(r_{12}) r_{12}^2 = -\frac{6}{S} \frac{\partial q_v}{\partial \beta \mu_{\sigma_1}} \quad . \tag{4.3}$$

This relation has been derived before [2]. Summation over σ_1, with weights e_{σ_1}, leads to the identity

$$\sum_{\sigma_1,\sigma_2} n_{\sigma_1} n_{\sigma_2} e_{\sigma_1} e_{\sigma_2} \int d\mathbf{r}_{12} h^{(2)}_{\sigma_1\sigma_2}(r_{12}) r_{12}^2 = -\frac{6}{\beta} \quad , \tag{4.4}$$

which is the well-known Stillinger-Lovett relation [1] for a mixture. It should be noted that (4.3) is valid for each species σ_1 separately, so that it is more general than (4.4). Likewise, (4.2) contains k independent species labels.

A useful alternative form for the sum rule (4.2) is obtained by adopting a different set of independent thermodynamic variables, namely the charge density q_v and a set of reduced thermodynamic potentials defined as [6]

$$\bar{\mu}_\sigma = \mu_\sigma - e_\sigma \frac{\sum_{\sigma'} e_{\sigma'} \mu_{\sigma'}}{\sum_{\sigma'} e_{\sigma'}^2} \quad , \tag{4.5}$$

which satisfy the constraint $\sum_\sigma e_\sigma \bar{\mu}_\sigma = 0$. In terms of these variables the combination of partial derivatives occurring in (4.2) gets a simple form, since one has

$$\frac{1}{S} \sum_\sigma e_\sigma \left(\frac{\partial}{\partial \beta \mu_\sigma} \right)_{\beta, \{\beta \mu_{\sigma'}\}} = \frac{1}{\beta} \left(\frac{\partial}{\partial q_v} \right)_{\beta, \{\beta \bar{\mu}_\sigma\}} \quad . \tag{4.6}$$

For the special case of the one-component plasma (with $s = 1$) the zeroth-moment sum rule (4.1) is completely equivalent to the standard charge sum rules for the correlation or Ursell functions. The second-moment sum rule (4.2) reduces then to the sum rule found before by Alastuey [4] and by Vieillefosse and Brajon [5] for the one-component plasma.

The sum rules derived here are useful in studying the behaviour of large-scale fluctuations in an ionic mixture. In particular, they can be employed to establish the small-wavenumber behaviour of the Fourier transform of cross-fluctuation expressions containing the product of the local partial density $n_\sigma(\mathbf{r}) = \sum_\alpha \delta(\mathbf{r} - \mathbf{q}_{\sigma\alpha})$ of species σ or of the local charge density $q_v(\mathbf{r}) = \sum_\sigma e_\sigma n_\sigma(\mathbf{r})$ of the particles on one hand, and an arbitrary localized microscopic configurational quantity $f(\mathbf{r})$ on the other hand. In general, the latter may be written as a linear combination of k-particle sum variables of the form

$$f(\mathbf{r}) = \sum_{\sigma_1\alpha_1,\ldots,\sigma_k\alpha_k}{}' f_{\sigma_1\ldots\sigma_k}(\mathbf{q}_{\sigma_1\alpha_1},\ldots,\mathbf{q}_{\sigma_k\alpha_k};\mathbf{r}) \quad . \tag{4.7}$$

The Fourier-transformed fluctuation expressions we are interested in have the form

$$\frac{1}{V} \langle [n_\sigma(\mathbf{k})]^* f(\mathbf{k}) \rangle \quad , \tag{4.8}$$

and

$$\frac{1}{V} \langle [q_v(\mathbf{k})]^* f(\mathbf{k}) \rangle \quad , \tag{4.9}$$

with the Fourier transform of the microscopic quantity $f(\mathbf{r})$ defined as $f(\mathbf{k}) \equiv \int^V d\mathbf{r} \exp(-i\mathbf{k}\cdot\mathbf{r}) f(\mathbf{r})$.

Writing the fluctuation expression (4.8) in terms of correlation functions and using the zeroth-order sum rule (4.1) one easily derives the identity

$$\lim_{\mathbf{k}\to 0} \frac{1}{V} \langle [n_\sigma(\mathbf{k})]^* f(\mathbf{k}) \rangle = \frac{\partial \langle f \rangle}{\partial \beta \mu_\sigma} - \frac{\beta}{S} \frac{\partial q_v}{\partial \beta \mu_\sigma} \sum_{\sigma'} e_{\sigma'} \frac{\partial \langle f \rangle}{\partial \beta \mu_{\sigma'}} \quad . \tag{4.10}$$

Summing over σ, with weights e_σ, one finds that all cross-fluctuation expressions containing the Fourier transform of the local charge density $q_v(\mathbf{r})$ of the particles as a factor vanish in the long-wavelength limit:

$$\lim_{\mathbf{k}\to 0} \frac{1}{V} \langle [q_v(\mathbf{k})]^* f(\mathbf{k}) \rangle = 0 \quad . \tag{4.11}$$

This result is equivalent to the charge sum rule for the k-point correlation function.

The first non-vanishing contribution in the small-wavenumber expansion of a cross-fluctuation expression with the charge density is of second order in k. The coefficient of the second-order term can be evaluated by employing the second-moment sum rule (4.2). One finds

$$\lim_{\mathbf{k}\to 0} \frac{1}{k^2}\frac{1}{V} \langle [q_v(\mathbf{k})]^* f(\mathbf{k}) \rangle = \frac{1}{S} \sum_\sigma e_\sigma \frac{\partial \langle f \rangle}{\partial \beta \mu_\sigma} \quad . \tag{4.12}$$

In particular, putting $f(\mathbf{k})$ equal to the Fourier transform of the local partial density $n_\sigma(\mathbf{r})$ or of the local charge density $q_v(\mathbf{r})$ of the particles one finds identities that are equivalent to (4.3) and (4.4). A suggestive form for (4.12) is obtained by using a different set of variables as in (4.6):

$$\lim_{\mathbf{k}\to 0} \frac{1}{k^2}\frac{1}{V} \langle [q_v(\mathbf{k})]^* f(\mathbf{k}) \rangle = \frac{1}{\beta} \left(\frac{\partial \langle f \rangle}{\partial q_v} \right)_{\beta,\{\beta\bar\mu_\sigma\}} \quad . \tag{4.13}$$

This identity is somewhat similar to the differentiation formulas that have been discussed in section 2. However, the similarity is superficial only: the present identity determines the second-order contribution in the long-wavelength expansion of a fluctuation expression, whereas the identities of section 2 fix the leading zeroth-order contributions of fluctuation expressions.

References

[1] F. H. Stillinger Jr. and R. Lovett, J. Chem. Phys. **49**, 1991 (1968)

[2] L. G. Suttorp and A. J. van Wonderen, Physica **145A**, 533 (1987); A. J. van Wonderen and L. G. Suttorp, Physica **145A**, 557 (1987)

[3] L. G. Suttorp, Contrib. Plasma Phys. **29**, 335 (1989)

[4] A. Alastuey, J. Phys. France **49**, 1507 (1988)

[5] P. Vieillefosse and M. Brajon, J. Stat. Phys. **55**, 1169 (1989)

[6] E. H. Lieb and J. L. Lebowitz, Adv. Mathem. **9**, 316 (1972)

THE SAHA EQUATION FOR A TWO–TEMPERATURE PLASMA

M.C.M. van de Sanden and P.P.J.M. Schram
Eindhoven University of Technology
Department of Technical Physics
P.O. Box 513, 5600 MB Eindhoven, The Netherlands

Abstract

A generalization of the Saha equation for the case of a two–temperature plasma is derived from thermodynamic considerations. The resulting equation depends in first approximation on the electron temperature only. A more rigorous approach concerns the application of the Zubarev formalism. This is done for arbitrary heat flow between the electrons and the internal states of the heavy particles on one hand and the kinetic degrees of freedom of the heavy particles on the other hand. In the case of zero heat flow the equation mentioned with electron temperature only is recovered. In general corrections to this are found which depend on the ratio of electron and heavy particle masses, the temperature difference and the specific atomic structure. For hydrogen plasmas the results indicate that corrections may be of the order of 10%.

1.   Introduction

We consider a chemical reaction

$$\sum_i \nu_i \, A_i = 0 \qquad\qquad (1.1)$$

taking place under conditions of thermodynamic equilibrium. The ν_i are the so–called stoichiometric coefficients. If the temperature T and the Volume V of the system are constant, the equilibrium condition is the stationarity of the free energy $F = E - TS$, where E is the internal energy and S the entropy, i.e. $dF = 0$. From the thermodynamic identity

$$TdS - dE = pdV - \sum_i \mu_i \, dN_i , \qquad\qquad (1.2)$$

where p is the pressure, N_i the number of particles of species i and μ_i the corresponding chemical potential, we conclude that $\mu_i = (\partial F / \partial N_i)_{T,V}$, so that $dF = 0$ leads to

$$\sum_i \nu_i \, \mu_i = 0 . \qquad\qquad (1.3)$$

Applied to the ionization–recombination reaction,

81

$$A_0^p + e \overset{\leftharpoonup}{\to} A_+^q + e + e, \qquad (1.4)$$

where e, A_0 , A_+ represent electron, neutral atom and ion respectively, whereas p and q are internal states, we have $\nu_+ = 1$, $\nu_e = 1$, $\nu_0 = -1$. Then eq. (1.3) leads to the equilibrium Saha equation[1] :

$$\frac{n_e n_+^p}{n_0^{\,q}} = \frac{g_e g_+^p}{g_0^{\,q}} \left[\frac{2\pi m_e k_B T}{h^2} \right]^{3/2} \exp\left[-\frac{E_+^p - E_0^q}{k_B T} \right], \qquad (1.5)$$

where n's are particle densities, g's are degeneracies, the subscript e, 0, + refer to electrons, neutrals and ions respectively, m_e is the electron mass, k_B Boltzmann's constant, h Planck's constant and $E_+^p - E_0^q$ is an energy difference between levels.

The problem under consideration concerns the generalization of eq. (1.5) for the case of different temperatures T_e and $T_0 = T_+ (\equiv T_h)$.

2. Discussion in the litterature

If the mass ratio is small, $m_e/m_h \ll 1$, then situations with $T_e \neq T_h$ have a long life time. Therefore we may assume in first approximation T_e and T_h to be constant. In the litterature two points of view about the generalization of (1.5) for $T_e \neq T_h$ are found:

(1)   Replace T by T_e in eq. (1.5). This result is obtained from
* Kinetic considerations[2,3,4] in which the electrons play a dominant role
* irreversible thermodynamics: principle of minimal entropy production[5].
(2)   A generalized Saha equation containing both T_e and T_h is derived by several
authors [6,7,8,9,10] on basis of attempts to generalize equilibrium thermodynamics.

We analyze the chain of reasoning in point of view (2). We have two subsystems:
—   electrons plus internal states – of both ions and neutrals – are described with T_e
—   translations of ions and neutrals are described with T_h.
If the systems are independent, then it is obvious that $T_i\, dS_i \geq dQ_i$, where dQ_i is an infinitesimal amount of heat transfered to subsystem i. More generally it is claimed that

$$\sum_i T_i\, dS_i \geq dQ \qquad (2.1)$$

with $dQ = \sum_i d\,Q_i$. Then it follows, as in equilibrium, that $dF = \sum_i dF_i = 0$ and again eq. (1.3) follows. The consequences of eq. (1.3) are, however, different. Prigogine[6] and Potapov[7] derived the following equation:

$$n_e \left[\frac{n_+^q}{n_0^p}\right]^{T_h/T_e} = g_e \left[\frac{q_+^q}{g_0^p}\right]^{T_h/T_e} \left[\frac{2\pi k_B T_e}{h^2}\right]^{3/2} \exp\left[-\frac{E_+^q - E_0^p}{k_B T}\right] . \tag{2.2}$$

Our criticism on this chain of reasoning concerns the fact that the subsystems are <u>not</u> independent, so that eq. (2.1) need not be true. During recombination at constant T_e , e.g., the entropy of the electrons decreases.

3. Modification of thermodynamic principles. Generalized Saha equation

The correct generalization of the second law of thermodynamics should be a statement about the total entropy $S = \sum_i S_i$. Our proposal[11] is:

$$dS \geq \sum_i T_i^{-1}\, dQ_i . \tag{3.1}$$

Consider, e.g., two subsystems which do not exchange heat (but, maybe, do exchange mass — as in the recombination — ionization process). The Potapov principle (2.1) leads to $T_2\, dS \geq (T_2 - T_1)dS_1$. If $dS_1 > 0$ and $T_2 < T$, then $dS < 0$ is allowed. This is not the case with our principle (3.1). If the volume V and the temperatures T_i ar constant, we have $dQ_i = dE_i$ and eq. (3.1) becomes

$$dF^* = d\sum_i \left(\frac{E_i}{T_i} - S_i\right) = 0 , \tag{3.2}$$

where F^* is the proper generalization of the free energy. The consequence now is <u>not</u> eq. (1.3), but

$$\sum_i \frac{\nu_i \mu_i}{T_i} = 0 . \tag{3.3}$$

Substitution of the proper expressions for the μ_i leads to the correct generalization of the Saha equation, i.e. eq. (1.5) with T_e instead of T.

4. Derivation on basis of the Zubarev formalism

The Zubarev approach to Nonequilibrium Statistical Mechanics[12] has been applied to the problem under consideration by the present authors[13] . The result is a generalization of eq. (3.3):

$$\sum_i \frac{\nu_i \mu_i}{T_i} = \ln\left(\frac{L}{R}\right) , \tag{4.1}$$

where L and R are expressions (summations) containing Boltzmann factors and transition probabilities $|\Phi_{ijmq}^{klp}|^2$, where i, j, k refer to the kinetic degrees of freedom of the electrons, l and m to the kinetic degrees of freedom of the heavy particles and q and p to the internal degrees of freedom. We distinguish three cases:

—   Thermal equilibrium: $T_e = T_h$. Then $L = R$ and the equilibrium Saha equation (1.5) is recovered.

—   The linear case, defined by

$$k_B^{-1} (T_e^{-1} - T_h^{-1})(E_l^{kin\ 0} - E_m^{kin\ +}) \ll 1. \tag{4.2}$$

Then we find:

$$\sum_i \nu_i\, \mu_i\, T_i^{-1} = k_B^{-1}(T_e^{-1} - T_h^{-1})\, \Delta_{he}\,, \tag{4.3}$$

where Δ_{he} is the heat exchange between electrons and heavy particles. It is of order m_e/m_h. If the right hand side of eq. (4.3) is neglected, then eq. (3.3) is recovered and therefore also the generalized Saha equation (1.5) with T_e instead of T. This is in perfect agreement with the results of Irreversible Thermodynamics, derived by Ecker and Kröll[5].

—   The general nonlinear case treated in the next section.

5.   Corrections to the generalized Saha equations

A model was introduced by the present authors[13] to calculate the right hand side of eq. (4.1) by means of an expansion in powers of $(m_e/m_h)^{\frac{1}{2}}$. The basis of the model is the assumption that the transition probabilities $|\Phi_{ijmq}^{klp}|^2$ are constant in the ranges of $E_m^{kin\ +}$ which are allowed by conservation of energy and momentum in the reaction, and zero outside these ranges. These ranges and approximations for L and R in eq. (4.1) are obtained by expansion in powers of $(m_e/m_h)^{\frac{1}{2}}$. The basic assumption of the model may seem an extreme simplification, but it should be noted that the transition probabilities occur in such similar ways in L and R that the influence of their detailed behaviour may be expected to be relatively small. The result of the calculation up to order m_e/m_h can be expressed by

$$\sum_i \frac{\nu_i \mu_i}{T_i} = -\frac{5}{3}\frac{m_e}{m_h}\left(1 - \frac{T_h}{T_e}\right)\frac{B}{0} \tag{5.1}$$

with

$$\left\{{}^{B}_{0}\right\} = \sum_{p,q} \exp\left[-\frac{E_p^{\,int0}}{k_B T_e}\right] \begin{bmatrix} T^*(\frac{\Delta_{pq}}{k_B T_e}) \\[2mm] N^*(\frac{\Delta_{pq}}{k_B T_e}) \end{bmatrix} \tag{5.2}$$

where $T^*(x)$ and $N^*(x)$ are known functions[13] and

$$\Delta_{pq} = E_q^{\,int\,+} - E_p^{\,int0} \tag{5.3}$$

(ionization energy from neutral state p to ionized state q). We define:

$$\delta b_p = n_p/n_p^{\,saha} - 1 \tag{5.4}$$

Then eq. (5.1) leads to

$$\delta b_p = -\frac{5}{3}\frac{m_e}{m_h}\left(1 - \frac{T_h}{T_e}\right)\frac{B}{0}\,. \tag{5.5}$$

In the case of hydrogen some numerical consequences of eq. (5.5) are presented in Figure 1.

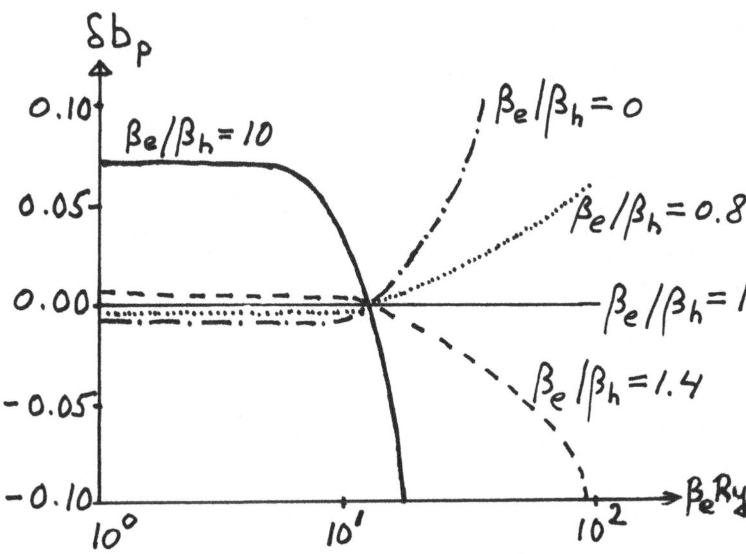

Figure 1

6. Conclusions

*   A simple thermodynamic argument leads in the case of two–temperature plasmas to the same generalization of the Saha equation as kinetic models. It is the equilibrium Saha equation with temperature T replaced by electron temperature T_e.
*   The Zubarev formalism[12] is suitable to calculate the generalized Law of Mass Action (Saha equation) in two–temperature plasmas.
*   For zero heat flow between electrons and heavy particles the Saha equation with T_e instead of T is obtained again.
*   For small heat flow the linear case is in agreement with the results of Ecker and Kröll[5].
*   The nonlinear case leads to new results. The dependence of the Law of Mass Action on T_h is of order m_e/m_h. For a hydrogen plasma the deviations from the (generalized) Saha equation may be significant for low $T_e (T_e < 0.5$ eV) and for $T_h/T_e < 0.8$ (up to 10%).

References

1. M.N. Saha, Proc.Roy.Soc. A 99 (1921) 135.
2. M. Mitcher and C. Kruger: Partially Ionized Gases, Wiley, New York, 1973.
3. L.M. Biberman, V.S. Vorob'ev and I.T. Yakubov: Kinetics of Non–equilibrium Plasmas, Consultants Bureau New York, 1987.
4. J.A.M. van der Mullen, Phys.Rep. 191 (1990) 109.
5. G. Ecker and W. Kröll, Proc. 5th Int. Conf. on Phenomena in Ionized Gases, Vol. 2, p. 64, Paris, 1964.
6. I. Prigogine, Bull.Cl.Acad.R.Belg. 26 (1940) 53.
7. A.V. Potapov, High Temp. 5 (1966) 48.
8. S. Veis, Czech. Conf. on Electronics and Vacuum Phys., Ed. L. Paty, Charles Univ. Prague, 1968.
9. H.G. Thiel, Wiss.Ber. AEG–Telefunken 44 (1971) 123.
10. K.C. Hsu, Ph.D. Thesis, Univ. of Minnesota, Minneapolis, 1982.
11. M.C.M. van de Sanden, P.P.J.M. Schram, A.G. Peeters, J.A.M. van der Mullen and G.M.W. Kroesen, Phys.Rev. A40 (1989) 5273.
12. D.N. Zubarev: Nonequilibrium Statistical Thermodynamics, Plenum, New York, 1965.
13. M.C.M. van de Sanden and P.P.J.M. Schram, Phys.Rev. A44 (1991) 5150.

PHASE TRANSITION IN SIMPLEST PLASMA MODELS

I.L.Iosilevski, A.Yu.Chigvintsev
Moscow Institute of Physics and Technology
Dolgoprudny, Moscow Region, 141700, USSR

The well-known simplest plasma model One-Component Plasma (OCP) is
the system of free moving charges of the same sign in the uniform compensating background of opposite sign. This model is studied carefully
nowadays [1÷3]. It should be emphasized that this model is not the single one but it is the family of models in fact. The difference may be
in the additional short-range interaction and in the type of statistics,
but the main subject for the following discussion is the difference in
the nature and the thermodynamic role of background.

In the ordinary version of the model – OCP with the rigid
background – the volume variations are not defined. The system can not
collapse or explode spontaneously and so the well-known negativity [1]
of formally defined pressure and compressibility does not signify the
thermodynamic instability. The only phase transition –crystallization –
occurs without change of density in this version of OCP .

Now the main subject of our interest is another variant of this
model. It is OCP with the compressible but still uniform background. It
means that the background does not screen the moving charges individually but does it in average only. More precisely, this version of model
may be defined by means of two variants of definitions:

A) Through the overall electroneutral grand canonical ensemble [4]

B) Through the density functional with the strong gradient correction at the Helmholtz free energy of background so that this term would
tend to infinity ($A\to\infty$) after the thermodynamic limit ($N\to\infty$, N/V=const)

$$F_{ocp+backgr}\left[n_{ocp}(x);n_b(y)\right] = F_{ocp} + F^{(local)}_{backgr} + A\int[\nabla n_b(\bar{x})]^2 d\bar{x} \qquad (1)$$

(The Coulomb interaction contribution in (1) is included in "OCP"-term).

It is known that the crystallization occurs with the small density
variation in this model [5]. But the main statement of this report is
the appearance of a new first-order phase transition [6], with the properties strongly depending on the precise definition of a background
thermodynamic contribution. It should be noted that the existence of
this phase transition may be proved independently of a calculation of
phase transition parameters. The simple arguments [7] with the use of
the Gibbs-Bogolubov unequality, dimensionless analysis and the well

known lower bound for the pressure correction of OCP [4] tend us to the form of equation of state of this model in the high density limit ($n \Rightarrow \infty$).

$$P = P_{backgr} + P_{ocp}^{(id.gas)} - n^{4/3} \text{const.} \qquad (2)$$

Two conclusion may be done:

1. The OCP of classical point charges in the Boltzmann ideal gas uniform background is thermodynamically unstable against a collapse at any temperature and density. This statement is in agreement, firstly, with the result of Lieb and Narnhofer [4] on a divergence of the over-all electroneutral grand canonical ensemble and, secondly, with the conclusion of Dyson [8] that thermodynamic stability of Coulomb system requires at least one sort of particles to be fermions.

2. In the case of ideal Fermi-gas background the phase transition of gas-liquid type with upper critical point appears at the sufficiently low temperature or sufficiently large value of charge.

Classical Point Charges in the Ideal Fermi-Gas Background

The equation of state of both these subsystems (OCP and background) are known almost exactly, so we can carry out direct calculations of the parameters of this phase transition. In this calculations we have used the analytical fits from [1,5] (OCP) and [9,10] (ideal Fermi-gas). The results are presented in Tab.I and Fig.1÷3 in standard notations.

$$\Gamma \equiv (4\pi n/3)^{1/3}(Z^2 e^2/kT) \qquad \Lambda_e^2 \equiv 2\pi \hbar^2/m_e kT \qquad r_s^{-3} \equiv 4\pi n_e a_B^3/3 \qquad a_B \equiv \hbar^2/m_e e^2$$

	Z	1	2	3	10	30	100	1000
I	T_c (Ry)	0.0393	0.152	0.314	2.17	10.6	55.8	1232
	Γ_c	8.86	12.6	16.3	42.7	117	376	3690
	$(r_s)_c$	5.73	3.32	2.43	1.00	0.467	0.206	0.044
	$(n_e \Lambda_e^3)_c$	7.22	4.90	4.22	3.30	3.03	2.94	2.89
	$\{P/(n_z + n_e)/kT\}_c$	0.125	0.129	0.130	0.130	0.129	0.128	0.127
II	T_{tr} (Ry)	0.0044	0.0225	0.0557	0.906	9.37	41.35	–
	$(\Delta n/n)_{tr}$	0.019	0.019	0.019	0.021	0.069	–	–
III	$(r_s)_{bin}$	2.47	1.55	1.19	0.531	0.255	0.115	0.025
	$(r_s)_{spin}$	3.08	1.94	1.48	0.664	0.319	0.143	0.031
	Γ_{spin}	5.76	8.39	11.0	28.9	79.6	256	2518
	$\{\Delta H_f/N_z\}_0$ (Ry)	0.3633	1.831	4.715	78.27	1016	16860	3.63+6

Table I. The parameters of phase transitions in OCP of classical point charges in the ideal Fermi-gas background for the different values of charge number Z: I,II - critical and triple points; III - the heat of sublimation and the parameters of condensed (r_s) and gaseous (Γ) binodals and spinodals in the limit $T \Rightarrow 0$.

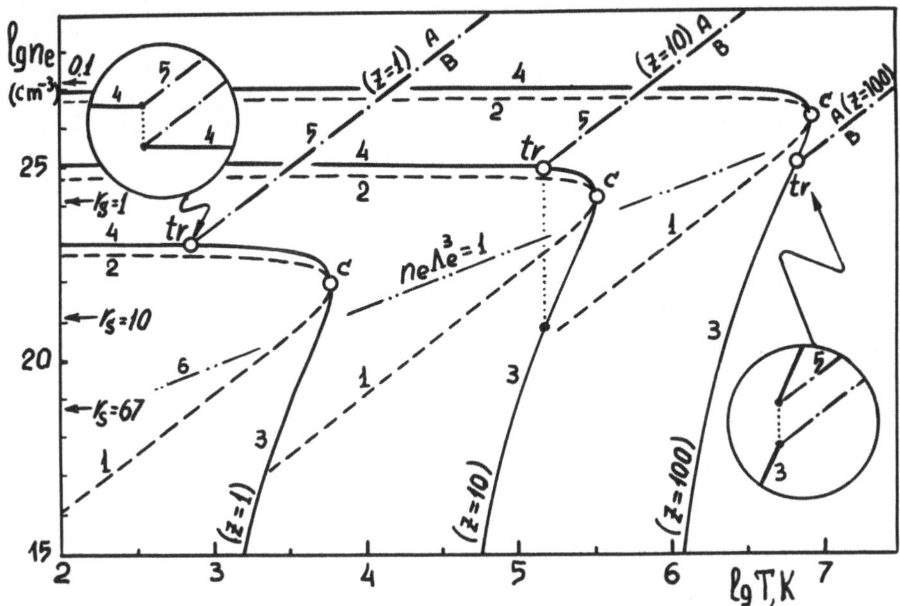

Fig.1. Phase diagram of OCP classical point charges in the ideal Fermi-
-gas background in temperature ↔ background electron density coordinates for different values of charge number Z. Spinodals (1,2) and
binodals (two-phase coexistence curve (3,4)) of condensed (2,4) and
gaseous(1,3) phases, melting (5) of crystalline (A) to fluid (B) phases
(Γ≈178), critical (c) and triple (tr) points, the position of electron
degeneracy (6) and constant Brueckner parameter r = 0.1, 1, 10 and 67
(the last value - cold melting of electron Wigner crystal [11]) are
remarked.

Fig.2.(a,b).Density ↔ temperature
diagram of coexistence curve and
its diameter: $d \equiv (n_l + n_g)/2$

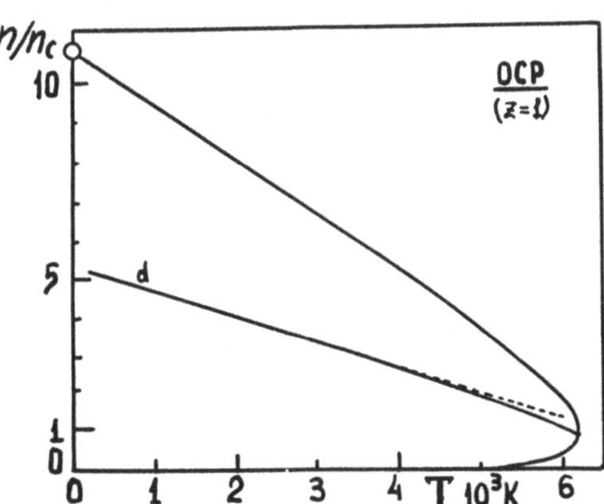

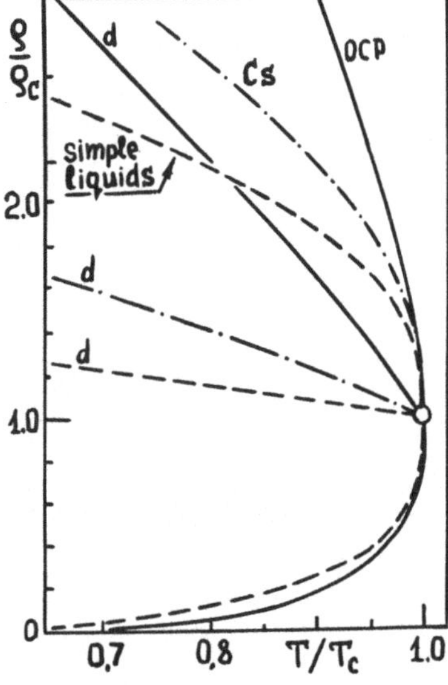

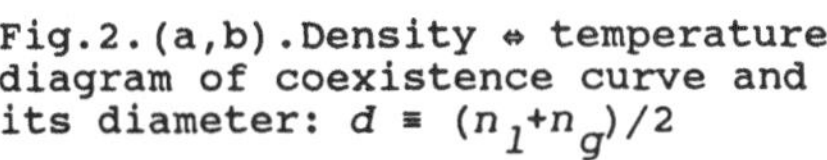

Some remarks to these results:

1) The special compressibility factor $Z_c \equiv P_c/kT_c(n_z + n_e)_c$ in critical point has almost the same rather low value for all charge numbers.[1]

2) For $Z > Z^* \simeq 46$ the melting curve (line $\Gamma \approx 178$) crosses the gaseous part of binodal and so it forms a specific picture of the phase diagram.

3) The density variation between crystalline and fluid phases in the triple point is fairly small (see Tab.I) when this point does not coincide with the critical point, in this latter case ($Z = Z^*$) the value of this variation increases remarkably. Besides that, this variation is much greater than the value estimated in [5] ($\Delta V/V \approx 0.0003$).

4) For the present phase transition the deviation from the well-known semi-empirical rule of rectangular diameter is similar to those for alkali metals (Fig.2), but the slope of this diameter and a relation of normal to critical densities is rather great in comparison with the real substances. For alkali metals $n_0/n_c \simeq 4 \div 5$.

5) The critical point of this phase transition is not the genuine one because the density fluctuations in the vicinity of the critical point are absent in the model definition "A" and are suppressed by the gradient term in the definition "B". In the limit of the infinite gradient term the critical exponents must tend to their classical values. In the case of strong but finite gradient term in (1) the critical point position must slightly deviate from its limiting value.

6) The remarkable feature of any nonuniformity in equilibrium system of Coulomb particles (in particular of inter-phase boundary) is a finite difference at the average electrostatic potential through this nonuniformity. The equality of two electrochemical potentials for every kind of Coulomb particles at both sides of nonuniformity leads to inter-phase potential drop depending on temperature (Fig.3)

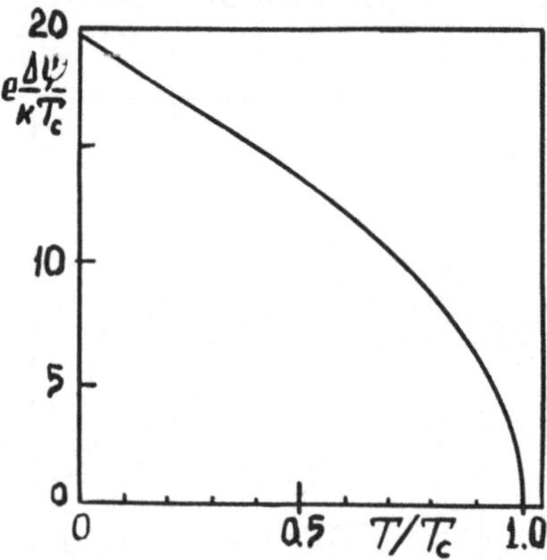

$$\mu_i^{(1)} + Z_i e\varphi^{(1)} = \mu_i^{(2)} + Z_i e\varphi^{(2)} \quad \Rightarrow$$

$$\varphi^{(2)} - \varphi^{(1)} = (Z_i e)^{-1}[\mu_i^{(1)} - \mu_i^{(2)}] \quad (3)$$

Fig.3 Potential drop between two phases ($\Delta\varphi \equiv \varphi_l - \varphi_g$)

In some sense the zero-temperature limit of this drop is the supplementary thermoelectrophysical constant of a substance. The high value of the drop for this model ought to be noted . Estimation for some metals [13] gives a value $\Delta\varphi/kT_c \simeq 2 \div 3$.

1) It is quite close to the estimation [12] with the use of equation of state of the van der-Waals type ($Z_c = 0.145$) with the Coulomb correction.

Quantum Electron Gas

In this variant of OCP the moving charges are the electrons. When the background is rigid the melting curve changes its shape in the high density limit because of cold melting appearance. The estimations [14] gives the Wigner crystallization inside the boundaries: $\Gamma \geq 178$; $r_s \leq 67$ [11] (classical and quantum melting); $T \leq 10^{-5}$ Ry [14]. A transition to the electron gas in the compressible but uniform background ought to change this phase diagram essentially because of the appearance of a vast phase transition of gas-liquid type. If we suppose that the parameters of this phase transition for Quantum Electron Gas are close to those of Classical OCP in the ideal Fermi-gas background, we must conclude that this new phase transition completely excludes the Wigner crystallization because of its melting curve placed deeply inside not only the coexistence domain but just inside the spinodal curve (Fig.1). So at non-zero temperatures this crystal is absolutely unstable against the phase decomposition into two fluid weakly coupled plasma - a dense phase and a rare one. For $T \approx 0$ this supposition was declared previously in [15].

Double OCP model.

The background thermodynamic properties may be define more meaningfully with the use of the variational principle of statistical mechanics. The choice of the N-particle distribution function in a multiplicative form as a product for nucleus and electrons $\rho_N = \rho_{(+)} \rho_{(-)}$ is equivalent to the complete switching off individual correlations between two subsystems of charges while (++) and (--) correlations are completely permitted. So the nucleus and electrons turn for each other into the uniform and compressible compensating background. This combination of two inserted one into another OCP-models will be called by the term "Double OCP". By the switching off another kinds of dynamic correlations inside each subsystems we can obtain the hierarchy of simplest plasma models with the decreasing free energy. Each of them presents the upper bound for a free energy of real plasma, the Double OCP being the best.

$$F \leq F_{ocp}^{(+)} + F_{ocp}^{(-)} \leq \left\{ \begin{array}{l} F_{ocp}^{(+)} + F_{HF}^{(-)} \leq F_{ocp}^{(+)} + F_{id.g}^{(-)} \\ F_{ocp}^{(-)} + F_{HF}^{(+)} \leq F_{ocp}^{(-)} + F_{id.g}^{(+)} \end{array} \right\} \leq F_{id.g}^{(+)} + F_{id.g}^{(-)} \qquad (4)$$

Being applied to the real substances, the Double OCP model gives the phase diagram depending on the masses of positive charges. For heavy particles the phase diagram is similar to those in Fig.1. The main difference is additional cold melting in the nuclear subsystem in the high density limit (Fig.4). For the mass-symmetrical plasmas (electron-positron and electron-holes plasmas) the phase diagram is similar to this

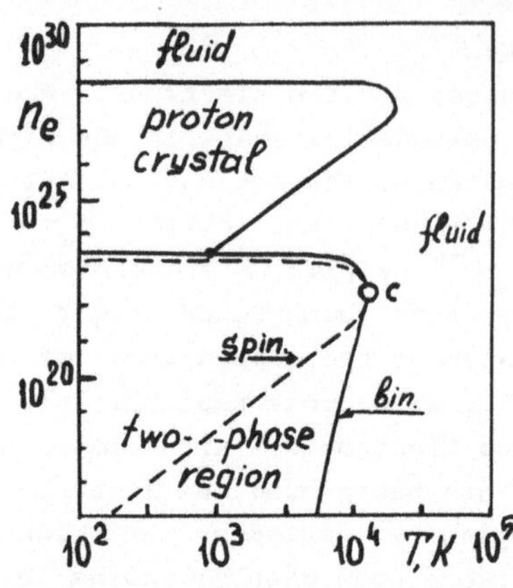

for electron gas with the only phase transition of gas-liquid type. The Double OCP model gives a simple estimation for the critical temperature of this phase transition. For example in electron-positron plasma $kT_c \approx 0.6$ eV.

Fig.4. Phase diagram of the Double OCP model of electron-proton system. Estimation with the data [14] and modified results Fig.1 .

OCP Phase Transition and Anomalies in the Equilibrium Spatial Charge Distribution

The discussed OCP phase transition seems to be practically unknown. Let us consider the situation where our knowledge of the properties of this phase transition may be useful. It is so in every case of a thermodynamically equilibrium spatial charge distribution in nonuniform plasma. An example is the quantum electron distribution in atomic cell. This problem may be formulated in terms of the Density Functional (for example [16]). The Free Energy Functional $F[n(\cdot)]$ is minimized over the electron density $n(r)$ with the additional normalization condition

$$F[n(\cdot)] = e\int \varphi_{ext}(\bar{x})n(\bar{x})d\bar{x} \ + \ \frac{e}{2}\int \frac{n(\bar{x})n(\bar{y})}{(\bar{x}-\bar{y})} \ d\bar{x}d\bar{y} \ + \ F^*[n(\cdot)] \qquad (5)$$

$$\int n(\bar{x})d\bar{x} = Z$$

In the local approximation the "Exchange-Correlation-Kinetic" term F^* is

$$F^*[n(\cdot)] = \int f(\bar{x})n(\bar{x})d\bar{x} \ ; \qquad f(\bar{x}) = \lim \left\{ \frac{F(N,V,T)}{N} \right\} \qquad (6)$$

$$(N \to \infty \ ; N/V = n)$$

When $F(N,V,T)$ is the free energy of Boltzmann or Fermi ideal gas we deal with the well known Poisson-Boltzmann or Thomas-Fermi approximations [16]. It ought to be emphasized that in frames of approximation (5, 6) any attempt to take into account the electron correlations by using the electron gas exact free energy $F(N,V,T)$ will require the equation of state of OCP with the compressible background. As a result the discussed OCP phase transition appears in the case of sufficiently low density and temperature ($T \le T^*$) in the form of discontinuity in a spatial charge distribution inside the cell. So the smooth Thomas-Fermi (or Poisson-Boltzmann in classical case) distribution will be subdivided into the condensed "drop" around an attractive center and the diffuse "atmosphere" at the cell periphery. In terms of the OCP phase transition the boundary temperature T^* is just the critical temperature (for electron gas $T_c \sim 1$ eV), and two values of local densities (Fig.5)

are the densities of coexisting condensed and gaseous phases depending
on temperature only. It ought to be noted especially that this   phase
decomposition occurs in the system of particles with entirely repulsive
interaction.

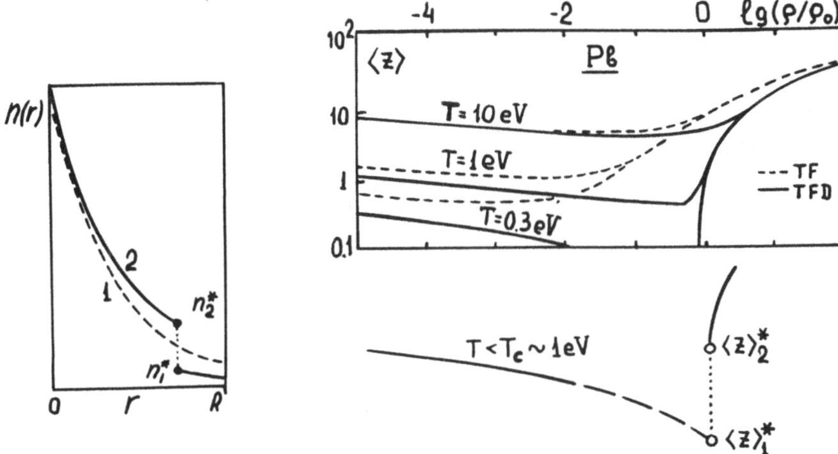

Fig.5. Electron gas profile in atomic cell at $0<T<T_c^{OCP}$. Qualitative comparison of two local approximation (5,6): 1- Thomas-Fermi; 2 - additional exchange and correlation correction in frames of exact equation of
state of strongly coupled OCP.

Fig.6. The "Ionization degree" of Pb atomic cell in TF and TFD approximations (Fig.2 from [19]). The curve under the figure represents the
supposed profile at low temperatures $0< T < T_c \sim 1$ eV.

This example allows to predict the appearance of similar phase
stratifications in all the cases when the equation of state of strongly
coupled OCP being used to calculate the nonuniform charge distribution
in local approximation (6). For example, it occurs in the cases of
equilibrium charges near charged hard wall or the edge of background,
the same-sign charges in macroscopic shell or cavity and so on [7,17].

It should be noted that it is not necessary to use in (6) the exact
equation of state of strongly coupled OCP for the appearance of discussed discontinuity. For this it is sufficient to use any local attractive correction (exchange or correlation). Therefore this discontinuity
must appear as a consequence of the Thomas-Fermi-Dirac approximations
or other local modifications. It is really so, just the same discontinuities were shown in the figures (12÷ ÷15) at the famous book of
P.Gombash [18]. Note that the density of a "gaseous" part of the profile in the Fig.5 is equal to zero for $T=0$. For $T \neq 0$ calculations in
TFD-approximation of electron distribution in atomic cell was made in
[19]. There the degree of ionization was defined to be proportional to

the boundary electron density. We can conclude this value to be discontinuous for $T < T_c^{OCP} \sim 1,0$ eV . Unfortunately this part of results was omitted in the figures 1,2 of [19]. The supposed structure of omitted part of this curves at $T \leq T_c$ are shown schematically under Fig.6. It ought to be noted that the range of this discontinuity must increase when $T \to 0$ so that $<z>_1^* \to 0$ and $<z>_2^* \to$ const≈ 0.035 (it corresponds to the boundary electron density $r_s \approx 4.2$ and $P_e = 0$).

Conclusion

The next models of OCP-family mentioned above (with the compressible but uniform background) seem to be interesting for future study:

1. Double OCP plasmas (exact parameters calculation)

2. Binary Ionic Mixture (BIM). Interrelation of present phase transition with the well-known phase decomposition in BIM-model [2,3].

3. Classical Charged Hard and Soft Spheres.

4. Classical Point Charges (+background) with the repulsion $\sim 1/r^S$ ($s \leq 2$), and Hard Spheres with the repulsion $\sim 1/r^S$ ($s \leq 3$).

References

[1] J. P. Hansen, Phys. Rev. **A**, **8**, 3096, (1973)

[2] M. Baus, J. P. Hansen, Phys. Reports, **59**, 1, (1980)

[3] S. Ishimaru, H. Yyetomi, S. Tanaka, Phys. Reports, **149**, 91, (1987)

[4] E.H.Lieb, H.Narnhofer, J.Stat.Phys.**12**,291,(1975);Ibid.**14**,465,(1976)

[5] E. L. Pollock, J. P. Hansen, Phys. Rev. **A**, **8**, 3110, (1973)

[6] I.L.Iosilevski, in "Equation of State under the Extreme Conditions" (in russian) /ed.by G. Gadiyak/, (Novosibirsk, 1981), p. 20

[7] I.L.Iosilevski, Teplofiz. Vysok. Temp. (in russ.), **23**, 1041, (1985)

[8] F. J. Dyson, A. Lenard, Journ. Math. Phys. **9**, 698, (1968)

[9] N.N.Kalitkin,L.V.Kuzmina,J.Vych.Met.Mat.Fiz.(in russ.)**15**,768,(1975)

[10] R.Zimmerman, Many-Particle Theory of Highly Exited Semiconductors, (Tubner Verlag, Leipzig, 1988)

[11] D. Ceperley, Phys. Rev. **B**, **18**, 3126, (1978)

[12] A.A. Likalter, Dokl.Akad.Nauk (ДАН СССР,in russ.), **259**, 96, (1981)

[13] I. L. Iosilevski, Proc. XIII All-Union Conference on Physics of Aerodisperse Systems, (in russ.), (Odessa, 1986), v1, p.69

[14] R. Mochkovitch, J. P. Hansen, Phys. Let., **73A**, 35, (1979)

[15] N. Wiser, M. Cohen, J. Phys. Ser.C, **2**, 193, (1969)

[16] D. A. Kirznitz, Y. E. Lozovik, G. V. Shpatakovskaia, Usp.Fiz.Nauk (in russ.), **117**, 3, (1975)

[17] I. L. Iosilevski, Proc. XIII All-Union Conference on the Equation of State (in russ.) /ed. by V.Fortov/, (IVTAN, Moscow, 1990), p.10

[18] H. Gombas, Die Statistische Theorie des Atoms. Vien, (1949)

[19] H. Szichman, S. Eliazer, D. Salzman, JQSRT, **38**, 281, (1987)

MEAN SPHERICAL APPROXIMATION FOR THE THERMODYNAMICS OF PARTIALLY IONIZED AND STRONGLY COUPLED PLASMAS

ANDREAS FÖRSTER AND WERNER EBELING

Institut für Theoretische Physik, Humboldt-Universität zu Berlin,
Invalidenstraße 110, D-1040 Berlin, Germany

Abstract

The combined action of Coulomb attraction and short-range repulsion between the atoms and ions in a dense plasma is described within the mean spherical approximation (MSA). We propose modifications of the original MSA results to account for the reduced volume effect and further electronic contributions to the thermodynamic functions. A numerical example is provided for the composition and the equation of state of Xe plasma.

INTRODUCTION

Within the chemical picture different bound states of electrons and nuclei are introduced as individual "chemical species". These species are distinguished by their total charge and mass, the internal energy spectrum, and by other parameters like, e.g., an effective radius, and are considered as the elementary constituents of the plasma. Following this concept, we regard the plasma as composed of atoms, ions, and free electrons, which interact through pair potentials. At long distance, the interaction between all charged particles is governed by pure Coulomb forces. The complex structure of the finite-size ions plays a minor role and give rise only to corrections of higher order, originating, e.g., from polarisation. At short separation the bound-electron shells appear to be essential and, due to the Pauli exclusion principle, lead to a strong repulsion between all ions and atoms. These potential branches are not well known. Generally, serious problems arise for a strong theoretical treatment of the short-range interaction between the finite-size plasma particles. The theoretical determination of the structure factor is a tedious task and has been carried out only for a small number of ions. As an alternative, we suggest in this article a rather crude approach by modelling the ions as charged hard spheres and utilizing the concept of the mean spherical approximation (MSA). The MSA was originally developed to analyse electrolytic solutions and molten salts, which naturally involve extended ions, and works very well up to high densities. One of the convenient features of the MSA is that it gives explicit, yet general results for the thermodynamic functions. We believe that these results may be successfully applied to dense plasmas.

THE FREE ENERGY FOR A MIXTURE OF CHARGED HARD SPHERES

We restrict our consideration to a plasma of only one chemical element and neglect the formation of molecules, clusters, and negative ions. Consequently, the plasma is described as consisting of atoms with particle density n_0, ions with charges $+ z e$ and densities n_z, and free electrons with density n_e. The total heavy particle density n is defined by

$$n = \sum_z n_z \; . \tag{1}$$

z denotes the charge numbers of the heavy particles with Z_{max} beeing the maximum charge number of the model, $0 \le z \le Z_{max} \le Z_{nucl}$. If not otherwise stated, summation over z runs from 0 to Z_{max}. As usual, the electro-neutrality condition should be fulfilled,

$$n_e = \sum_z z \, n_z \; . \tag{2}$$

The detailed plasma composition is conveniently expressed by the relative fractions of the heavy particles,

$$\alpha_z = n_z/n \; . \tag{3}$$

A reduced description of the ionisation state is based upon the mean charge of the heavy particles $\bar{z}$,

$$\bar{z} = \sum_z z \, \alpha_z \; . \tag{4}$$

For fixed volume V, temperature T, and heavy particle density n, the free energy density f is usually chosen to be the basic thermodynamic potential,

$$F/V = f(n_0, n_1, \ldots, n_{Z_{max}}, n_e, T) \; . \tag{5}$$

In thermodynamic equilibrium f assumes a minimum with respect to the individual heavy particle densities $\{n_z\}$.

Debye and Hückel (1923) were the first who payed attention to the influence of the finite ionic radii on the thermodynamic properties. They considered an electrolytic solution of ionic species with different charge numbers $\{z\}$ and diameters $\{d_z\}$ and solved the Poisson equation for the electrical microfield in the case of weak interaction. The influence of the finite radii appears as additional multipliers τ_{DH} in the well-known Debye-Hückel law,

$$f = f_{id} - \frac{k_B T \kappa}{12 \pi} \sum_z \frac{n_z z^2 e^2}{\varepsilon k_B T} \tau_{DH}(x_z) \; ,$$

$$x_z = \kappa d_z \; , \tag{6a-c}$$

$$\tau_{DH}(x) = \frac{1}{x^3} \left(\frac{9}{2} + 3 \ln(1 + x) - 6 (1 + x) + \frac{3}{2} (1 + x)^2 \right) \; ,$$

with

$$\kappa^2 = \sum_z \frac{n_z z^2 e^2}{\varepsilon k_B T} \; . \tag{7}$$

Here, f_{id} denotes the ideal part of the free energy density, κ the inverse Debye-Hückel length, and ε the electrical permeability of the solution. For plasmas, the summations over z in Eqs. (6a) and (7) include additionally the electrons with $z_e = -1$ and $d_e = 0$. The pioneering article of Debye and Hückel provided the fundament of forthcoming reseach in the theory of electrolytes and dense plasmas. However, their theory works well only for very diluted systems.

The mean spherical approximation was proposed by Lebowitz and Percus (1966) as a generalization of the spherical model for Ising spin systems to classical fluids and has exact solutions for several interesting cases. The basic MSA integral equation consists of the Ornstein-Zernike equation,

$$h_{zz'}(r) = c_{zz'}(r) + \sum_{z''} n_{z''} \int d^3r' \, h_{zz''}(|r'|) \, c_{z''z'}(|r - r'|) \; , \tag{8}$$

together with the closure relations for the total correlation functions $h_{zz'}(r)$ and the direct correlation functions $c_{zz'}(r)$:

$$h_{zz'}(r) = -1 \; , \quad \text{for} \quad r < \frac{1}{2} (d_z + d_{z'}) \; , \tag{9}$$

$$c_{zz'}(r) = - \frac{z \, z' \, e^2}{4 \pi \varepsilon r k_B T} \; , \quad \text{for} \quad r > \frac{1}{2} (d_z + d_{z'}) \; . \tag{10}$$

The first condition, Eq. (9), is exact for hard-sphere particles and the second one, Eq. (10), is the specific assumption of the MSA. Waisman and Lebowitz (1970, 1972) solved this equations for the primitive model of a electrolyte with ions of arbitrary charges $\{z\}$ and equal diameters $d_z \equiv d$ and found

$$f = f_{id} + f_{hs} - \frac{k_B T \kappa^3}{12 \pi} \tau_{WL}(x) \; ,$$

$$x = \kappa d \; , \tag{11a-b}$$

$$\tau_{WL}(x) = \frac{1}{x^3} \left(2 + 6 \, x - 2 \, (1 + 2 \, x)^{3/2} + 3 \, x^2 \right) . \tag{11c}$$

Here, f_{hs} denotes the excess contribution of the uncharged hard-sphere system. Note, that for small κ and/or small diameters the series expansion of the Waisman-Lebowitz result coincides with that of the Debye-Hückel theory. The correction multipliers $\tau_{DH}(x)$ and $\tau_{WL}(x)$ have the same limit for $x \to 0$:

$$\tau_{DH}(x) = 1 - \frac{3}{4} \, x + \frac{3}{5} \, x^2 + O(x^3) , \tag{12}$$

$$\tau_{WL}(x) = 1 - \frac{3}{4} \, x + \frac{3}{4} \, x^2 + O(x^3) . \tag{13}$$

The exact solution for the unrestricted general case of the MSA which allows for arbitrary charges $\{z\}$ and arbitrary diameters $\{d_z\}$ was constructed by Blum (1975) and Blum and Høye (1977).

Another class of MSA model systems involves besides the charged hard spheres a rigid, uniform, and neutralizing charge background with density ρ (Palmer and Weeks, 1973). The most general case of arbitray charges and diameters was treated by Parrinello and Tosi (1979). Their resulting formulae for the free energy density read

$$f = f_{id} + f_{hs} + f_{PT} , \tag{14a}$$

$$f_{PT} = k_B \, T \, \frac{K^3}{3 \, \pi} - \frac{e^2}{4 \, \pi \, \varepsilon} \left\{ K \sum_z \frac{n_z \, \tilde{z}^2}{1 + K \, d_z} + \frac{\pi}{2 \, \Delta} \, \Omega \, P_n^2 + \right.$$

$$\left. + \frac{\pi}{2} \, \rho \sum_z n_z \, z \, d_z^2 - \frac{\pi^2}{30} \, \rho^2 \sum_z n_z \, d_z^5 \right\} , \tag{14b}$$

with

$$\Delta = 1 - \frac{\pi}{6} \sum_z n_z \, d_z^3 , \qquad \tilde{z} = z - \frac{\pi}{6} \, \rho \, d_z^3 ,$$

$$\tag{14c-f}$$

$$\Omega = 1 + \frac{\pi}{2 \, \Delta} \sum_z \frac{n_z \, d_z^3}{1 + K \, d_z} , \qquad P_n = \frac{1}{\Omega} \sum_z \frac{n_z \, \tilde{z} \, d_z}{1 + K \, d_z} .$$

The electro-neutrality condition reads now

$$\rho = \sum_z z \, n_z . \tag{15}$$

This solution is expressed in terms of a single parameter K which can be obtained from the nonlinear equation

$$(2 \, K)^2 = \frac{e^2}{\varepsilon \, k_B \, T} \sum_z n_z \left(\frac{\tilde{z} - \frac{\pi}{2 \, \Delta} \, d_z^2 \, P_n}{1 + K \, d_z} \right)^2 . \tag{16}$$

The first two terms of f_{PT} correspond to the Debye-Hückel result, Eq. (6a), with $2K$ beeing the analogue of the Debye-Hückel parameter κ. For small diameters $\{d_z\}$ we obtain again in first order of $(\kappa\, d_z)$

$$f_{PT} = -\frac{k_B T \kappa}{12\,\pi} \sum_z \frac{n_z\, z^2\, e^2}{\varepsilon\, k_B\, T} \left(1 - \frac{3}{4}\,\kappa\, d_z + o(\kappa\, d_z)\right). \qquad (17)$$

in full correspondence with Eqs. (6)+(12) and (11)+(13), respectively.

APPLICATION TO DENSE PLASMAS

Now, we will use the solution of Parrinello and Tosi as starting point to model a partially ionised and strongly coupled plasma. There are two principally different ways to apply the PT solution to plasmas (EBELING et al., 1991):

(a)   The electrons are considered as point charges with $z_e = -1$ and $d_e = 0$. Then the sums over z in Eqs. (14)-(16) cover all species including the electrons and we set $\rho = 0$.

(b)   We consider the electrons as a uniform charge background. Then their density n_e is related with the density ρ and the sums over z cover only the heavy particles.

The first case is related to high temperatures and/or low densities when the electrons form a classical gas and the screening capabilities of electrons and ions are comparable. The second case is related to low-temperatures and/or high densities when the electrons are degenerated. Then, the polarizability of the electron gas is small and its screening capability is low. For the remainder of this article we will deal only with the second case.

In the PT version of the MSA, the charge background homogeneously penetrates the extended particles. This situation has no adequate correspondence in real plasmas: the free electrons are not allowed to penetrate into the atoms and ions due to the Pauli exclusion principle. Therefore, we propose certain modifications of the PT result:

(i)   we substitude z by $z + \frac{\pi}{6}\,\rho\, d_z^3$ in Eqs. (14) and (15) to compensate the background charge contained inside the extended particles,

(ii)   we cancel the last two terms in Eq. (14b) which represent the selfenergy of the background charge included inside the hard spheres and its interaction with the ionic charge,

(iii) we put $\rho = n_e^* = n_e/\Delta$ in order to fulfill the electro-neutrality condition of Eq. (15).

The modifications (i) and (ii) formally reduce the solution of

Parrinello and Tosi to that of Blum and Høye with the only difference that the electro-neutrality condition still involves the electronic background ρ. The modification (iii) has an obvious physical meaning: since the free electrons are not allowed to penetrate into the volume which is filled with atoms and ions the volume accessible to the electrons is reduced and their local density n_e^* is higher than the average density with respect to the whole system n_e.

Our total model of the free energy density reads as follows:

$$f = f_{id,i} + f_{id,e}^* + f_{PT,mod} + f_{hs} + f_{xc,e}^* \ . \tag{18}$$

The first two terms denote the usual ideal contributions of atoms and ions (i) and of electrons (e) following from the Boltzmann statistics and from the Fermi-Dirac statistics, respectively. The third term stands for the modified MSA, Eqs. (14) with modifications (i)-(iii). The excess contribution of the uncharged hard-sphere system (hs) is modeled by the quasi-exact formula of Mansoori et al. (1971). The last term of the free energy density stands for the exchange-correlation contribution of the electron gas (xc,e). Here, we take adventage of a Padé-fit formula given by Tanaka et al. (1985) which reproduce numerical data obtained from the Singwi-Tosi-Land-Sjölander theory. The asterisk * means, that the expression is rescaled according to the generalized concept of the reduced volume for the electrons (Förster, 1991; Förster et al., 1992), i.e., throughout the marked expressions the *local* density of the free electrons is replaced by $n_e^* = n_e/\Delta$. In this article, we extend this principle to the interaction contributions of the electrons, in contrast to our previous work, where only the ideal contribution $f_{id,e}$ has been rescaled. Note, that in this procedure the average electron density with respect to the whole system, i.e. with respect to the system volume V, always remains unchanged, e.g., if $f = n_e \phi(n_e)$ so $f^* = n_e \phi(n_e^*)$. Finally, the charge interaction between the electrons and the heavy particles is assumed to be small and has been neglected in our model.

NUMERICAL RESULTS AND DISCUSSION

In continuation of our previous efforts (Ebeling et al., 1988; Kahlbaum and Förster, 1990) we present a numerical example for xenon. In the ideal contribution of the heavy particles, the statistical sums are restricted to the ground state only. A consequent account of the excited levels of the atoms and ions would require to consider a system of at least hundreds of species with different radii and charges (Ebeling, 1990) what is outside the scope of this article. The

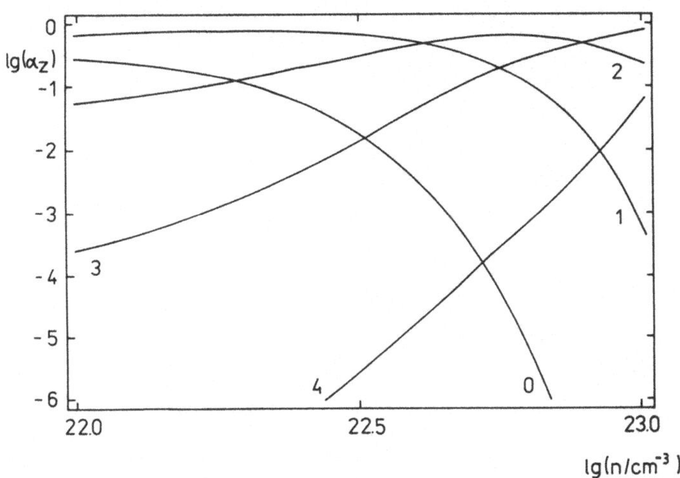

Figure 1   Relative fractions of the heavy particles $\{\alpha_z\}$ vs. the heavy particle density n for T = 30000 K. The curves are labeled by the charge numbers z.

hard-sphere radii $\{d_z\}$ are generally not well known, especially for the higher charged ions and, therefore, we use the simple fit of Bespalov and Polishchuk (1989), $d_z = a_o/(0.043\ (I_z/eV)^{1/2})$, where $\{I_z\}$ denote the unperturbed ionisation energies and a_o the Bohr radius. The equilibrium state of the plasma was determined by solving the coupled set of nonideal Saha equations,

$$\mu_z = \mu_e + \mu_{z+1}\ , \qquad z = 0,\ \ldots,\ z_{max}-1\ , \tag{19}$$

supplemented by the electro-neutrality condition of Eq. (2), with $\{\mu_\nu,\ \nu=z,e\}$ beeing the chemical potentials. Fig. 1 shows the detailed composition for one particular temperature and the most interesting density range where the pressure ionisation occurs. At these densities the electrons are highly degenerated, $n_e^*\ \Lambda_e^3 \geq 1$ (Λ_e - thermal de Broglie wavelength of the electrons). With increasing density, we observe the subsequent interchange of dominant species starting from a low ionised plasma towards full ionisation. Simultaneously, the mean charge of heavy particles (Fig. 2) monotonously increases with more or less pronounced steps at integers of $\bar{z}$ (Förster et al., 1991). We suppose that these steps disappear after a complete account of the excited levels. Finally, we calculated the pressure (thermal equation of state) according to

$$p = - f + n_e\ \mu_e + \sum_z n_z\ \mu_z\ . \tag{20}$$

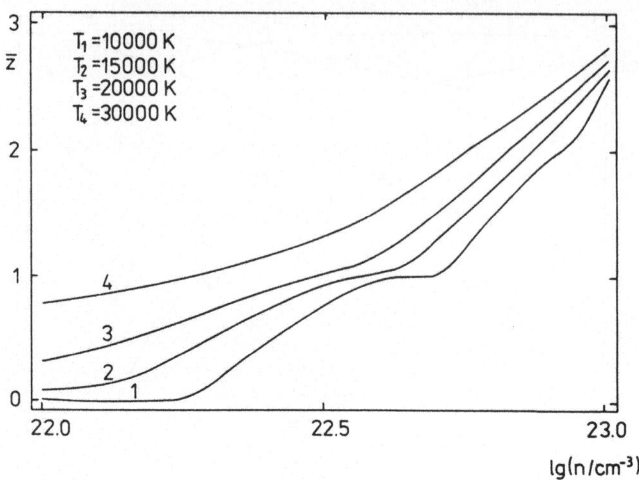

Figure 2   Mean charge of heavy particles $\bar{z}$ vs. the heavy particle density n for four different temperatures.

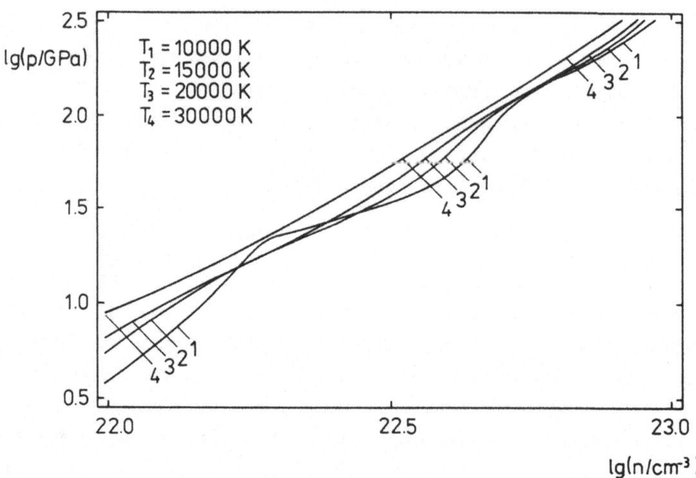

Figure 3   Pressure p vs. the heavy particle density n for four different temperatures.

and present in Fig. 3 several isotherms. The isotherms show a local flattening at the same densities where the interchanges of dominant species in the plasma composition take place. With lowering of the temperature, the flattening transforms into branches of thermodynamic instability what can be interpreted as the occurance of the plasma phase transition (Ebeling and Grigo, 1980; Ebeling et al., 1991). We observe first a phase transition which is mainly connected with the interchange $Xe^0 \rightarrow Xe^+$. The corresponding coexistence line shows a negative slope and ends up in a critical point at $T_c = 8680$ K, $p_c = 25.5$ GPa, $n_c = 2.1\ 10^{22}$ cm^{-3}, and $\bar{z} = 0.09$. At temperatures around 5000 K we detect another phase transition which is linked with the ionisation step $Xe^+ \rightarrow Xe^{++}$. As a substantial improvement of our previous results (Kahlbaum and Förster, 1990), these findings not depend artificially on the maximum charge number Z_{max} and the two-phase regions lie entirely inside the window of validity. The relatively low critical temperature of the first transition, as compared with the phase diagram of xenon by Ebeling et al. (1988), is due to the neglection of the charge interaction between the electrons and the heavy particles. We mention the forthcoming work of Kahlbaum and Schneidenbach (1992), who adopted our model, completed it by a proposal for the electron-ion interaction, and found higher critical temperatures. However, the poor model for the neutral fluid and the unphysically high packing fraction in the transition regions, $\Delta \leq 0.5$, do not allow to present a reliable recalculation of the phase diagram.

Instead of a summary we mention the necessary improvements of a MSA-based theory for partially ionised and strongly coupled plasmas:
- the link of the MSA with elaborated theories for the neutral fluid, like the fluid perturbation theory, what is necessary for a description of low ionisation,
- the transition to *effective* radii for the bound states of electrons and nuclei, i.e. of the ground and excited states of atoms and ions, which are included into the model,
- the derivation of quantum-theoretical expressions for the electron-ion interaction,
- the link of the MSA with models for point charges at high densities to describe the transition to the full ionised state.

Following this line of successive improvements we may hope to arrive at a reliable combination of MSA and quantum theory of plasmas.

REFERENCES

I.M. Bespalov, A.Ya. Polishchuk, 1989, Pisma v Zh. Tekhn. Fiz. (USSR) 15, no. 2, 4 (in Russian)

L. Blum, 1975, Mol. Phys. 30, 1529

L. Blum, J.S. Høye, 1977, J. Phys. Chem. 81, 1311

P. Debye, E. Hückel, 1923, Physik. Z. 24, 185

W. Ebeling, 1990, Z. phys. Chem. **271**, 233

W. Ebeling, A. Förster, V.E. Fortov, V.K. Gryaznov, A.Ya. Polishchuk, 1991, *Thermophysical Properties of Hot Dense Plasmas* (TEUBNER-TEXTE zur Physik, Vol. 25), Teubner Stuttgart Leipzig

W. Ebeling, A. Förster, W. Richert, H. Hess, 1988, Physica A **150**, 159

W. Ebeling, M. Grigo, 1980, Ann. Physik (Leipzig) **37**, 21

A. Förster, 1991, *Composition, Equation of State, Phase Diagram, and Ionization Kinetics of Strongly Coupled and Partially Ionized Plasmas* (Dr.rer.nat. Thesis), Humboldt-Universität zu Berlin (in German)

A. Förster, T. Kahlbaum, W. Ebeling, 1992, Laser Part. Beams **10**, 215

A. Förster, T. Kahlbaum, A. Rickert, 1991, Suppl. Z. Phys. D **21**, S171

T. Kahlbaum, A. Förster, 1990, Laser Part. Beams **8**, 753

T. Kahlbaum, H. Schneidenbach, 1992, in: Contributions to the 1992 International Conference on Plasma Physics, Innsbruck

J.L. Lebowitz, J.K. Percus, 1966, Phys. Rev. **144**, 251

G.A. Mansoori, N.F. Carnahan, K.E. Starling, T.W. Leland, Jr., 1971, J. Chem. Phys. **54**, 1523

R.G. Palmer, J.D. Weeks, 1973, J. Chem. Phys. **58**, 4171

M. Parrinello, M.P. Tosi, 1979, Chem. Phys. Lett. **64**, 579

S. Tanaka, S. Mitake, S. Ichimaru, 1985, Phys. Rev. A **32**, 1896

E. Waisman, J.L. Lebowitz, 1970, J. Chem. Phys. **52**, 4307

E. Waisman, J.L. Lebowitz, 1972, J. Chem. Phys. **56**, 3086 and 3093

QUASICLASSICAL SHELL MODEL AND THERMODYNAMICAL

FUNCTIONS OF DENSE PLASMA

E.A.Kuzmenkov

Thermophysics Intense Actions Scientific Center,
Izhorskaya 13/19, Moscow, 127412, Russia

G.V.Shpatakovskaya

National Center of Mathematical Modelling,
Miusskaya 4, Moscow, 105047, Russia

Abstract

A quasiclassical shell model (QM) is developed to
describe the thermodynamical functions of nonideal
dense plasma. QM model can be employed over a wide
range of densities and temperatures from the Saha model
region of application to the corrected Thomas-Fermi
model area of use.

INTRODUCTION

The quasiclassical shell model is proposed for the matter with a
high energy concentration. The statistical models [1] commonly
employed in this range of parameters are the Thomas-Fermi (TF) model
[2], the TF model with gradient and exchange corrections (TFC)
[3,4], and the quantum-statistical model [5-7] treating these
corrections self-consistently. The statistical TF model is a
dominant term of the Hartree-Fock model expansion in a semiclassical
parameter $\xi = \dfrac{d\lambda}{dx}$ (λ is de Broiglie wavelength). It averages all the
physical quantities ignoring the atomic shell structure. The
gradient and exchange corrections have a second order in ξ and do
not change the averaged behaviour of physical functions. However, a
Taylor series in ξ can't describe all the atomic system features
because the functions are nonanalytical in ξ. There are terms
$\sim\xi^n\exp(1/\xi)$ and a separation of the dominant such term enables us to
consider the shell effects [8] and to construct the equation-of-
state(EOS) quasiclassical model. It should be noted that some

semiclassical aspects (wave functions, quantization condition) one can find in different publications, for example in [9-11]. But if one uses a succesive semiclassical approach all the resulting physical quantities are the terms of an expansion series in a semiclassical parameter ξ, the dominant TF term, the gradient, exchange and shell corrections being included. A similar approach has been used independently in [12] to describe the total binding energy of an isolated atom at zero temperature.

GENERAL PRINCIPLES

The TF model in a Wigner-Seitz spherical cell is the basis of our theory. Electron density, particle number, free energy, internal energy, and pressure in the TF model are (in the following equations the atomic units are used):

$$\rho_{TF}(r) = \frac{\sqrt{2}}{\pi^2} T^{3/2} I_{1/2}\left(\frac{\mu-V(r)}{T}\right) \quad , \quad (1)$$

$$N_{TF} = \int d\bar{r}\, \rho_{TF}(r) \quad , \quad (2)$$

$$F_{TF} = -\frac{2\sqrt{2}}{3\pi^2} T^{5/2}\int d\bar{r}\, I_{3/2}\left(\frac{\mu-V(r)}{T}\right) - \frac{1}{2}\int d\bar{r}\, \rho_{TF}(r)\left(V(r)+\frac{Z}{r}\right) + \mu N \quad , \quad (3)$$

$$E_{TF} = \frac{\sqrt{2}}{\pi^2} T^{5/2}\int d\bar{r}\, I_{3/2}\left(\frac{\mu-V(r)}{T}\right) + \frac{1}{2}\int d\bar{r}\, \rho_{TF}(r)\left(V(r)-\frac{2Z}{r}\right) \quad , \quad (4)$$

$$P_{TF} = \frac{2\sqrt{2}}{\pi^2} T^{5/2} I_{3/2}(\mu/T) \quad , \quad (5)$$

where T, μ, Z, and $I_n(X)$ respectively are the temperature, chemical potential, nuclear charge, and Fermi-Dirac integral, and $V(r)$ is a potential electron energy. If one wants to account for some other effect in addition to the TF approach there is a general expression for the correction to the TF free energy through some effect [1]:

$$dF = -\int_{-\infty}^{\mu} d\mu'\, dN(\mu') \quad , \quad (6)$$

where $dN(\mu)$ is the particle number correction, the correction to the TF electron density $\delta\rho$ through the effect being small. In the treated nonrelativistic self-consistent potential approach the TF method does not consider the effects of the electron density inhomogeneity, exchange interaction, and atomic shell structure. The use of Eq.(6) leads us to construct the additive expression for the free energy with regard to all the above effects on the TF model basis. The result is a quasiclassical EOS model (QM):

$$F_{QM} = F_{TF} + \delta F_{qu-ex} + \delta F_{sh} \quad , \tag{7}$$

$$E_{QM} = E_{TF} + \delta E_{qu-ex} + \delta E_{sh} \quad , \tag{8}$$

$$P_{QM} = P_{TF} + \delta P_{qu-ex} + \delta P_{sh} \quad , \tag{9}$$

where subscript $qu-ex$ and sh denote quantum-exchange and shell. Since quantum-exchange corrections are calculated as in the TFC model [4], in this paper we consider the shell corrections only. One can derive from Eq.(6) the pressure correction in a general form

$$\delta P = - \left.\frac{\partial \delta F}{\partial W}\right|_T = \rho_{TF}(R_0)\,\delta\mu + \int_{-\infty}^{\mu} d\mu'\delta\rho(\mu',R_0) \quad , \tag{10}$$

where R_0 is the spherical cell radius, and W is the atomic cell volume. To obtain from Eq.(6) the relation for a shell correction to the internal energy one needs the form of the shell density correction. When using the spherically symmetric semiclassical wave function, one can obtain the electron density and particle number expressions:

$$\rho(r,\mu) = \sum_{(n)} |\Psi_{(n)}(\bar{r})|^2 f_F\left(\frac{E_{(n)} - \mu}{T}\right) =$$

$$= \frac{1}{2\pi^2} \int d\bar{\ae} \sum_{n,l} f_F\left(\frac{E_{nl}(\ae)-\mu}{T}\right) \frac{2l+1}{r^2 p_{nl}(r)} \frac{dE_{nl}}{dn} \quad , \tag{11}$$

$$N(\mu) = 2 \int d\bar{\ae} \sum_{n,l} (2l+1)\, f_F\left(\frac{E_{nl}(\ae)-\mu}{T}\right) \quad , \tag{12}$$

where $\ae$ is quasiimpulse, $p_{nl}(r) = \sqrt{2(E_{nl}-V(r)-(l+1/2)^2/2r^2)}$, n and l are the principal and angular quantum numbers, $f_F(X)$ is Fermi function. If the sums $\sum_{n,l}$ in Eqs.(11) and (12) are replaced by integrals $\iint dn\,dl$, all the information about shell effects is lost and we obtain the TF density expression. The use of the Poisson equation

$$\sum_{n=a}^{b} \mathfrak{Z}_n = \sum_{k=-\infty}^{\infty} \int_{a-e}^{b+e} dn\,\mathfrak{Z}(n)\,\cos(2\pi kn), \qquad 0 < e < 1$$

enables us to make the proper substitution $\sum_{n\,l}\ldots \rightarrow \int dn\,dl\ldots$. Subtracting the TF term ($k=s=0$), we obtain the density shell correction and the internal energy shell correction :

$$\delta E_{sh} = \delta F_{sh} - T\,\frac{\partial \delta F_{sh}}{\partial T} = \int d\bar{r}\left(\frac{3}{2}\,\rho(r) + \frac{\partial\rho}{\partial\mu}\,V(r)\right)(\delta\mu - \delta V) + \delta E_2 \quad , \quad (13)$$

$$\delta E_2 = \frac{1}{2\pi^2}\int d\bar{r}\sum_{k,s}{}'\int d\bar{k}\int dE\ E\ f_F\left(\frac{E-\mu}{T}\right)\int dl\,\frac{2l+1}{r^2 p_{El}(r)}\,\cos(2\pi kn)\cos(2\pi sl) \quad , \quad (14)$$

where $\sum_{k,s}'$ denotes the sum without term $k=s=0$.

To calculate the integrals in Eqs.(10-14) we need the spectrum that links E with n and l. There are both discrete and band spectra for the compressed atom. The Bohr-Sommerfeld quantization condition

$$S_{El} = \pi(n-l-1/2)$$

is inapplicable in this case ($S_{El} = \int dr\, p_{nl}(r)$ is an action). So in order to describe such spectrum features, we need a generalized quantization condition for which a number of requirements must be :

1. For strongly bound electrons this quantization condition transforms to Bohr-Sommerfeld one.

2. For free electron states it describes a continuous spectrum.

3. For $\mathfrak{æ}=0$:

$$\mathfrak{R}_l(E_n(\mathfrak{æ}=0),R_0) = 0, \qquad l \text{ is even}$$

$$\mathfrak{R}_l(E_n(\mathfrak{æ}=0),R_0) = 0, \qquad l \text{ is odd}$$

where $\mathfrak{R}_l(E,r)$ is a radial wave function.

4. The function $E(\mathfrak{æ})$ near the point $\mathfrak{æ}=0$ is quadratic:

$$E(\mathfrak{æ}) = E(0) + ak^2 + \ldots$$

Proper quantization condition for $\tilde{S}_{El}$ was constructed in [13].

SHELL CORRECTION TO THE PARTICLE NUMBER

Let us calculate the particle number shell correction using quantization condition [6]. When assuming for simplicity the distribution of electron energy states within the band to be one-dimensional

$$\int d\bar{\mathfrak{x}} \ldots = \frac{1}{\mathfrak{x}_0} \int_0^{\mathfrak{x}_0} d\mathfrak{x} \ldots \tag{15}$$

one can obtain the relation for the particle number:

$$N(\mu) = \frac{2}{\pi} \sum_k \sum_l (2l+1) \int dE\, \tilde{\tau}_{El}\, f_F\left(\frac{E-\mu}{T}\right) \cos[2k(\tilde{S}_{El} + p(l+1/2))]\, I_k(\varphi_{El})\ , \tag{16}$$

$$\tilde{\tau}_{El} = \frac{\partial \tilde{S}_{El}}{\partial E}\ .$$

Function $I_k(\varphi_{El})$ is a result of the integration over k:

$$I_k(\varphi_{El}) = \frac{(-1)^k}{2}\left[P_k(2\varphi_{El}^2 - 1) - P_{k-1}(2\varphi_{El}^2 - 1)\right], \qquad k \neq 0$$

$$I_0(\varphi_{El}) = 1, \qquad \varphi_{El} = \tan \Delta_{El},$$

$P_k(X)$ is a Legendre polynomial, Δ_{El} describes bandwidth. For strongly bound electrons $\varphi_{El} \to 0$, $I_k(\varphi) \to 1$, and the corresponding energy levels are discrete. So the discrete level contribution to the shell correction to the TF model spectrum is maximal. For sufficiently large energies (free electron states) $\Delta_{El} \to \frac{\pi}{4}$, $\varphi_{El} \to 1$, $I_k(\varphi_{El}) \to 0$ and the corresponding energy state contribution to the shell correction equals zero. Such a spectrum is ccntinuous. There is an intermediate energy region between discrete and continuous spectra in which the shell effect role diminishes. So function $I_k(\varphi_{El})$ describes shell effect damping when passing from a discrete spectrum to a continuous one. The atomic spectrum analysis indicates a rather smooth dependence of $E_n(l)$ on l within the shell with the principal quantum number n. But the shells with different n are separated by wide energy intervals. So in the low order X-expansion series we can replace a summing over discrete l by an integral over l. Then, an accurate summing over n by means of the Poisson formula enables us to take into account dominant effects of the atomic shell structure. The term with k = 0 in Eq.(16) is the TF part. The rest of the sum in Eq.(16) is a shell correction

$$\delta N_{sh}(\mu) = \frac{2}{\pi} \sum_k{}' \int d\lambda^2 \int dE \; \tilde{\tau}_{E\lambda} \; f_F\!\left(\frac{E-\mu}{T}\right) \; \cos[2k(\tilde{S}_{E\lambda}+\pi\lambda)] \; I_k(\varphi_{E\lambda}) \quad , \qquad (17)$$

where $\lambda=1+1/2$, and $\sum_k{}'$ denotes the sum without term $k=0$.

NONDEGENERATE PLASMA

Now let us calculate from Eq.(17) the particle number shell correction for the nondegenerate plasma. The chemical potential μ is negative in this case and $|\mu|/T \gg 1$. So the factor $f_F\!\left(\frac{E-\mu}{T}\right)$ in Eq.(17) limits the integration region by discrete spectrum for which $\tilde{S}_{E\lambda} = S_{E\lambda}$, the function $I_k(\varphi_{E\lambda})\approx 1$ and varies weakly . After integrating by parts over E one obtains

$$\delta N_{sh}(\mu) = - \frac{1}{\pi} \sum_k{}' \frac{1}{k} \int dE \frac{\partial f_F}{\partial E} \int d\lambda^2 \sin[2k(S_{E\lambda}+\pi\lambda)] \; I_k(\varphi_{E\lambda}) \quad . \qquad (18)$$

To evaluate the integral over λ^2 we use a squared λ-expansion series

$$S_{E\lambda} = S_E - \pi\lambda - \frac{\delta_E \lambda^2}{2} \; , \qquad \delta_E = - \left.\frac{\partial^2 S_{E\lambda}}{\partial \lambda^2}\right|_{\lambda=0} \quad . \qquad (19)$$

That expansion is obvious for small λ. The integral over E is easily calculated because of the derivative $f_F'\!\left(\frac{E-\mu}{T}\right)$ behaves like a delta function. Expanding all the integrand functions in the neighbourhood of the point $E=\mu$ one can obtain finally

$$\delta N_{sh}(\mu) = \frac{2}{\pi\delta_{\mu 0}} \sum_{k=1}^{\infty} \left\{ \frac{2\pi k \frac{\delta\pi\lambda_{max}}{\delta\mu} T}{\sinh(2\pi k \frac{\partial\pi\lambda_{max}}{\delta\mu} T)} \cos(2\pi k\lambda_{max}(\mu) - \right.$$

$$\left. - \frac{2\pi k \tau_{\mu 0} T}{\sinh(2\pi k \tau_{\mu 0} T)} \cos(2k S_{\mu 0}) \right\} \frac{I_k(\varphi_{\mu 0})}{k^2} \quad . \qquad (20)$$

Equation (20) describes the "temperature" oscillations because of the electron shell ionization with increased temperature. The function $I_k(\varphi_{\mu 0})=1$ for the nondegenerate plasma and tends to zero when the chemical potential becomes positive and increases. Hence it describes "temperature" oscillation damping when passing from nondegenerate plasma to condensed matter. A similar computation for the density and energy corrections $\delta\rho_{sh}$ and δE_2 gives the next relation

$$\delta E_2 = \mu \int d\bar{r}\, \delta\rho_{sh} = \mu\, \delta N_{sh}(\mu) \quad . \qquad (21)$$

Qualitative evaluation and numerical calculations show that the shell correction δV_{sh} on the average is signficantly smaller than the chemical potential correction $\delta\mu_{sh}$. So $\delta\mu_{sh}$ may be calculated approximately from the normalization condition

$$\delta\mu_{sh} = -\,\delta N_{sh} / \int d\bar{r}\, \frac{\partial\rho}{\partial\mu} \qquad (22)$$

and the internal energy correction δE_{sh} may be determined from Eqs.(13),(21),(22)

$$\delta E_{sh} = \left(\frac{3}{2}\, Z - \int d\bar{r}\, \frac{\partial\rho}{\partial\mu}\, (\mu - V(r))\right) \delta\mu_{sh} \quad . \qquad (23)$$

Analysis of the pressure correction, expressed by Eq.(10), discloses that the second term is exponentially small for the nondegenerate plasma. The reason is that the cell boundary region is classically forbidden for the electron states with negative energies. So for the nondegenerate plasma one can use

$$\delta P_{sh} = \rho(R_0)\, \delta\mu_{sh} \quad . \qquad (24)$$

Equations (20),(22-24) allow us to calculate the shell corrections to the nondegenerate plasma equation of state. The comparison of the suggested model results with both the plasma [14] and the statistical [4] model calculations is good.

QUASICLASSICAL TEMPERATURE MODEL (QMT)

An extension of Eqs.(20),(22-24) over all ranges of the parameters with Eqs.(4),(5),(8),(9) enables us to construct the quasiclassical "temperature" model (QMT) which turns into TFC model for the degenerate matter and checks well with Saha model for the Boltzmann plasma. Indeed all the "temperature" shell corrections vanish for the degenerate matter because of the factor $I_k(\varphi_{\mu 0})$. Of course in the intervening region of the parameters QMT is a reasonable physical interpolation only. Thus, one can use this model instead of the mathematical union of the TFC and Saha models without any

joining problems or any empirical data. The other great virtue of
the QMT is its Z-scaling: a single set of calculations for one Z
suffices for all Z.

CONCLUSIONS

A simple quasiclassical model is proposed to calculate the equation
of state for the matter with a high energy concentration. This model
may be employed over a wide range of densities and temperatures from
the Saha model region of application to the improved Thomas-Fermi
model (TFC) area of use. The proposed model describes ab initio
typical step behaviour of the ionization state and energy as a
result of successive shell ionization with increased temperature.
The model naturally includes the effects of electron-ion interaction
with increased density.
The authors thank Academician V.E.Fortov for fruitful discussions.

REFERENCES

[1] D.A.Kirzhnits, Yu.E.Lozovik and G.V.Shpatakovskaya,
    Sov.Phys.Usp. **18**, 649 (1976)

[2] R.P.Feynman, N.Metropolis and E.Teller, Phys.Rev. **75**, 1561
    (1949)

[3] D.A.Kirzhnits, Sov.Phys.JETP. **5**, 64 (1957)

[4] N.N.Kalitkin, Sov.Phys.JETP. **11**, 1106 (1960)

[5] N.N.Kalitkin and L.V.Kuz'mina, Sov.Phys.Solid State. **13**, 1938
    (1972)

[6] R.More, Phys.Rev.A. **19**, 1234 (1979)

[7] F.Perrot Phys.Rev.A. **20**, 586 (1979)

[8] D.A.Kirzhnits and G.V.Shpatakovskaya, Sov.Phys.JETP. **35**, 1088
    (1972)

[9] J.W.Zink, Astroph.J. **162**, 145 (1970)

[10] C.M.Lee and E.I.Thorsos, Phys.Rev.A. **17**, 2073 (1978)

[11] A.V.Andriyash and V.A.Simonenko, Plasma Phys.(USSR). **14**, 1201
    (1988)

[12] B.-G.Englert and J.Schwinger, Phys.Rev.A. **32**, 26 (1985)

[13] G.V.Shpatakovskaya and E.A.Kuzmenkov, High Pressure Res. **1**, 345
    (1989)

[14] B.N.Bazylev, F.N.Borovik, G.A.Vergunova, S.I.Kaskova,G.S.Romanov,
    V.B.Rozanov, L.K.Stanchits, K.L.Stepanov and A.V.Teterev,
    Laser and Particle Beams. **6**, 709 (1988)

A MODEL FOR THE IONIZATION EQUILIBRIUM
OF A VERY DENSE HYDROGEN PLASMA.

J. Wallenborn

Physique Statistique, Plasmas et Optique Non Linéaire
Association Euratom-Etat Belge
Université Libre de Bruxelles, CP 231
Boulevard du Triomphe, 1050 Bruxelles, Belgique

B. Bernu

Laboratoire de Physique Théorique des Liquides
Université Pierre et Marie Curie, Tour 16
4, Place Jussieu, 75230 Paris Cedex 05, France

Abstract

A method to calculate ionization properties of very dense correlated plasmas is proposed. A mixture composed of hydrogen atoms in their ground state and unbound protons and electrons is studied. All species are treated as classical point particles interacting through pseudo-potentials. As an effect of the correlation between hydrogen atoms it is found that the ionization can decrease as the temperature increases. All the correlation functions and excess chemical potentials are evaluated in the HNC approximation.

1. Introduction.

Up to now, the transport properties of a strongly coupled plasma have been essentially studied, both theoretically [1] and numerically [2], in the case of fully ionized systems. However, it is known that it is experimentally difficult to reach a domain of temperature and density where no neutral particles are present [3],[4]. These neutral particles could deeply modify the value of the transport coefficients.

As an example, let us mention the measurement by Shepherd *et al.* [4] of the electrical conductivity of a plasma, created by means of a capillary discharge in polyurethane, at a temperature of about 10 eV and an electronic density of $6 \cdot 10^{22}$ cm^{-3}. The measured value of the electrical conductivity was by a factor of 10 smaller than the value predicted by the kinetic theory of fully ionized plasmas [1]. It is thus desirable that the transport theory of dense plasmas takes into account the degree of ionization of the system.

We have developped a model [5] to treat on an equal footing the equilibrium and transport properties of a partially ionized dense hydrogen plasma.

So far, the diverse attempts to study the ionization equilibrium of a strongly coupled plasma don't consider consistently the correlations between particles, either they are based on the average-atom model [6] or on the generalization of the Saha equation [7]. Here, we go beyond this deficiency.

2. The model.

We have modelized a dense hydrogen plasma by a mixture of unbound electrons (e), unbound protons (p) and hydrogen atoms (H), which are all treated as classical point particles interacting through pseudo-potentials.

The total electron number density is $\rho = \rho_e + \rho_H$, where ρ_e is the unbound electron density. The electroneutrality of the system imposes $\rho_p = \rho_e$.

The degree of ionisation $\alpha = \rho_e / \rho$ is evaluated as a function of ρ and T when the chemical reaction

$$H \rightleftharpoons p + e \tag{1}$$

is at equilibrium. Practically, the chemical potentials of each species are evaluated within the HNC approximation for fixed α and T. Then ρ is varied in order to fulfil the equilibrium condition :

$$\mu_H (\rho, T, \alpha) = \mu_e (\rho, T, \alpha) + \mu_p (\rho, T, \alpha) \tag{2}$$

In order to obtain a tractable solution of the problem we made the assumption that the hydrogen atoms are frozen in their ground state. This assumption, which drastically simplify the internal partition function of the hydrogen atom, can be seen as a screening effect of the medium on the excited states, as has been verified at the end of the calculations [5]. In a very dense plasma, the screening length is of the same order as the mean distance between charges. We thus consider a density ρ such that

$$2.5 \cdot 10^{22} \, \mathrm{cm}^{-3} = \frac{3}{4 \pi (4 \, a_0)^3} < \rho < \frac{1}{4 \pi a_0^3} = 1.6 \cdot 10^{24} \, \mathrm{cm}^{-3} \tag{3}$$

where a_0 is the Bohr radius. These bounds ensure that the ground state exists and that the excited ones are screened out. At lower density our assumption could be modified in order to include a finite number of atomic excited states, each one corresponding to a new chemical species. More details can be found in ref. 5.

We consider only temperature $T > 5$ eV. In the range of density given by eq. (3), this means that the electrons de Broglie wavelength is shorter than their mean separation. This justifies that all the particles are treated as classical point particles. On the other hand, in this range of temperature there is no formation of H_2 , H_2^+ and H^-, nor can appear the so-called plasma phase transition [8].

The interaction potentials are taken from the quantum two body problem. When necessary they are properly modified to avoid the Coulomb singularity at short separation by accounting for the quantum diffraction effect [9]. Their explicit expession can be found in ref. 5. Let us only mention that hydrogen atoms interact via different potentials according to the spin of their respective bound electron : parallel spins result in a pure repulsion and thus an enhancement of the degree of ionization; antiparallel spins produce in addition an attractive part in the interaction potential. In our semi-classical description, that means that we consider two kinds of hydrogen atoms (H↑, H↓) that is, with unbound electrons and protons, a four-component system.

3. Results and discussion.

We have computed the seven different pair correlation functions within the HNC approximation. The excess chemical potentials are directly obtained [10]. The ideal parts are :

$$\beta \mu_\alpha^0 = \ln (\rho_\alpha \lambda_\alpha^3 / g_\alpha) + \beta \varepsilon_\alpha^0 \qquad (\alpha = e, p, H) \qquad (4)$$

where λ_α, g_α, ε_α^0 are respectively the thermal de Broglie wavelength, the degeneracy factor and the internal energy of each particle. For the hydrogen atom, ε_H^0 is the ground state energy of the bound electron. We used $\varepsilon_H^0 = -13.6$ eV which is the value corresponding to an isolated atom. This leads to a lower bound of the ionization. When the ideal value of the chemical potential (4) is used in eq. (2), the Saha equation [11] is recovered.

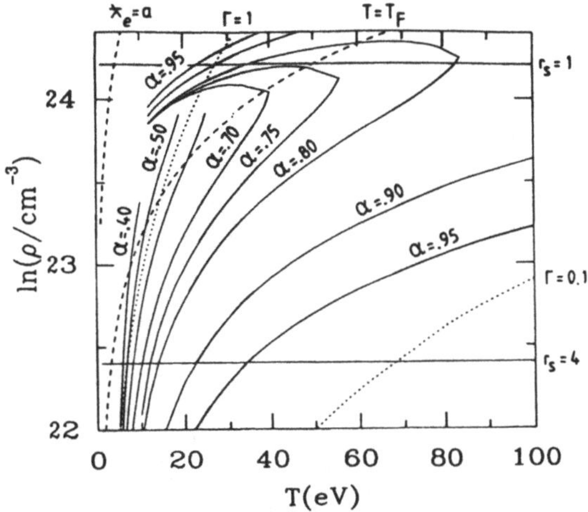

Fig. 1. Curves of constant ionization . The validity domain is approximately comprised between $r_s = 1$ and $r_s = 4$ [$r_s = a/a_0$, $a = (3/4\pi\rho)^{1/3}$]. $\Gamma = e^2/k_b Ta$ is the coupling parameter. For $T > T_F$ (electron Fermi temperature) the quantum symmetry effects are negligible, whereas for $\lambdabar_e = \hbar/(2\pi m_e k_b T)^{1/2} > a$ the system is fully quantum.

The results are given on fig.1 which shows curves of constant ionization. For a fixed temperature, when the density ρ is increased from 10^{22} cm^{-3} to 10^{24} cm^{-3}, the degree of ionization α first decreases (perfect gas behaviour) and then increases abruptly (Debye-Hückel [DH] - like behaviour). For $\rho < 10^{23}$ cm^{-3}, α increases monotonically with temperature at fixed density. At higher densities however, α shows a minimum with respect to T. This unexpected behaviour can be interpreted as caused by the tendency of H-atoms to form clusters at low temperature and high density. The H-atoms with antiparallel electronic spins tend to fall in the minimum of their interaction potential. The volume occupied by the clusters is less than the volume occupied by the same number of uncorrelated atoms; this favours ionization. When the temperature increases, the correlations decrease, the H-atoms occupy more volume and the ionization decreases. At higher temperature, there are no more clusters. One recovers the same qualitative behaviour as in DH approximation, that is ionization increases with temperature.

This interpretation is based on the fact that the pair correlation function $g_{H\uparrow H\downarrow}(r)$ shows a small peak in the domain where α decreases with temperature. This, however, could be an artefact of the HNC approximation, which could be inacurate for neutral particles. For, a Monte Carlo simulation is now in progress to test the validity of the HNC approximation for the correlations as well as for the degree of ionization α as a function of ρ and T.

On the other hand, the adaptation of the classical kinetic theory of dense systems to mixtures of neutral and charged particles is being considered. Indeed, the knowledge of the degree of ionization is an intermediate step for the evaluation of the electrical conductivity and other transport coefficients.

References.

[1] Boercker D.B., Rogers F.J. and DeWitt H.E., Phys. Rev. A **25** (1982) 1623; Ichimaru S.I. and Tanaka S., Phys. Rev. A **32** (1985) 1790; Wallenborn J., Zehnlé V. and Bernu B., Europhys. Lett. **3** (1987) 661; Zehnlé V., Bernu B. and Wallenborn J., J. Phys. France **49** (1988) 1147; Bernu B., Wallenborn J. and Zehnlé V., ibid **49** (1988) 1161.

[2] Hansen J.P. and McDonald I.R., Phys. Rev. A **23** (1981) 2041; Sjögren L., Hansen J.P. and Pollock E.L., Phys. Rev. A **24** (1981) 1544; Bernu B. and Hansen J.P., Phys. Rev. Lett. **48** (1982) 1375; Physica **122A** (1983) 129.

[3] see e.g. these proceedings.

[4] Shepherd R.L., Kania D.R. and Jones L.A., Phys. Rev. Lett. **61** (1988) 1278.

[5] Bernu B. and Wallenborn J., Europhys. Lett. **14** (1991) 203.

[6] Perrot F. and Dharma-wardana M.W.C., Phys .Rev. A **26** (1982) 2096; ibid. **29** (1984) 1378; ibid **36** (1987) 1378; Ying R. and Kalman G., Phys. Rev. A **40** (1989) 3927 and references quoted therein

[7] Kraeft W.D., Kremp D., Ebeling W. and Röpke G., Quantum Statistics of Charged Particle Systems, Plenum (1986)

[8] Ebeling W. and Richert W., Phys.Lett. **108A** (1985) 80; Saumon D. and Chabrier G., Phys. Rev. Lett. **62** (1989) 2397.

[9] Kelbg G., Ann. Phys. (Leipzig) **12** (1963) 219.

[10] Verlet L. and Levesque D., Physica **28** (1962) 1124; Hansen J.P. and Vieillefosse P., Phys. Rev. Lett. **34** (1976) 391.

[11] Landau L. and Lifchitz E., Physique Statistique, MIR (Moscow), chap. 10.

MICROFIELD, QUASI-ZONES AND PLASMAS NONIDEALITY

I.O.Golosnoy, N.N.Kalitkin, V.S.Volokitin

National Centre for Mathematical Modelling

Miusskaya sq.,4-A, Moscow, 125047 USSR

Abstract

It is shown that microfield is the main factor determing plasmas non-ideality. It explains 1) quasi-zonic structure of electron spectra in dense plasmas, 2) transforming of ionization equilibrium model into quantum-statistical model at ultrahigh densities, 3) thermodynamical and optical properties of matter in a huge region of densities.

The simple models were developed for 1) microfield in nonideal plasmas and 2) plasmas nonideality, based self-consistently on microfield.

INTRODUCTION

A problem of reliable Equation of State (EOS) constructing is a very important for physics and industry. It has a lot of applications such as shock waves in gases and solids, strong current discharges (for powerful lasers), magneto-hydrodynamical generators, hypersonic aircrafts, etc. These various processes need EOS for medias from gas and plasma to compressed solids. No one model can describe EOS with good accuracy in such wide region. So we have to use different models and "sew a blanket from pieces", as it was done in the famous library SESAME [1-3] (Fig.1).

The next models were developed by our department from 1960 till now [4-10]: 1) ionization and chemical equilibrium model (ICEM) for gases and plasma with different nonideality corrections; the last one is based on microfield; 2) quantum-statistical model (QSM) for highly compressed matter; 3) quasi-zonic model (QZM) joins these cases; 4) quasi-zonic interpolation (QZI) for "gap" between ICEM and QSM.

Here we present our results concerning ICEM and QZM. All formulas below are written in atomic system of units.

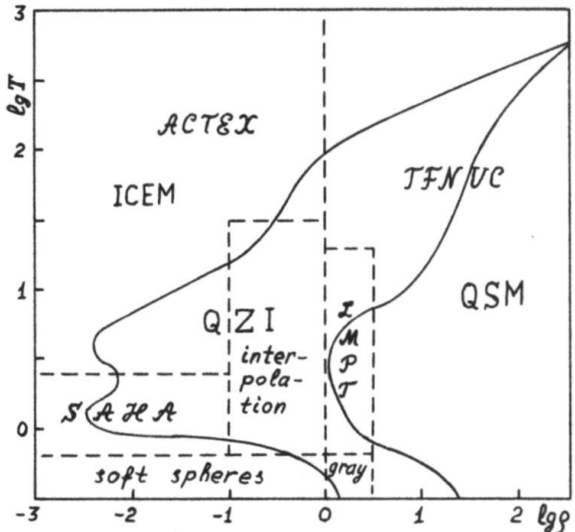

Fig. 1. Wide-range EOS for Al. Solid lines – real boundaries for different models (normal letters); dashed lines - boundaries used in SESAME library (models are written with italic, their description see in [1-3]).

1. WIDE-RANGE MICROFIELD MODEL

A calculation of EOS tables in a wide-range of temperatures and densities requires a lot of computations of the microfield distribution function. That's why we need a simple, wide-range microfield model. We didn't develope any new model, but improved essentially numerical algorithm for a well known APEX model (see [11,12]).

Self-consistent treatment of ions and electrons is a very difficult problem. So one ussually considers ions immersed in a uniform neutralizing background. The physical error this model (up to 30%) is caused by two facts: 1) electronic neutralizing background isn't uniform, 2) electrons bear their own microfield. Ions are contained in a volume Ω at a temperature T. Let us denote x_{jk} for a concentration of ions with charges k of j sort of atoms respectively, and V for an average atomic volume. Then we can define the mean radius of atomic cell R and parameter of nonideality Γ from formulas: $(4\pi/3)R^3=V$ and $\Gamma=1/(RT)$.

Monte-Carlo (MC) simulations consider to be the most accurate calculations for such mixture. They are applicable as for an ideal so for a strong coupled plasmas. But MC simulations are not simple, they demand a large amount of calculations.

Today APEX is the most convenient method for computations of microfield distribution function. Calculations over APEX and MC give close results, but APEX is easier. APEX is a complicated model which takes into account correlations between plasmas particles. An influence of electrons can be taken into account by effective screening of ions and considering electronic gas in neutralizing background of ions. This consideration made the problem more difficult but results obtained differs slightly. Furthemore, the influence of rapid electronic microfield on ionizing balance is rather complex problem.

The probability density for microfield intensity ε near a test particle with charge k_0 is

$$P(\varepsilon) = \frac{2\varepsilon}{\pi} \int_0^\infty \sin(l\varepsilon)T(l)l\,dl \ . \tag{1}$$

where the Fourier transform of $P(\varepsilon)$ is

$$T(l) = < \exp(il\mathbf{E}) >, \tag{2}$$

The angular brackets in (2) denote a canonical ensemble average over configurations weighted with a Boltzmann factor, $\mathbf{E}$ denotes the electric field on the test particle due to any configuration.

We construct a simple approximation for [10]

$$T(l) = \exp\left(\sum_k x_k F_k(l) \right), \quad x_k = \sum_j x_{jk}, \tag{3}$$

where $F_k(l)$ is an analitical function which depends on Γ and charges k and k_0. This

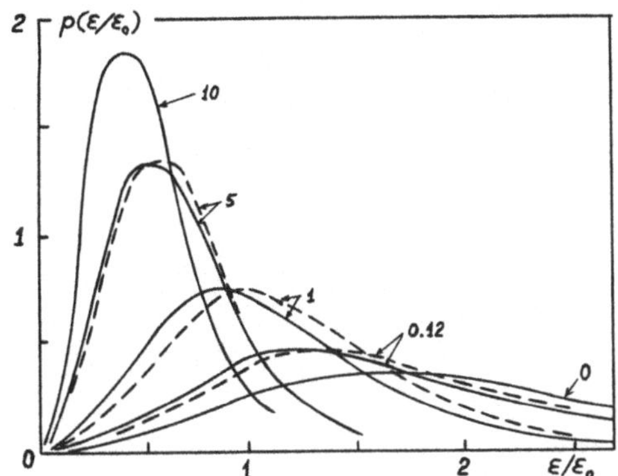

Fig. 2. Microfield distribution function $p(\varepsilon)$ of one-component plasmas; figures near curves - Γ; (solid line - APEX, dashed line - our computations).

approximation allows to compute the microfield distribution function of multicomponent plasmas with complicated composition and an arbitrary degree of nonideality. Its physical accuracy and region of validity don't yield to the usual models. This approximation is very simple and it is suitable for numerical calculations of EOS.

Fig. 2 illustrates our calculations of $P(\varepsilon)$ compared with APEX results for one component plasmas. We consider four values of parameter Γ; difference between two methods is less than 30%.

It is known there exist ionization potential decreasing (IPD) and statistical sums cutting in plasmas. For a long time there was assumed that these phenomena can be described by statistic models (Debye screening, temperature cut off etc.). However, the experiments of 70th showed that the fluctuating microfield is the main factor causing IPD. We consider IPD of an ion or an atom with charge $(k-1)$ as a random value with probability density

$$f(\varphi_k) = 2\varphi_k P(\varphi_k^2), \tag{4}$$

where φ_k is instantaneous IPD connected with instantaneous microfield $\varphi_k = -2(k\varepsilon)^{1/2}$. By means of the approximation (3) and formula (4) it is possible to derive the mean potential decreasing $\Delta\varphi_k^{(mf)}$ [10]:

$$\Delta\varphi_k^{(mf)} = - \frac{2 (k\, k_m /R^2)^{1/2}}{\left(0.027(k_m/k_h)^4 + (k_m^4 k_s^2 k/k_h^5)\Gamma + 0.18(k_m k\Gamma)^2\right)^{1/8}}, \tag{5}$$

where

$$k_m = \sum_k kx_k, \quad k_h = \left(\sum_k k^{3/2}x_k\right)^{2/3}, \quad k_s^2 = \sum_k k^2 x_k.$$

In the case of low-dense ideal plasmas $(\Gamma \to 0)$ $\Delta\varphi_k^{(mf)} = - 2(k\, k_h)/R$, i.e. IPD limit is not of Debye type.

2. MICROFIELD MODEL OF PLASMAS NONIDEALITY

To find composition and thermodynamical functions of plasma system one can minimize free energy:

$$F = F_e + \sum_j \sum_{k=0}^{z_j} x_{jk} F_{jk} + \Delta F,$$

where F_e – electron free energy for the arbitrary degeneration case, F_{jk} – free energy of each heavy component, depending also on statistical sums G_{jk} [4,5].

Minimization lead us to ionic equilibrium equations:

$$\mu + T \ln(G_{j,k-1} x_{jk} / (G_{jk} x_{j,k-1})) + \varphi_{jk} + \Delta\varphi_{jk} + \delta\varphi_{jk} = 0, \tag{6}$$

where μ is is electron chemical potential, connected with x_e, φ_{jk} are potentials of ionization,

$$\Delta\varphi_{jk} = (\partial/\partial x_{jk} - \partial/\partial x_{j,k-1} + \partial/\partial x_e)\Delta F \tag{7}$$

are IPD due to interaction between particles, $\delta\varphi_{jk}$ are negligible additional IPD due to the cutted statistic sum's differentiating.

Many different models of plasmas nonideality were proposed. Among them are: Debye model (DM); One-component plasmas (OCP), generalized for mixtures; Debye in grand canonical ensemble (GDM); Smooth gap (SG); Thomas–Fermi correction (TFC) for nonhomogeneouty of electron gas;

All these models are based on a priori nonideality term ΔF constructed from some model principles. First four models have Debye limit $\Delta F \approx k_s^2/r_D$ for low nonideality $\Gamma{\to}0$ (r_D – is Debye radius); model TFC has cell limit $\Delta F \approx k_s^2/R$. DM and OCP have serious defects: non-unique decision of equation (6), prediction plasmas phase transition which is not confirmed by experiment etc. All Debye-type models have also some internal contradictions. We omit the discussion about them in this short paper.

We offered here quite new self-consistent approach for plasmas nonideality model's obtaining. If we know any model of IPD we can derive free energy correction term

$$\Delta F = \sum_{k=1}^{z_m} \int_0^{x_k} \sum_{j=1}^{k} \Delta\varphi_j^{(k)}(x, x_{k-1}..x_1) \, dx,$$

where $\Delta\varphi_j^{(k)}$ is potential shift for $(j-1)$ ion in system, which contains ions not more k-ionized. We can find also statistic sum's formfactor

$$\omega(E) = 1 - \exp(- |\Gamma(1+1/n)(\varphi_k - E)/\Delta\varphi_k|^n),$$

where E is energy of ion level and n is "stiffness" of formfactor shape. However, one can deduce $\Delta\varphi$ from formfactor chosen:

$$\Delta\varphi = \int_0^\infty \xi \left.\frac{d\omega(x)}{dx}\right|_{x=\varphi-\xi} d\xi \ .$$

Thus we finally connected all values in (6) together. The way $\Delta\varphi\to\Delta F$ is more preferable than $\Delta F\to\Delta\varphi$, because we can check reliably validity of $\Delta\varphi$ models according to spectral data, but accuracy of thermodynamical experiments is insufficient for checking ΔF.

We chose the microfield ionization shift (5) for construction the microfield nonideality model (MFN). But we have to consider also the influence of neutral particles, which may be essential at large densities [6]:

$$\Delta\varphi_k^{(np)} \approx -1.07kx_0^{1/3}/R,$$

where x_0 is concentration of neutral particles. The total shift is [9]:

$$\Delta\varphi_k = \Delta\varphi_k^{(mf)} + \Delta\varphi_k^{(np)} - \Delta\varphi_k^{(mf)}\Delta\varphi_k^{(np)}/(\Delta\varphi_k^{(mf)} + \Delta\varphi_k^{(np)})$$

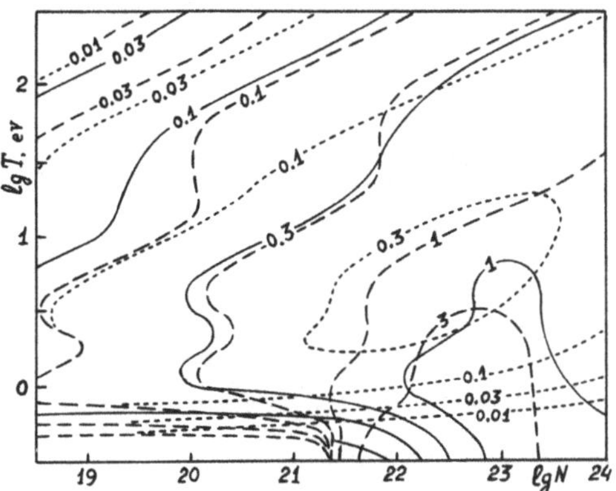

Fig. 3. Parameter of nonideality for Al - different models (MFN - solid line, OCP - dashed line, GDM - dotted line); figures near curves - γ_E.

This model has cell limit $\Delta F \approx k_s^2/R$ for low nonideality.

We've checked the correspondence of some well known nonideality models (MFN among them) and formfactors to the experimental spectral data on Hydrogen (Gavrilov, USSR, 1986; $T=20000^{\circ}$K and $N_e=2,4,8\times10^{18}$cm^{-3}) and Laser produced plasmas (Rochester, 1988 [13]; $T=1.1$ keV, $\rho=0.5$ g/cm^3, complex composition with $k_m\approx10$). The occupancy of energy levels was compared with numbers of spectral lines observed in experiments.

The Table 1 show the results of this comparison. Here the numbers of visual lines predicted by different models placed. "Star" to the right of figure means that upper theoretical lines are much more intensive than experimental ones, and "star" to the left means vice versa. One can see, that MFN describes both experiments correctly though temperatures and densities differ each other more than 1000 times in these cases, and all other models contradict experimental data.

The EOS for Al was computed by PLAZMA-5 code considering different models of nonideality. Some isolines of thermodynamical nonideality parameter $\gamma_E=-E_{pot}/E_{kin}$ are shown on Fig.3 (solid line - MFN, dashed line - OCP, dotted line - GDM). GDM gives smaller theoretical γ_E at large densities than one expects. On the contrary OCP gives γ_E more than 2 and even 3, which lead to non-physical plasmas phase transition, negative pressure etc. Simple Debye (DM) has the same defects at even smaller densities. MFN has essentially better qualitative behaviour of energy, pressure and nonideality, provides ionization by pressure, i.e. gives satisfactory description of plasmas properties at high densities.

Table 1. Numbers of observed spectral lines.

model	discharge in H Ba line, $N_e \times 10^{-18} =$			Rochester laser plasma Ly lines	
	= 2	4	8	Al^{+12}	Ar^{+17}
cutting with T	0	0	0	0	0
Plank-Larkin	* 1	* 1	* 1	* 1	* 1
experiment	3	2	1	2	3
MFN	3	2	1	2	3
Debye	4 *	3 *	2 *	3 *	4 *
OCP	5	4	3	4	5
turning point	6-7	5-6	4-5	5	6
GDM	10 *	9 *	8 *	10 *	11 *

3. QUASI-ZONES

Let us consider a structure of electron spectra under different conditions. An isolated atom has a discrete spectrum; that correspons $\rho = 0$. Condensed matter at $T = 0$ is a crystal, and its electron spectrum has a zonic structure (Fig. 4). But what will be at high temperatures and $\rho \neq 0$, when matter is a dense plasmas?

Each atom in plasmas at a given moment is situated in certain instant microfield created by it's neighbourhood; so it has a discrete spectrum, but with levels shifted (and splitted) according their values for the isolated atom. This shift is a random value, and at a given moment it is different for different atoms. If we take all atoms of matter, each level of the isolated atom will be replaced with totality of shifted levels, which we name quasi-zone (Fig. 4). The distribution function of quasi-zone will be just the same one as for IPD (5).

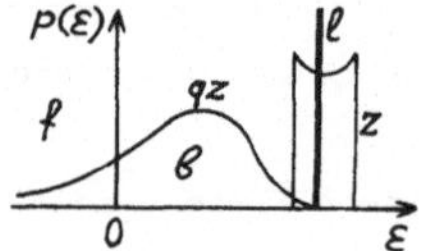

Fig. 4. Types of electron spectrum: l - level, z - zone, qz - quasi-zone; f - free electrons, b - bounded ones.

There are some essential differences between zones and quasi-zones. 1) Zonic energy distribution has sharp edges (see Fig. 4). Quasi-zone has bell-shape form with infinite "tail"; so some part of electrons lies up Fermi surface, i.e. will be ionized. 2) Widths of quasi-zones are of the same order as IPD (5) and practically don't depend on number of quasi-zone. Width of lower zones are much less than for upper ones. 3) Zones for non-compressed crystal are rather narrow: even upper zones width are ~ 1 eV. Quasi-zones at the same densities are much wider; for example, in Rochester laser produced plasmas [13] it is ~ 200 eV.

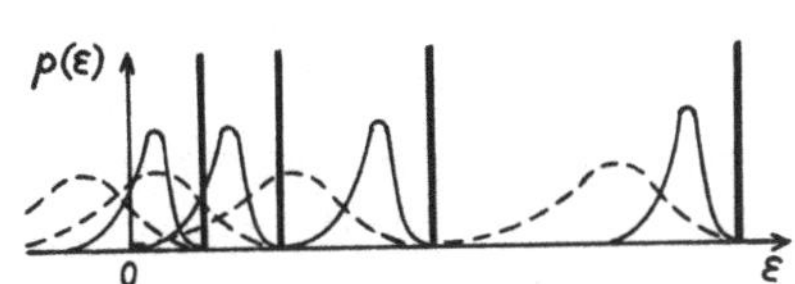

Fig. 5. Quasi-zones at small (solid lines) and large (dashed lines) densities. Bold lines are levels for zero density.

Fig.5 shows a system of atomic levels; lower levels are remote, upper levels are close. For gas densities quasi-zones are very narrow, and quasi-zones for different levels don't overlap practically. When density enlarges, outer quasi-zones begin to overlap. At large densities even lower quasi-zones may overlap.

Essential overlapping of quasi-zones means that electron spectrum under Fermi surface becomes not only continious, but it has almost constant distribution. The last situation means that Thomas-Fermi statistical approach is valid. So quasi-zonic conception explains how ICEM turns into QSM at large densities. Detailed theory joining ICEM and QSM isn't easy to develope, but a simple method of quasi-zonic interpolation (QZI) is described below.

4. SHELL EFFECTS IN DENSE PLASMAS

It is understandable that shell structure of atoms leads to certain oscillations in thermodynamical and optical properties of matter. For example, dependence of ionization on temperature has almost "jumps" when one shell is just totally ionized and the next shell must be touched. These oscillations are distinct at small densities and smooth gradually when densities enlarge. This effect exist even in ICEM with level structure of spectrum, but quasi-zones strengthen it.

Let's consider turning ICEM into QSM. It is clear that as totally ionized so inner shells don't influence on thermodynamical properties of matter; only partially ionized shell is essential. Therefore we must compare quasi-zone

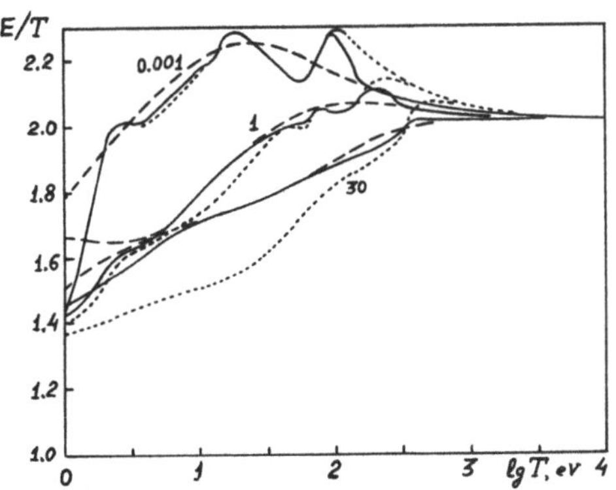

Fig. 6. EOS for H_2O: solid lines – QZI, dashed lines – QSM, points – ICEM; figures near curves – densities (g/cm^3).

width $\Delta\varphi$ with distance between ionization potentials ΔI of this shell. For $\Delta\varphi \ll \Delta I$ usual ICEM is valid; if $\Delta\varphi \gg \Delta I$, QSM is applicable. These boundaries of applicability are shown on Fig.1; they differ essentially from boundaries of SESAME library (the last one have no serious background). Intermediate case $\Delta\varphi \sim \Delta I$ may be calculated reliably with special quasi-zonic interpolation QZI between ICEM tables and QSM ones. Fig.6 shows an example of such interpolation. Results are close to ICEM at small densities and to QSM at large ones; for intermediate densities they lie between these two models.

These results permit to achieve an excellent quantitative agreement with unique experiments on shock compression of solids up to 500 MBar [14]. For Al the mean difference of experimental and theoretical (from the first principles) densities was $\approx 2\%$; when the theory was fitted a little to some experimental data at small pressures, the difference lessens to $\approx 1\%$. It is essentially better than another theories give.

So the theory described in this paper is now confirmed by optical as well as thermodynamical experiments. It predicts an essential smoothing of shell effects at large densities and permits to construct EOS the most precise for today.

REFERENCES

[1] R.M. More, J.F. Barnes, R.D. Cowan, Bull. Am. Soc. **II 21**, 1153 (1976)

[2] K. Holian, LosAlamos Nat. Lab. rep. LA-10160-MS, (1984)

[3] D.A. Young et. al., Phys. Rev. Lett. **108A**, 157 (1985)

[4] N.N. Kalitkin, in: Mathematical Modelling; Physical and Chemical Properties of Matter, Ed. (Nauka, Moscow, 1989), p.114

[5] N.N. Kalitkin, Sov. J. Mathem. Model. 1, N2 (1989)

[6] N.N. Kalitkin, V.S. Volokitin, ibid., 3, N5 (1991)

[7] P.D. Shirkov, ibid., 3, N6 (1991)

[8] V.S. Volokitin, ibid., 3, N7 (1991)

[9] V.S. Volokitin, ibid., 3, N8 (1991)

[10] I.O. Golosnoy, ibid., 3, N9 (1991)

[11] C.F. Hooper, Electric microfield distribution function: past and present, Ed. (Univ. Florida, 1986)

[12] C. Iglesias et. al., Phys. Rev., **A30**, 2001 (1984)

[13] C.F. Hooper jr, R.C. Mancini, D.P. Kilcrease et al, SPIE. **913**, N 129 (1988)

[14] E.N. Avrorin et.al., Sov. JETPH, **93**, 613 (1987)

BEHAVIOR OF THE ELECTRICAL MICROFIELD
IN A PLASMA, FOR A STRONG FIELD

Marie-Madeleine GOMBERT and TU KHIET
Laboratoire de Physique des Gaz et des Plasmas[1]
Bât. 212, Université de Paris-Sud, F91405 ORSAY cedex, France

Abstract

We develop a method to study asymptotic expansions of microfields in a plasma for a large field. This is based on the Baranger and Mozer formalism. We give the first results and show how to take into account the quantum corrections by means of pseudo-potentials.

1   Introduction

In astrophysics and in fusion by inertial confinement as well, a precise knowledge of the atomic and ionic line shapes emitted by a plasma is of great importance. A lot of works were done on this subject. The aim of our study is the behavior of $W\left(\vec{E}\right)$ for large $\left|\vec{E}\right|$. Calculations of wings of spectral lines need the knowledge of asymptotic behavior of $W\left(\vec{E}\right)$ and, more precisely, its high frequency component (due to the electrons).

2   Formalism

We use the Baranger and Mozer formalism [1]. The Fourier transform of $W\left(\vec{E}\right)$, denoted by $F\left(\vec{k}\right)$ can be expressed in terms of correlation functions :

$$F\left(\vec{k}\right) = \exp\left[\sum_{p=1}^{\infty} \frac{\rho^p}{p!} h_p\left(\vec{k}\right)\right] \tag{1}$$

where ρ is the density of the particles which create the field (electronic density in the case of high frequency component) and $h_p\left(\vec{k}\right)$ is related to g_p , the correlation function of p particles producing electrical field :

$$h_p\left(\vec{k}\right) = \int \varphi_1 ... \varphi_p \, g_p\left(\vec{r_1}, ..., \vec{r_p}\right) \mathrm{d}^3 r_1 ... \mathrm{d}^3 r_p \quad . \tag{2}$$

Here the origin is chosen on the point where the microfield distribution is calculated. It can be either neutral or charged. φ_j is :

$$\varphi_j = \exp\left(\mathrm{i}\,\vec{k}\,.\,\vec{E_j}\right) - 1 \tag{3}$$

where $\vec{E_j}$ is the field created by the particle labelled j and located at $\vec{r_j}$. If the correlations between electrons are neglected, the high frequency microfield at a neutral point is the Holtsmark microfield :

$$F_{Hol.}\left(\vec{k}\right) = \exp\left[\rho h_1^N(k)\right] \qquad \text{with} \; : \; h_1^N = -\frac{4}{15}(2\pi e k)^{3/2} \tag{4}$$

---

[1]Unité de Recherche associée au C. N. R. S.

superscript N means neutral. With the usual notations :

$$\frac{4}{15}(2\pi)^{3/2}r_0\rho = 1 \quad , \quad E_0 = \frac{e}{r_0} \quad , \quad \beta = \frac{E}{E_0} \quad \text{and} : \quad u = kE_O \quad , \tag{5}$$

we have :
$$\rho h_1^N(u) = -u^{3/2} \quad . \tag{6}$$

r_0 is very close the mean distance between particles. The microfield distribution can be written :

$$H(\beta) = 4\pi\beta^2 W(\beta) = \frac{2\beta}{\pi}\int_0^\infty du\, u \sin(\beta u) F(u) \quad . \tag{7}$$

In the following, the distances are measured in Debye length :

$$x = r/\lambda_D \qquad \text{with} : \lambda_D = \left(4\pi\rho^2/k_BT\right)^{-1/2} \tag{8}$$

and the correlation parameter : $\qquad\qquad y = r_0/\lambda_D \tag{9}$

is related to the usual plasma parameter, Λ, through :

$$\Lambda = \left(4\pi\rho\lambda_D^3\right)^{-1} = \frac{\sqrt{8\pi}}{15}y^3 \quad . \tag{10}$$

To study the asymptotic behavior of $h(\beta)$ (Eq.(7)), we have to evaluate expansions of $F(u)$ (Eq.(1)) and $h_p(u)$ (Eq.(2)) for small u. For small correlations, h_p (Eq.(2)) can be written as a graph expansion exactly as it was done for g_p evaluation [2]. The order of h_p is Λ^{p-1}. So, at the second order with respect to Λ, Eq.(1) becomes :

$$F(u) = \exp\left[\rho h_1(u) + \frac{\rho^2}{2!}h_2(u) + \frac{\rho^3}{3!}h_3(u)\right] \quad . \tag{11}$$

In the case of high frequency microfield at a neutral point :

$$\rho h_1^N(u) = -u^{3/2} \quad , \tag{12}$$

$$\frac{\rho^2}{2!}h_2^N(u) = \frac{1}{2!(4\pi\Lambda)^2}\int \varphi_1\varphi_2\left(g\left(x_{12}\right) - 1\right)d^3x_1 d^3x_2 \quad , \tag{13}$$

$$\frac{\rho^3}{3!}h_3^N(u) = \frac{1}{3!(4\pi\Lambda)^3}\int \varphi_1\varphi_2\varphi_3\, g_3\left(\vec{x_1},\vec{x_2},\vec{x_3}\right)d^3x_1 d^3x_2 d^3x_3 \tag{14}$$

with : $\qquad\qquad x_{12} = \left|\vec{x_1} - \vec{x_2}\right| \quad . \tag{15}$

Using the well known nodal expansion for the correlation functions, the $\dfrac{\rho^p}{p!}h_p(u)$ can be represented as following :

$$\rho h_1^N(u) = \quad \circ\!\!-\!\!-\!\!-\!\!\bullet\; -e \, , \tag{16}$$

$$\frac{\rho^2}{2!}h_2^N(u) = \quad \cdots\;, \tag{17}$$

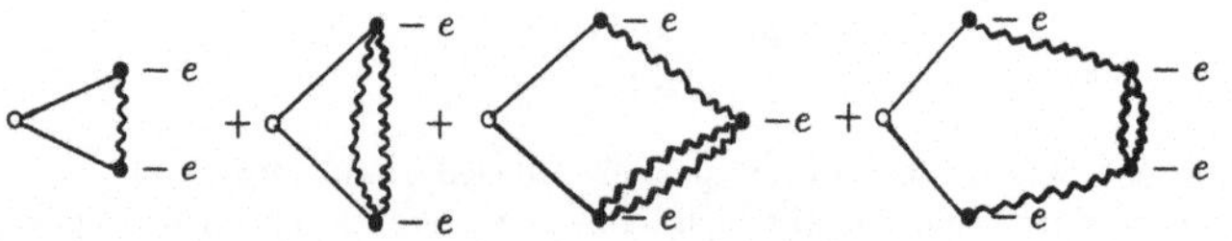

$$\frac{\rho^3}{3!}h_3^N(u) = \quad \cdots \; . \tag{18}$$

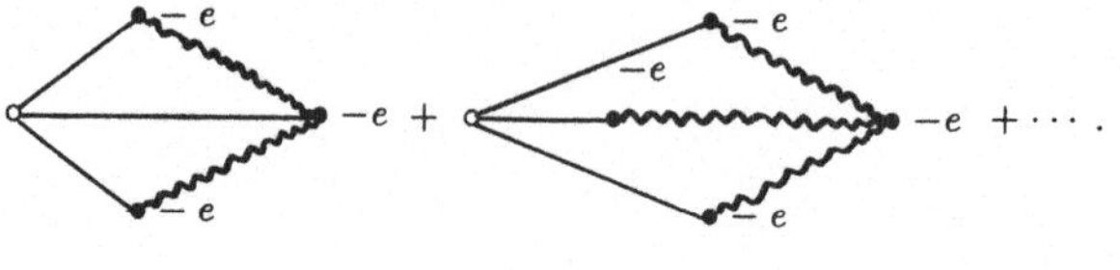

The correlation functions are expanded until the order Λ^2. A straight line represents a function φ and a waved line a Debye chain :

$$-\beta V(x) = -\frac{\Lambda}{x}\exp(-x) \quad .$$

(19)

The open point is the radiator and black points are electrons of plasma. For instance :

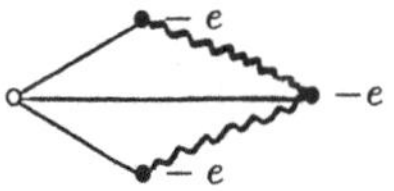

$$\cdots \quad -e \quad = \quad \frac{15}{32(2\pi)^{7/2}y^3}\int \mathrm{d}^3 x_1\varphi_1\left[\int \mathrm{d}^3 x_2\varphi_2\frac{\exp(-x_{12})}{x_{12}}\right]^2$$

(20)

with :

$$\varphi_j = \exp\left(iy^2\frac{\vec{u}.\vec{x}_j}{x_j^3}\right) - 1.$$

(21)

If the radiator is an ion with an electrical charge $+Ze$, we have to take into account the correlations between the electrons and the charge $+Ze$. The correlation functions, g_p, are modified. So :

$$\rho h_1^C = \frac{1}{4\pi\Lambda}\int \mathrm{d}^3 x\varphi g_{ei}(x) \,,$$

(22)

$$\rho h_1^C = \rho h_1^N + \cdots + \cdots + \cdots$$

(23)

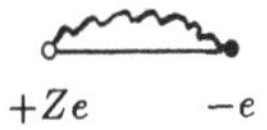

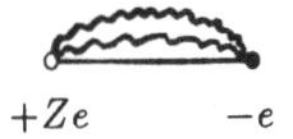

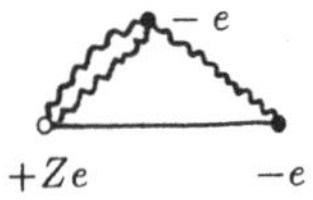

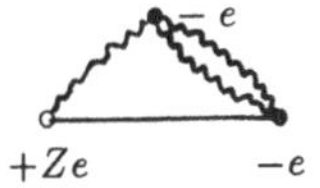

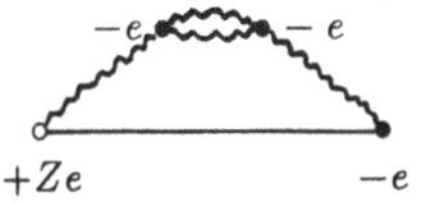

$$\frac{\rho^2}{2!}h_2^C = \frac{\rho^2}{2!}h_2^N + \cdots + \cdots + \cdots \quad .$$

(24)

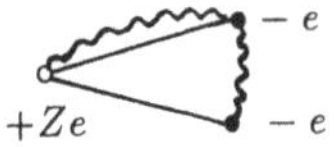

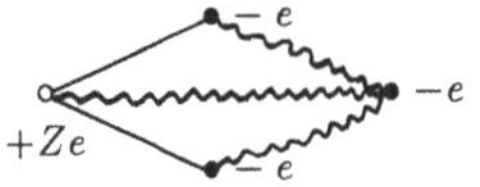

In the last equations, superscript C stands for charged . g_{ei} is the pair correlation function between the charge $+Ze$ and an electron. In g_{ei} , the screening is due only to the electrons. Calculations are made in a one component plasma.

3   First results

For $h_1^C(u)$:

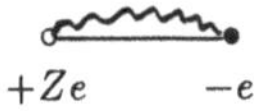

$$\cdots = Z\int_0^\infty \mathrm{d}x\,x\exp(-x)\left[\mathrm{j_0}\left(\frac{y^2 u}{x^2}\right) - 1\right]$$

(25)

with :

$$\mathrm{j_0}\left(\frac{y^2 u}{x^2}\right) = \frac{x^2}{y^2 u}\,\sin\left(\frac{y^2 u}{x^2}\right) \quad .$$

(26)

The integration (26) is performed by dividing the domain of integration into two parts. $\exp(-x)$ is expanded for small x, and j_0 for large x. So :

$$
\begin{aligned}
\text{[graph]} \quad = \quad & -\frac{\pi Z}{8}y^2 u + \frac{2\sqrt{2\pi}}{15}Zy^3 u^{3/2} + \frac{Z}{24}y^4 u^2 \ln\left(y^2 u\right) + \frac{Z}{8}y^4 u^2 \left(\frac{C-29}{18}\right) \\
& -\frac{2\sqrt{2\pi}}{315}Zy^5 u^{5/2} + \frac{\pi}{2304}Zy^6 u^3 - \frac{\sqrt{2\pi}}{14175}Zy^7 u^{7/2} \\
& -\frac{Z}{172000}y^8 u^4 \ln\left(y^2 u\right) - \frac{Z}{57600}y^8 u^4\left(C-\frac{431}{30}\right) + \mathcal{O}\left(y^9 u^{9/2}\right)
\end{aligned}
\tag{27}
$$

where C is the Euler's constant.

$$
\text{[graph]} \quad = \quad \frac{Z^2\sqrt{2\pi}}{15}y^3\int_0^\infty \mathrm{d}x\,\exp(-2x)\left[j_0\left(\frac{y^2 u}{x^2}\right)-1\right] .
\tag{28}
$$

Using the same method to evaluate integration (28) :

$$
\begin{aligned}
\text{[graph]} \quad = \quad & -\frac{2\pi}{45}Z^2 y^4 u^{1/2} + \frac{\pi}{30}\sqrt{\frac{\pi}{2}}Z^2 y^5 u - \frac{8\pi}{225}Z^2 y^6 u^{3/2} - \frac{2\sqrt{2\pi}}{135}Z^2 y^7 u^2 \ln\left(2yu^{1/2}\right) \\
& -\frac{\sqrt{2\pi}}{45}Z^2 y^7 u^2\left(C-\frac{14}{3}\right) + \frac{16\pi}{4725}Z^2 y^8 u^{5/2} + \mathcal{O}\left(y^9 u^3\right) .
\end{aligned}
\tag{29}
$$

The order of the graph : $\quad$ [graph] $\quad$ with n Debye chains is : $u^{3/2}\left(y^2\big/u^{1/2}\right)^n$. Thus, the criteriom of validity of our expansion is : $y^2 < u^{1/2}$ or : $\beta < y^{-4}$.

For $h_2^N(u)$:

Each integrand is expanded in spherical harmonics, as usual [1]. It follows :

$$
\begin{aligned}
\text{[graph]} \quad = \quad & -\frac{15}{(8\pi)^{3/2}y^3}\sum_{\ell=0}^\infty (-1)^\ell(2\ell+1)\int_0^\infty \mathrm{d}x_1 x_1^2\left[j_\ell\left(\frac{y^3 u}{x_1^2}\right)-\delta_{\ell 0}\right] \\
& \int_0^\infty \mathrm{d}x_2 x_2^2\left[j_\ell\left(\frac{y^2 u}{x_2^2}\right)-\delta_{\ell 0}\right]f_\ell(x_1,x_2)
\end{aligned}
\tag{30}
$$

with : $\quad f_\ell(x_1,x_2) = 2\pi\int_{-1}^{+1}\mathrm{d}\mu P_\ell(\mu)\frac{\exp\left(-x_{12}\right)}{x_{12}}\quad$ and : $\quad \mu = \cos\left(\overrightarrow{x_1},\overrightarrow{x_2}\right)$ $\quad$ (31)

where j_ℓ are spherical Bessel functions and $P_\ell(\mu)$ are Legendre polynomials.

$$
\text{[graph]} \quad = \quad \frac{5}{4\sqrt{2\pi}}yu^2 + \mathcal{O}\left(y^2 u^{5/2}\right) .
\tag{32}
$$

To calculate the integral on x_1 and on x_2 (Eq.(30)), the same method as for h_1^C is used. The domains of integration are divided into two parts : small and large x_1 and small and large x_2. In this way, it is possible to evaluate the correction terms of the expansion (Eq.(32)).

4 Quantum corrections

We use the method of effective potentials. For the evaluation of the correlation functions the Coulomb potential is replaced by effective potentials [3]. For instance, in the high temperature limit :

$$u_{ei}(r) = Z\frac{e^2}{r}\left[\exp\left(-\frac{2r^2}{\pi\lambdabar^2}\right) - 1\right] + Z\sqrt{2}\frac{e^2}{\lambdabar}\left[\Phi\left(\frac{r}{\lambdabar}\sqrt{\frac{2}{\pi}}\right) - 1\right] \; , \tag{33}$$

$$\begin{aligned}
u_{ee}(r) \;=\;& -k_B T \ln\left[1 - \frac{1}{2}\exp\left(-\frac{r^2}{\pi\lambdabar^2}\right)\right] \\
&+ \left[1 - \frac{1}{2}\exp\left(-\frac{r^2}{\pi\lambdabar^2}\right)\right]^{-1} \cdot \left\{-\frac{e^2}{r}\left[\exp\left(-\frac{r^2}{\pi\lambdabar^2}\right) - 1\right] - \frac{e^2}{\lambdabar}\left[\Phi\left(\frac{r}{\lambdabar\sqrt{\pi}}\right) - 1\right]\right. \\
&\left. - \frac{e^2}{2\lambdabar}\,{}_1F_1\left(1, \frac{3}{2}; -\frac{r^2}{\pi\lambdabar^2}\right) - \frac{e^2}{2r}\sum_{\ell=1}^{\infty}\left(-\frac{r^2}{\pi\lambdabar^2}\right)^{\ell}\frac{1}{\ell!}\sum_{p=0}^{\infty}\frac{1}{2p+1}\right\}
\end{aligned} \tag{34}$$

where : $\lambdabar = \hbar\left(\pi m_e k_B T\right)^{-1/2}$ is the electronic De Broglie length, m_e the electron mass, ${}_1F_1$ the degenerate hypergeometric function and Φ the error function : $\Phi(x) = \frac{2}{\sqrt{\pi}}\int_0^x \exp\left(-x^2\right)\mathrm{d}x$. The potential between an electron and an ion, u_{ei}, is the well known Kelbg potential [4]. In Eq.(34)), the exchange effect, is taken into account.

With the same formalism, we are able to calculate g_p and further h_p and $H(\beta)$. The quantum corrections are depending on $\lambdabar/\lambda_D$ (diffraction effect) and on $\lambdabar^3\rho$ (symmetry effect). These calculations are valid if the quantum effects are small :

$$\lambda_D \gg \lambdabar \Rightarrow \gamma^{1/2} \gg y^3 \qquad \text{with :} \qquad \gamma = \frac{1}{\pi}\left(\frac{e^2}{\lambdabar k_B T}\right)^2 \; . \tag{35}$$

γ is a parameter due to Davies and Storer [5]. The simplest form of the screened potential between an electron and an ion is :

$$-\beta V_{ei}(x) \;=\; \frac{Z\Lambda}{x}\left(\exp(-x) - \exp\left(-x\frac{\lambda_D\sqrt{2}}{\lambdabar}\right)\right) \tag{36}$$

and h_1^C can be approximated :

$$\rho h_1^C(u) \;=\; \frac{15}{y^3\sqrt{8\pi}}\int_0^{\infty}\mathrm{d}x x^2 \exp\left(-\beta V_{ei}(x)\right)\cdot\left[\mathrm{j_0}\left(\frac{y^3 u}{x^2}\right) - 1\right] \tag{37}$$

which is diverging if there is no quantum effect. g_{ei} can be expanded :

$$\exp\left(-\beta V_{ei}(x)\right) \;\simeq\; \exp\left(Z(2\pi\gamma)^{1/2} - \frac{Zy^3\sqrt{8\pi}}{15} + \mathcal{O}\left(\frac{\gamma x}{y^3}\right)\right) , \tag{38}$$

$$\exp\left(-\beta V_{ei}(x)\right) \;\simeq\; g_{ei}(0) + \text{ expansion of } x \text{ , } \gamma \text{ and } y \; . \tag{39}$$

$g_{ei}(0)$ has been evaluated precisely by Davies and Storer [5] as an expansion with respect to $\gamma^{1/2}$. Thus, the large field limit is corrected :

$$\rho h_1^C(u) = -u^{3/2}g_{ei}(0) + \cdots \; . \tag{40}$$

References

[1] M. Baranger and B. Mozer, Phys. Rev. **115**,569(1959)
   B. Mozer and M. Baranger, Phys. Rev. **118**,626(1960)
[2] C. Deutsch, Y. Furutani and M.-M. Gombert, Phys. Reports **69**,85(1981)
[3] H. Minoo, M.-M. Gombert and C. Deutsch, Phys. Rev. **23A**, 924(1981)
   M.-M. Gombert and H. Minoo, Contrib. Plasma Phys. **29**,355(1989)
[4] G. Kelbg, Ann. Phys. (Leipzig) **12**,219(1963)
[5] B. Davies and R.G. Storer, Phys. Rev. **171**,150(1968)

CRITICAL POINT OF LITHIUM UNDER INFLUENCE OF COULOMB INTERACTION

H. HESS

Institute of Low-Temperature Plasma Physics Greifswald
Robert-Blum-Str. 8 - 10, O-2200 Greifswald, Germany

Abstract

A combination of the vapour pressure equation with equation of state data assuming only Coulomb interaction in the attractive term gives couples of critical data (p_c, T_c) for the alkali metals which are in good agreement with experimental data for heavier species. For lithium, where experimental data are not available so far, the data estimated here are not in agreement with data recommended earlier.

Introduction

Lithium is an alkali metal of considerable technological interest. It is used or planned to be used in inertial confinement fusion, solar power plants, electrochemical energy storage, magnetohydrodynamic power generators and in a lot of further applications. Therefore the knowledge of the properties of lithium is required in a wide range of parameters often beyond the experimental possibilities for determining them. Unlike the other alkali metals, the critical data, for example, of lithium cannot be obtained in a stationary experiment due to their high values. Pulsed experiments are in preparation but have not been performed yet. Therefore methods for estimation of critical values are of substantial interest.

Estimation of Critical Temperatures

One of the most frequently used methods for estimating critical temperatures T_c of metals is the Guldberg rule /1/ which relates them to the boiling temperatures T_b at atmospheric pressure:

$$T_c^{\ G} \approx 1.5 \, T_b. \tag{1}$$

As can be seen from Fig. 1, however, there is no unique factor in this
rule valid for all metals. There are rather different groups as, for
example, the alkalis, further a group with medium critical
temperatures, and a group with high critical temperatures. For each
group, however, the factor is greater than originally given by
Guldberg.

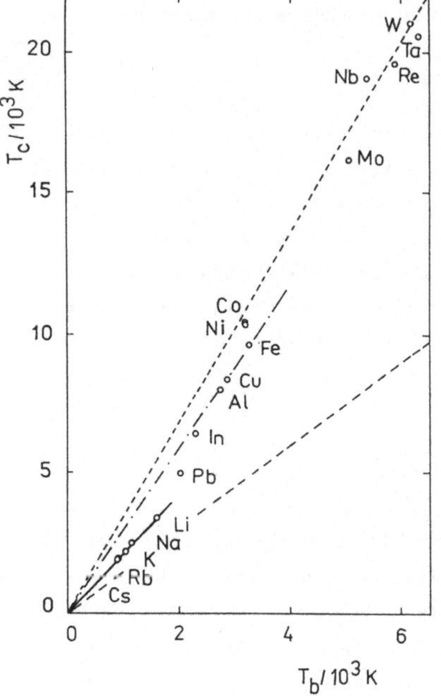

Fig. 1: Critical temperature versus normal boiling tempe-
rature for some metals.
– – – original Guldberg rule (1)
——— modified (for alkalis!)
Guldberg rule: $T_c^G \approx 2.1\ T_b$
– · – representing metals with
medium critical temperature
(5000 – 10 000 K)
– – – – representing metals with
high critical temperature
(15 000 – 20 000 K)

Another method for estimating critical temperatures is the Kopp-Lang
relationship /2/:

$$T_c^{KL} \approx 28.74\ \Delta H_V \qquad\qquad (2)$$

where ΔH_V is the vaporization enthalpy at the normal boiling point in
kJ cm^{-1} (temperature in K).

The Kopp-Lang relationship follows from the Guldberg rule by using the
well-known Trouton's rule that the vaporization entropy of metals is
approximately constant. Similarly as for the Guldberg rule there is
also no unique constant in Trouton's rule. In combining both rules to
the Kopp-Lang relationship (Fig. 2), the critical temperatures are now
much better represented as a function of the vaporization enthalpy

than as a function of the boiling temperature.

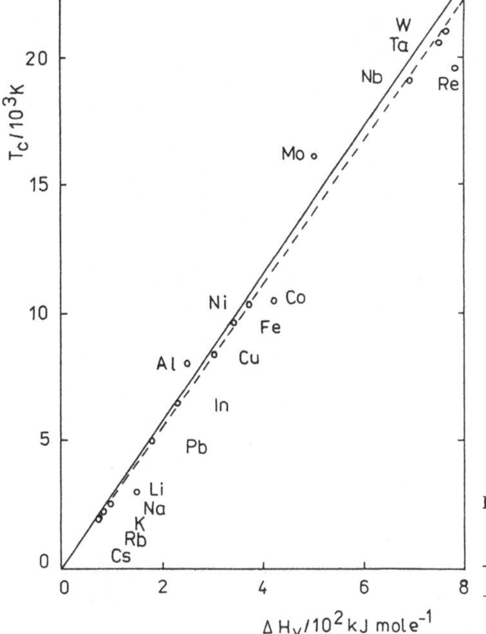

Fig. 2: Critical temperature versus vaporization enthalpy
——— Kopp-Lang relationship (2)
– – – modified Kopp-Lang rela-
tionship: $T_c^{KL} \approx 27.5 \; \Delta H_v$

Deviations from this relation, as for example in the case of lithium
or cobalt, should be seriously discussed. Both these metals should
possibly have higher critical temperatures than given in /3/ (the vaporization enthalpies are from /4/).
Now as an example for our further consideration, we will use the alkali metals. In Table I, critical temperatures are shown: in the first
column the "best" experimental values (selected by the author; the
lithium value in parentheses is the recommended value from /1/).

Table I Critical temperatures from different sources

Element	T_c/K	References	T_c^{G}/K	T_c^{KL}/K
Li	(3344)	/1/	3399	4126
Na	2485	/5/	2433	2763
K	2198	/6/	2172	2270
Rb	2017	/7/	2033	2044
Cs	1924	/7/	1984	1943

The Guldberg values seem to be too low, the Kopp-Lang values too high
(both especially for the lighter alkalis), therefore we expect for
lithium:

$$T_c^{\ G} \approx 3999 \ K < T_c < T_c^{\ KL} \approx 4126 \ K. \tag{3}$$

Estimation of Critical Pressure

If one knows the critical temperature, the critical pressure should be
calculated using a vapour pressure equation. For the alkalis, the
following empirical relation can be used /1/:

$$\ln(p/MPa) = A + B/(T/K) + C \ln(T/K) \tag{4}$$

The constants A, B, and C are given in /1/, and the equation appears
to be well-proved, in lithium at least in the low-pressure range. With
the "best" critical temperatures from Table I we can calculate critical pressures $p_c^{\ V}$ and compare them with the "best" experimental values
(the recommended lithium value - not measured - is given in parentheses).

Table II Critical pressures from vapour pressure equation

Element	T_c/K	p_c/MPa	References	$p_c^{\ V}$/MPa
Li	3399...4126	(30,4)	/1/	33.07...79.97
Na	2485	24.8	/5/	25.12
K	2198	15.5	/6/	14.31
Rb	2017	12.5	/7/	12.24
Cs	1924	9.25	/7/	9.10

The recommended lithium value appears to be too small (cf. (3)).

Coulomb Interaction around the Critical Point

There were some attempts to improve the van der Waals equation of
state with regard to its application to metals near their critical
point (/8/ - /13/). In two papers (1981, 1985), Likalter /10,11/ used
Coulomb interaction only in the attractive term of the van der Waals
equation for describing critical point properties, arguing that in a
certain neighbourhood of the critical point the metallic fluid should
be strongly ionized. In the same way, Chapman and March /12,13/
proceeded later (1986,1987).
Therefore the procedure of Likalter is not isolated, his conclusions,

however, go much further than those of other authors. A first experimental hint which may be considered as a justification of a purely
Coulombic attractive term was given by Hensel et al. in 1979 /14/.
Modelling their pVT measurements in expanded liquid rubidium with a
van der Waals-like equation of state, they got for the exponent of the
specific volume in the attractive term a value of 1.31 which has to be
compared with the Coulombic value of 4/3.

Proceeding so, Likalter /11/ got a relation

$$p_c \sim E_i^2 \, T_c^2 \qquad\qquad (5)$$

where E_i is the ionization energy of the undisturbed atom. As can be
seen from Fig. 3, this relation reflects at least in a qualitative
manner the behaviour of the critical data of metals.

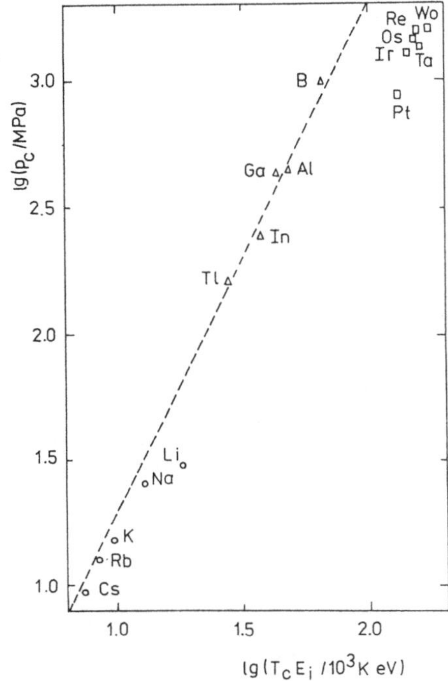

Fig. 3: Critical pressure versus $T_c \cdot E_i$, double-logarithmic plot.

o – group I A of the periodic system

Δ – group III B

□ – platinum group

The dashed line with the slope 2 corresponds to the relation (5).

Combined Method

Using the known alkali values from Tables I and II, one can make the
relation (5) to an equation:

$$P_c \approx (1.66 \pm 0.12)\ 10^{-7}\ (T_c \cdot E_i)^2\ MPa/(KeV)^2 \tag{6}$$

The corresponding curve can be seen in Fig. 4 together with the vapour
pressure curves for the different alkali vapours including lithium.
For each species there exists a point of intersection which is marked
by a cross. The open circles stand for the "best" experimental values
(the experimental scatter does not exceed the circle area!). The open
square is the lithium value as recommended by a recent standard book
authorized by the International Union of Pure and Applied Chemistry
/1/.

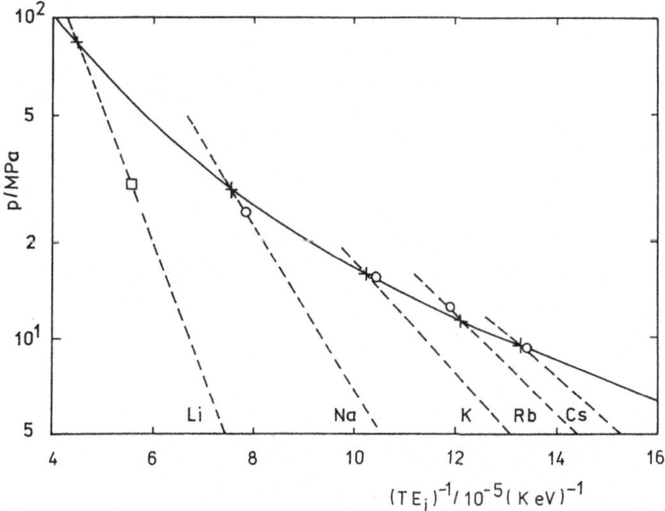

Fig. 4: Pressure versus $(T \cdot E_i)^{-1}$ for the determination of the critical
point (temperature and pressure) by the "combined" method.
  - - -   eq. (4); vapour pressure equation
  ------   eq. (6); "Coulomb" behaviour
  o - values from Table 1; "experimental" values
  + - values from the intersection of eq. (4) with eq. (6);
        "combined" method
  □ - values for lithium, recommended in /1/.

As can be seen, the p_c-T_c values from Cs to Na are sufficiently well described by the points of intersection. In Table III, the so calculated values are shown (p_c', T_c'; recommended values for lithium from /1/ in parentheses) and compared with the "best" experimental values from Tables I and II.

Table III Critical temperature and pressure

Element	T_c/K	T_c'/K	p_c/MPa	p_c'/MPa
Li	(3344)	4176	(30.4)	84.0
Na	2485	2556	24.8	29.0
K	2198	2254	15.5	15.7
Rb	2017	1969	12.45	11.2
Cs	1924	1941	9.25	9.44

For the critical temperatures, the differences are smaller than 3 % for all alkalis (maximum experimental error $\leq \pm$ 1.5 %) but lithium. The critical pressures in the case of K and Cs differs by 2 % only, whereas for Na and Rb these differences are higher - 11 % and 14 % (maximum experimental error $\leq$ 10 %).

Compared with the deviations in the case of lithium ($\Delta T_c \approx$ 25 %, $\Delta p_c \approx$ 176 %), the differences between the values for the heavier alkalis calculated by the "combined" method proposed here and the "best" experimental values are quite small. Supposed the method can be applied to lithium too, it should be expected that the deviations are of the same order of magnitude ($\Delta T_c \approx$ 3 %, $\Delta p \approx$ 10 %).

4. Concluding remarks

The method presented here gives a further estimation of critical point data for lithium (p_c, T_c) which is consistent with the vapour pressure equation and with the assumed Coulomb interaction around the critical point.

The values derived here ($p_c \approx$ 84.0 MPa, $T_c \approx$ 4176 K) are appreciably higher than the values recommended earlier ($p_c \approx$ 30.4 MPa, $T_c \approx$ 3344 K /1/) although they are well within the broad range of estimations given so far (17 MPa $\leq p_c \leq$ 242 MPa, 2994 K $\leq T_c \leq$ 4400 K /1/). The good agreement of our estimations with the experimental values of the heavier alkali metals within the experimental errors speaks for a

certain reliability of the lithium values too. There is further the
hope that the same procedure as used here will also give reliable
values for other metals.

Acknowledgment

Part of this work has been done during a stay at the Institute of
Experimental Physics of the Technical University Graz/Austria. The
author is indebted to Prof. H. Jäger and his colleagues, especially to
Dr. G. Pottlacher and Mr. E. Kaschnitz, for their kind hospitality and
the stimulating atmosphere in their institute. Financial support has
been given by the Austrian Ministry of Science and Research.

References

/1/ Magill, J., and R.W. Ohse, in: Handbook of Thermodynamic and
    Transport Properties of Alkali Metals, Ed.: R.W. Ohse, Blackwell
    Scientific Publications, Oxford 1985, p. 73

/2/ Lang, G., Z. Metallkde **68** (1977) 213

/3/ Fortov, V.E., and I.T. Yakubov, Fizika neidealnoi plazmy (in
    russian), Chernogolovka 1984, p. 261

/4/ Schulze, G.E.R. Metallphysik, Springer-Verlag, Wien, New York,
    1974

/5/ Binder, H., Experimentelle Bestimmung von pVT-Daten, kritischen
    Größen und Zustandsgleichung des Natriums bis 2600 K und 500
    bar, Doctoral Thesis, Karlsruhe 1984

/6/ Freyland, W.F., and F. Hensel, Ber. Bunsenges. Physik. Chemie **76**
    (1972) 16

/7/ Jüngst, S., B. Knuth, and F. Hensel, Phys. Rev. Lett. **55** (1985)
    2160

/8/ Young, D.A., and B.J. Alder, Phys. Rev. **A 3** (1971) 364

/9/ Blander, M., and H.R. Leribaux, in: Proc. 7th Symp. Thermophys.
    Prop., Ed.: A. Cezairliyan, Am. Soc. Mech. Eng., Gaithersburg
    1977, p.882

/10/ Likalter, A.L., Doklady Akad. Nauk **259** (1981) 96

/11/ Likalter, A.A., Teplofiz. vys. temp. **23** (1985) 465

/12/ Chapman, R.G., and N.H. March, Phys. Chem. Liq. **16** (1986) 77

/13/ Chapman, R.G., and N.H. March, Phys. Chem. Liq. **17** (1987) 165

/14/ Pfeiffer, H.P., W. Freyland, and F. Hensel, Ber. Bunsenges. Phys.
    Chem. **83** (1979) 204

HIGH PRESSURE - HIGH TEMPERATURE THERMOPHYSICAL MEASUREMENTS ON LIQUID METALS

E. Kaschnitz, G.Pottlacher

Technische Universtät Graz, Institut für Experimentalphysik

Petersgasse 16, 8010 Graz, Austria

Abstract

Two resistive pulse heating systems allow the determination of thermophysical properties of solid and liquid metals, such as heat capacity and the mutual dependences between enthalpy, electrical resistivity, temperature and volume up to temperatures of about 10.000 K and ambient pressures up to 0.5 GP. Investigations have been performed on the following elements: C, W, Re, Ta, Mo, Nb, Fe, Ni, Co, Cu, Pb, In, and Hg. From the investigations at elevated pressures critical point data of some metals can be estimated.

Introduction

The behaviour of metals is reasonably well known in the solid state. The liquid state of metals is more difficult to investigate. It covers the temperature range from the melting point up to the critical temperature which, for some metals, is more than 10.000 K. Traditional static steady-state techniques for the measurement of thermophysical properties are generally limited to temperatures below about 2000 K. Fast dynamic methods have been developed to avoid these difficulties and to permit the extension of the measurements to higher temperatures. Properties of matter at high temperatures are useful for many purposes, including applications which are subjected to high temperature/high pressure conditions, and for modelling and basic theory. A general survey of resistive pulse heating techniques and laser pulse heating techniques is given by Pottlacher [1].

Experimental

The submicrosecond discharge circuit stores the energy in a 5.4 µF capacitor, with charging voltages between 4000 and 8000 V. Heating rates of more than 10^9 K/s can be achieved. The recently installed microsecond pulse heating system [2] has a variable capacitor bank from 240 µF to 500 µF. This circuit is typically charged up to 10.000 V, the heating rates are in the range of 10^8 K/s. Both systems are built up coaxially.

The metal samples are resistively volume-heated by passing a large current pulse through them. In order to avoid peripheral discharges, water is chosen as the ambient medium. The experiment is performed in a high pressure vessel with sapphire windows and a maximum pressure capability of 0.5 GPa.

Measured quantities are:
a) The current through the wire, measured in the faster system using an induction coil. The coil signal is subsequently integrated. The slower system measures the current more accurate by the aid of a current transformer (Pearson Probe).
b) The voltage, using a coaxial ohmic voltage divider.

c) The surface radiance temperature of the sample, using a fast pyrometer operating at a certain wavelength (e.g. 850 nm or 1500 nm).
d) The black-body radiation from inside of the liquid metal tube.
e) The final volume of the sample, using a shadowgraph technique employing a camera with a 30 ns exposure time. This camera uses a Kerr-cell shutter. Currently a video imaging system for monitoring the sample volume is under construction.
f) The transmission time of a sound wave passing through the metal sample.
g) The vapor pressure of the liquid sample.

All measured quantities of the submicrosecond system have rise-times less than 10 ns, those of the microsecond system less than 100 ns. More experimental details are given in [2,3].

Datareduction and Results

The time-dependent current $I(t)$, the voltage $U(t)$, the surface radiation intensity $J_{SF}(t)$, as well as the blackbody radiation intensity $J_{BB}(t)$, are measured simultaneously with fast digitaloscilloscopes. Many additional thermophysical properties may be calculated:

Inductive contributions are subtracted from the measured voltage $U(t)$ to produce ohmic voltage $U_C(t)$. The specific enthalpy $H(t)$ in J kg^{-1} may be calculated from $U_C(t)$ and $I(t)$, starting at room temperature $[H = H(t) - H(298K)]$ from:

$$H(t) = \frac{1}{m} \int_0^T U_C(t)\, I(t)\, dt \tag{1}$$

where m is the mass of the sample.

The electrical resistivity $\rho_0(t)$ in Ω m, without a volume correction, can be calculated from:

$$\rho_0(t) = \frac{U_C(t)\, \pi\, r_0^2}{I(t)\, 1} \tag{2}$$

where r_0 is the initial radius in m and 1 the length in m of the specimen.

The measurement of the sample radius $r(t)$ allows determination of the volume change V/V_0 (V_0 initial volume). This allows the calculation of the specific electrical resistivity $\rho(t)$ in Ω m.

$$\rho(t) = \rho_0(t) \frac{r^2(t)}{r_0^2} \tag{3}$$

The temperature of the sample is calculated using the known melting temperature of the specimen as the calibration point. Temperatures are found using the KIRCHHOFF-PLANCK law and forming ratios of the radiance at temperature T to the radiance at the melting temperature T_m. This is possible due to the linear amplification of the pyrometer. At melt (index m) the pyrometer detects a constant radiance with the Intensity J_m :

$$J_m(T_m) = \frac{g\, \varepsilon_m\, (\lambda, T_m)\, c_1}{\lambda^5[(\exp(c_2/\lambda T_m) - 1)]} \tag{4}$$

Here $J_m(T_m)$ is the measured intensity in Volts at the melting point, T_m is the melting temperature, g is a geometric factor, $\varepsilon_m(\lambda, T_m)$ is the spectral emissivity at melting temperature, c_1 is the first radiation constant, λ is the wavelength in m, c_2 is the second radiation constant.

At any point in the liquid phase the pyrometer detects an Intensity J(t) in Volts, at a wavelength of 800 nm or 1500 nm:

$$J(t) = \frac{g\,\varepsilon(\lambda,T)\,c_1}{\lambda^5\{\exp[c_2/\lambda T(t)] - 1\}} \tag{5}$$

Here J(t) is the measured intensity at the instant t. The related unknown temperature T(t) we obtain from the ratio

$$\frac{J_m(T_m)}{J(t)} \tag{6}$$

and using equations (4) and (5).

$$T(t) = c_2/\lambda \ln\left[1 + \frac{\varepsilon(T)\,J_m(T_m)}{\varepsilon(T_m)\,J(t)}[\exp(c_2/\lambda T_m) - 1]\right] \tag{7}$$

These quantities are on the same time base, and by comparing values at common times we find the interdependences of enthalpy, resistivity, temperature and volume expansion.

The slope of the graph of enthalpy versus temperature allows the determination of c_P, the specific heat at constant pressure. From the same graph it is also possible to determine the enthalpy of fusion. These two quantities, obtained for the investigated elements, are listed in Table I.

As mentioned above, the sound speed in the metal sample and the vapor pressure are measured directly. The emissivity of liquid metals we obtain by comparing the surface radiation of the sample to the black-body radiation of the interior of the tube-shaped sample. More information is given in [3,4].

From different measurements under varying ambient pressure an experimental estimation of the critical point of metals can be obtained, as described by Pottlacher and Jäger [5]. The main items of the discussion are repeated in the following three paragraphs.

Fig.1 shows a schematic Van der Waals phase diagram in the p - V plane. The three main areas are the region of the solid state (s), of the liquid state (l) and of the gaseous state (g). Mixtures where two states can exist at the same time are the solid - liquid region (between the s- and the l- region), the region of liquid - vapor (l + v) and the region of solid - vapor (s + v). The line A - K is the normal boiling line, the binodal line is A - K - D; B - K - C represents the spinodal line; K is the critical point.

The binodal line is the equilibrium curve for liquid and vapor; in the region A - K - B a mixture of liquid and vapor as well as a superheated liquid state can exist. The spinodal line B - K is the boundary of thermodynamic stability of the superheated metastabile liquid. Using heating rates of more than 10^9 K/s, we can superheat liquid metals up to the spinodal line. Near this line volume fluctuations in the sample will occur due to the sudden growth of homogeneous vapor nuclei in the superheated liquid. The area within B - K - C is unstable, and therefore a 'jump' into the region K - C - D, the so called phase explosion, will occur with a large increase of volume. Crossing the spinodal line causes a sharp increase in the electrical resistivity for this kind of experiments, as the conducting cross-section of the sample is reduced. The rapid creation of vapor nuclei connected with a powerful expansion, the so called phase explosion, produces a shock wave in the surrounding medium. At that moment the surface

<u>Table I:</u> Experimentally determined values of the enthalpy of fusion (ΔH) and of the specific heat at constant pressure (c_p) for some elements. For c_p the valid temperature range is given.

Element	ΔH	c_p	Temperature range
	MJ kg⁻¹	J kg⁻¹ K⁻¹	K
C	>8.0	2641	3000 - 4950
W	0.256	305	3680 - 12000
Re	0.150	250	3453 - 7500
Ta	0.150	250	3500 - 8500
Mo	0.420	560	2890 - 10000
Nb	0.309	513	2750 - 6500
Fe	0.269	897	2250 - 4250
Ni	0.292	762	2000 - 7000
Co	0.273	780	1768 - 6000
Cu	0.230	500	1500 - 2750
Pb	-	156	2000 - 5000
In	-	390	2000 - 5000
Ni	0.292	762	2000 - 7000
Co	0.273	780	1768 - 6000
Cu	0.230	500	1500 - 2750
Pb	-	156	2000 - 5000
In	-	390	2000 - 5000

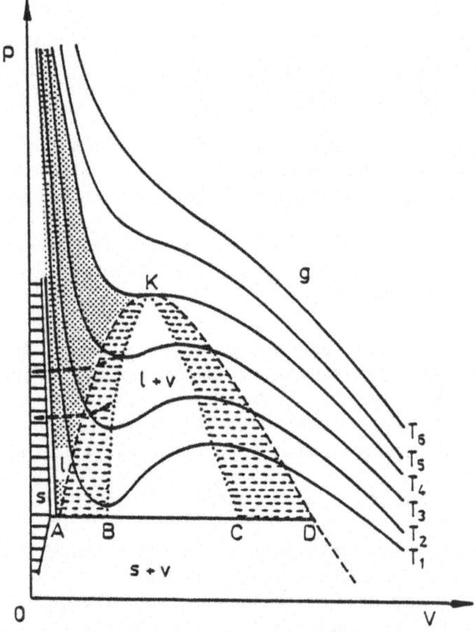

<u>Fig. 1.</u> Schematic Van der Waals phase diagram in the p-V plane. Isotherms $T_1 < T_2 < T_3 < T_4 < T_5 < T_6$; T_4 critical isotherm. A-K-D binodal line; B-K-C spinodal line; K critical point; s, solid; l, liquid; v, vapor; g, gas. Broken lines path of our experiment at two different static pressures.

temperature starts to decrease due to the cooling of the liquid phase, as it looses the more energetic atoms to the vapor phase, which expands adiabatically [6].

For the experimental determination of critical point data, we proceeded as follows: If the static pressure p_s is increased, the sharp rise of the electrical resistivity will occur at higher enthalpy values and the shock wave will become weaker. At static pressures higher than the critical pressure the shock wave and the sharp rise of the electrical resistivity will vanish. This seems to be a direct method to determine p_c. When reaching the spinodal line, the critical temperature T_c can be calculated from the corresponding enthalpy values. The volume expansion versus enthalpy curve gives the critical volume v_c.

Up to now we have performend experiments to obtain critical point data with the metals lead [5], indium [7] and cobald [4]. The elements mercury and lithium are the next materials to be investigated. To see the scatter of data, e.g. for lead, a summary of experimentally determined or theoretically estimated critical point data, as reported in literature, is given in Table II.

<u>Table II:</u> Summary of experimentally determined (∗) or theoretically estimated (#) critical point data of lead.

Reference	T_k	p_k	V_k	V_k/V_0	
	K	bar	$cm^3\,mol^{-1}$		
Bohdansky [8]	5500				#
Carlson [9]	6266	2530	68	3.7	#
Fortov [10]	5600	2400			∗
Gates [11]	3584	420	206	11.3	#
Grosse [12]	5400	850	96	5.1	#
Hodgson [13]	5300-6000	2000-3000	65	3.6-3.7	∗
Hornung [14]	5500	1010	82	4.5	#
Kopp [15]	4760				#
Lang [16]	5150				#
Morris [17]	5400	1400	101	5.5	#
Pottlacher [5]	5400±400	2500±300	65±5	3.6±0.4	∗
Young [18]	5158	2260	67	3.7	#
Zadumkin [19]	4200				#

Estimate of errors

Reliable estimates of the total experimental uncertainties are difficult to make for this kind of experiment. An uncertainty of 5% has to be assumed for the enthalpy of fusion; the uncertainty of the specific heat should not exceed 15 %. A detailed analysis of errors is given in [2,3].

References

[1] G. Pottlacher, "Thermophysical measurements, Subsecond", in Advances in Materials Sience and Engeneering, third Supplement, R. W. Cahn Editor, Pergamon Press, Oxford (1992)
[2] G. Pottlacher, E. Kaschnitz, H. Jäger, J. Phys.: Condens. Matter **3**, 5783, (1991)
[3] E. Kaschnitz, G. Pottlacher, H. Jäger, to be published in the Int. J. Thermophys.
[4] H. Heß, E. Kaschnitz, G. Pottlacher, in preparation.
[5] G. Pottlacher, H. Jäger, Int. J. Thermophys. **11**.4, 719, (1990)
[6] W. Fucke, U. Seydel, High Temp. High Press. **12**, 419, (1980)
[7] G. Pottlacher, T. Neger, H. Jäger, to be published in High Temp. High Press., (1992)

[8]   J.Bohdansky, L. Chem. Phys. **49**, 2982, (1968)

[9]   M. Carlson, H. Eyring, T. Ree, Proc. Nat. Acad. Sci. USA **46**, 649, (1960)

[10] V. Fortov, Moscow, personal communication, (1990)

[11] D. Gates, G. Thodos, Am. Inst. Chem. Eng. J. **6**, 50, (1960)

[12] A. Grosse, A. Kirshenbaum, J. Inorg. Nucl. Chem. **22**, 739, (1962)

[13] W. Hodgson, PhD Thesis, UCRL - 52498, (1978)

[14] K. Hornung, J. Appl. Phys. **46**, 2548, (1975)

[15] I. Kopp, Rus. J. Phys. Chem **41**, 782, (1967)

[16] G. Lang, Z. Metallkunde **68**, 213, (1977)

[17] E. Morris, AWRE Report 0-67/64 London UKAEA, (1964)

[18] D. Young, UCRL - 52352 LLNL, (1977)

[19] S. Zadumkin, Inzh.- Fiz. Zh. **3**, 63, (1960)

EQUATION OF MOTION FOR HIGHER ORDER DISTRIBUTION FUNCTIONS AND LINEAR RESPONSE THEORY

G. Röpke

University Rostock, Department of Physics

Universitätsplatz 3, 0 - 2500 Rostock

1 Aim of contribution

Within a quantum statistical approach to a many particle system such as dense plasmas, correlations have to be taken into acount in an appropriate way. In general we have to treat the nonequilibrium state and to derive equations of motion for the relevant observables of the system. The equilibrium state should result as a limiting case. The following problems are considered in detail: The concept of entropy in nonequilibrium is given in terms of relevant observables. The description of the system can be performed on different levels. As a relevant quantity of the system we can introduce the *single–particle Wigner distribution function*

$$f_1(p, r, t) = \mathrm{Tr}\left\{\rho(t) n_1(p, r)\right\},\tag{1}$$

with the single–particle Wigner density

$$n_1(p, r) = \int \frac{d^3 q}{(2\pi)^3} e^{iqr} a^+_{p-q/2} a_{p+q/2}.$$

In dense plasmas where many particle effects are of importance, higher order distribution functions have also to be included in order to describe correlations and bound state formation. We consider the *two–particle Wigner distribution function*

$$f_2\left(p_1 p_2, r_1 r_2, t\right) \;=\; \mathrm{Tr}\left\{\rho(t) n_2\left(p_1 p_2 r_1 r_2\right)\right\},$$

$$n_2\left(p_1 p_2, r_1 r_2\right) \;=\; \int \frac{d^3 q_1}{(2\pi)^3} \int \frac{d^3 q_2}{(2\pi)^3} e^{iq_1 r_1 + iq_2 r_2} a^+_{p_1-q_1/2} a^+_{p_2-q_2/2} a_{p_2+q_2/2} a_{p_1+q_1/2}.\tag{2}$$

The extension of the results given below to higher order distribution functions is straightforward.

2 Nonequilibrium statistical operator

The quantity of main interest in eqs. (1), (2) is the *statistical operator* $\rho(t)$. According to Zubarev [1] it is given by the expression

$$\rho(t) = \eta \int\limits_{-\infty}^{t} e^{\eta(t'-t)} U(t, t') \rho_{\mathrm{rel}}(t') U(t', t) dt',\tag{3}$$

where $U(t, t')$ is the time evolution operator; the limit $\eta \to 0$ has to be taken after the thermodynamic limit. The statistical operator $\rho(t)$ solves the von Neumann equation, where the initial conditions are incorporated in form of a source term. The relevant statistical operator

$$\rho_{\mathrm{rel}}(t) = \exp\left[-S(t)/k_{\mathrm{B}}\right] \tag{4}$$

corresponds to the maximum of the *entropy* $\langle S(t) \rangle$ at given normalization $\mathrm{Tr}\rho_{\mathrm{rel}}(t) = 1$ and given mean values of a set of relevant observables B_n:

$$\mathrm{Tr}\left\{\rho_{\mathrm{rel}}(t)B_n\right\} = \langle B_n \rangle^t \equiv \mathrm{Tr}\left\{\rho(t)B_n\right\}. \tag{5}$$

The solution of this variational principle is the generalized Gibbs state (4) with

$$S(t) = \sum_n F_n(t)B_n, \tag{6}$$

where the *Lagrange multipliers* $F_n(t)$ are determined by the conditions (5). Now, we are able to derive *equations of motions* for the relevant observables B_n in the form

$$\frac{d}{dt}\langle B_n \rangle^t = \mathrm{Tr}\left\{\rho(t)\frac{i}{\hbar}\left[H(t), B_n\right]\right\}. \tag{7}$$

$H(t)$ is the Hamiltonian of the system (an explicit dependence on time is given, for instance, by a time–dependent external field); the statistical operator $\rho(t)$ is related to the Lagrange multipliers $F_n(t')$ in the past $-\infty < t' < t$ according to eqs. (3), (4), (6). The time evolution of the Lagrange parameters is implicitly given by eqs. (5) and (7).

There are two essential problems in describing a many particle system:

(i) We have to evaluate the relations between the Lagrange multipliers $F_n(t)$ and the mean values $\langle B_n \rangle^t$ of the relevant observables according to eq. (5).

(ii) We have to evaluate the generalized kinetic equation (7).

In *thermodynamic equilibrium*, the solution of eq. (7) is trivial: B_n are constants of motion as the system energy H and the particle numbers N. The determination of the Lagrange parameters F_n is the main problem to be solved, they have the meaning of inverse temperature and chemical potential. The solution of eq. (5) which takes the special form

$$\mathrm{Tr}\left\{\rho_o H\right\} = U, \qquad \mathrm{Tr}\left\{\rho_o N\right\} = \Omega_o n, \tag{8}$$

with $\rho_o = Z_o^{-1}\exp\left[-\beta(H - \mu N)\right]$, results in the *equations of state* such as

$$\mu = \mu(\beta, n). \tag{9}$$

Presenting the density n as a function of the inverse temperature β and the chemical potential μ, the solution can be found by using the *thermodynamic Green function $G_1(p, z)$*:

$$n(\beta, \mu) = \frac{1}{\Omega_o} \sum_p \int \frac{d\omega}{2\pi} \frac{1}{e^{\beta(\omega-\mu)} + 1} \mathrm{Im}G_1(p, \omega + i\varepsilon) \tag{10}$$

(we consider a system of Fermions, p denotes the quantum numbers of single particle states as momentum, spin, species). We can employ the perturbation expansion and *diagram representation* to evaluate the thermodynamic Green function, see also [2]. Special partial summations have to be performed in order to take into account many particle effects:

(i) Dyson equation: The *self–energy* $\Sigma(p, z)$ and the quasiparticle concepts are introduced.

146

(ii) Screened potential: The *polarization function* $\Pi(q, z)$ is introduced.

(iii) Bethe–Salpeter equation: in the ladder approximation,

$$G_2^L(12, 34, z) = G_2^o(12, z) \left[\delta_{13}\delta_{24} + \sum_{1'2'} V(12, 1'2')G_2^L(1'2', 34, z) \right], \tag{11}$$

the properties of a *two–particle system* (correlations, bound states) imbedded in a plasma are introduced.

An important prescription to select out diagrams which are of physical relevance is the so–called *chemical picture*: Each diagram in the perturbation expansion is completed by further diagrams, where the single particle propagator G_1 is replaced by the two particle propagator G_2^L or higher order propagators, respectively. This approach realizes that bound states are treated on the same level as the "elementary" particles. A theory, where bound states are treated like a new species, is appropriate to describe a partially ionized plasma.

It should be mentioned that the Matsubara Green functions can also be introduced to evaluate averages with the relevant statistical operator (4). The evaluation of the equation of motion (7) needs, however, an additional perturbative expansion which applies to the time evolution operators $U(t, t')$.

Explicit relations to determine the Lagrange multipliers $F_n(t)$ according to eq. (5) can be given near the equilibrium where we can apply the theory of linear response. This special case is considered in Section 4.

3 Single–particle approximation for the entropy

For the entropy $S(t)$ (6), a cluster decomposition can be given:

$$S(t) = S_o^{(t)} + \sum_{pp'} S_1(p, p', t)a_p^+ a_{p'} + \sum_{p_1 p_2 p_1' p_2'} S_2 a_{p_1}^+ a_{p_2}^+ a_{p_2'}^+ a_{p_1'}^+ + \dots \tag{12}$$

We choose the set of relevant observables as the *single particle disribution function*, i.e. we restrict ourselves to only $S_o(t)$, $S_1(p, p', t)$ in the decomposition (12). Such a description of the system has the advantage that mean values with respect to $\rho_{\text{rel}}(t)$ are immediately evaluated in the following way:

(i) diagonalization

$$S_1(t) = \sum_p F_1(k, t)a_k^+ a_k, \tag{13}$$

(ii) Wick's theorem,

(iii) elimination of the Lagrange multipliers $F_1(p, t)$ according to eq. (5):

$$f_1(k, t) = \left\langle a_k^+ a_k \right\rangle^t = \{\exp[F_1(k, t)/k_{\text{B}}] + 1\}^{-1}. \tag{14}$$

The evaluation of the equation of motion for the single particle distribution function (*kinetic equation*, [3]) according to eq. (7) can be performed using a partial integration

$$\frac{\partial}{\partial t} f_1(p, r, t) = -D^{\text{Vlasov}}(p, r, t) + I^{\text{coll}}(p, r, t) \tag{15}$$

with

$$D^{\text{Vlasov}} = -\frac{i}{\hbar}\text{Tr}\left\{\rho_{\text{rel}}(t)\left[H, n_1(p, r)\right]\right\}$$

$$= \frac{p}{m}\frac{\partial}{\partial \vec{r}}f_1(p, r, t) - \frac{\partial U(r)}{\partial \vec{r}}\frac{\partial}{\partial \vec{p}}f_1(p, r, t). \tag{16}$$

The *collision term*

$$I^{\text{coll}} = -\int_{-\infty}^{t} e^{\eta(t'-t)}\text{Tr}\left[U(t, t')\left\{\frac{i}{\hbar}\left[H(t'), \rho_{\text{rel}}(t')\right]\right.\right.$$

$$\left.\left. + \frac{\partial}{\partial t'}\rho_{\text{rel}}(t')\right\}U(t', t)\right]dt' \tag{17}$$

takes the form of the Boltzmann–Nordheim collision integral after the approximations $\rho_{\text{rel}}(t') \approx \rho_{\text{rel}}(t)$, and considering only binary collisions from the time evolution operator $U(t, t')$. It is well known that such an approach (15) to the many Fermion system leads to an incorrect equilibrium distribution. The Fermi distribution function as equilibrium solution of the Boltzmann collision integral and the Boltzmann entropy are correct for an *ideal Fermion system*. For interacting Fermion systems, we have also to include the formation of two–particle correlations (S_2 in eq. (12)) what will be discussed in sections 5, 6.

4   The linear response theory

In order to realize the correct thermal equilibrium ρ_o (8) of the system, the entropy $S(t)$ (6), (12) should contain two–particle contribution at least of the form of the potential energy in the Hamiltonian H. Before giving the general treatment of higher order correlations, we consider the special case of near equilibrium where the Lagrange multiplies $F_n(t)$ can also explicitly evaluated from eq. (5) and eliminated from the equation of motion (7), see Refs. [4,5]. The deviation from equilibrium is considered to be caused by an *external potential*, $H_{\text{tot}} = H + H_{\text{ext}}^t$,

$$H_{\text{ext}}^t = \sum_q \int \frac{d\omega}{2\pi}e^{i\omega t}U_{\text{ext}}(q, \omega)\sum_{c,p}e_c n_{p,-q}^c. \tag{18}$$

The set of relevant observables to describe the nonequilibrium state of the system of charged particles is divided into the constants of motion H, N_c, and the induced "currents" B_ν^t (possible explicitly time dependent)

$$\begin{array}{ccccc} \{B_n\} & \to & \{H, & N_c, & B_\nu^t & \}; \\ \{F_n(t)\} & \to & \{\beta, & -\beta\mu_c, & -\beta\varphi_\nu(t) & \} \end{array} \tag{19}$$

are the corresponding Lagrange multipliers.

We assume that the *response parameters* $\varphi_\nu(t)$ are small quantities as long as the perturbation H_{ext}^t (18) is also small. Expanding the difference $\rho(t) - \rho_{\text{rel}}(t)$ (eqs. (3), (4)) with respect to $U_{\text{ext}}(q, \omega)$ and $\varphi_\nu(t)$ up to the first order, we have from eq. (5) the linear response equations

$$- U_{\text{ext}}(q_o, \omega_o)N_n^c(\omega) = \sum_{m d\omega'} D_{nm}^{cd}(\omega, \omega')\varphi_m^d(q_o, \omega_o - \omega') \tag{20}$$

Here, we consider only a single mode of the external field because the linear response to an arbitrary field H_{ext}^t can be given in form of a superposition. The response equations (20) represent a system of coupled linear equations to determine the response parameters $\varphi_n^c(\omega)$ (Fourier–transform of $\varphi_\nu(t)$). The coefficients $N_m^c(\omega)$, $D_{nm}^{cd}(\omega,\omega')$ are *equilibrium correlation functions*.

Describing the nonequilibrium state only by single–particle distribution functions as in section 3, the B_ν^t in eq. (19) are given by the Wigner densities $n_{p,-q}^c$. In order to have a finite set of response equations (20), we introduce moments of the single particle densities

$$B_\nu^t = B_n^{c,t}(q) = \sum_p \int \frac{d\omega}{2\pi} b_n^c(p,q,\omega) e^{-i\omega t} n_{p,-q}^c. \tag{21}$$

Whereas the response equations (20) for the relevant observables $n_{p,-q}^c$ correspond to generalized linear Boltzmann equations, the use of a finite set of moments (21) corresponds to a *variational solution* of these equations. The correlation functions occuring in (20) are moments of the single–particle equilibrium correlations:

$$\begin{aligned}
D_{nm}^{cd} &\leftrightarrow \left\langle \dot{n}_{k,-q_o}^d(\eta - i\omega_o)\,; n_{p,q_o}^c \right\rangle - i\omega_o \left\langle n_{k,-q_o}^d(\eta - i\omega)\,; n_{p,q_o}^c \right\rangle \\
N_n^c &\leftrightarrow \left\langle \dot{n}_{k,-q_o}^d(\eta - i\omega_o)\,; n_{p,q_o}^c \right\rangle
\end{aligned} \tag{22}$$

with

$$\begin{aligned}
<A(z); B> &= \int_{-\infty}^{0} e^{izt}(A(t),B)\,dt, \\
(A,B) &= \int_{o}^{\beta} d\tau\, \mathrm{Tr}\left\{\rho_o A(-i\hbar\tau)B\right\}
\end{aligned} \tag{23}$$

The evaluation of the equilibrium correlation functions can be performed using the technique of thermodynamic Green functions.

Solving the set of linear equations (20) using Cramer's rule we find for the mean values of the relevant observables

$$\langle B_i^a(-q_o,\omega)\rangle = \beta U_{\mathrm{ext}}(q_o\omega_o) \begin{vmatrix} 0 & M_{im}^{ad}(\omega,\omega'') \\ N_n^c(\omega') & D_{nm}^{cd}(\omega',\omega'') \end{vmatrix} \left| D_{nm}^{cd}(\omega',\omega'') \right|^{-1} \tag{24}$$

with M_{im}^{cd} being moments of $\left(n_{k,-q_o}^d\,; n_{p,q_o}^c\right)$. Especially, we have for the *dielectric function*

$$\begin{aligned}
\epsilon(q,\omega) &= 1 + \frac{i}{\epsilon_o\omega}\sigma(q,\omega) = 1 - \frac{1}{\epsilon_o q^2}\Pi(q,\omega); \\
\Pi(q,\omega) &= \sum_c e_c \frac{1}{\Omega_o} \sum_p f_c(p,q,\omega) U_{\mathrm{eff}}^{-1}(q,\omega) \\
&= \frac{\beta}{i\omega\Omega_o} \begin{vmatrix} 0 & M_m^d(\omega') \\ N_n^c(\omega) & D_{nm}^{cd}(\omega,\omega') \end{vmatrix} \left| D_{nm}^{cd}(\omega,\omega') \right|^{-1}
\end{aligned} \tag{25}$$

The following set of moments (21) is appropriate to obtain the correct Vlasov (RPA) limit of the collisionless plasma as well as the static conductitivity:

$$\begin{aligned}
b_s^c(p) &= (k_{\mathrm{B}} T m_c)^{1/2} \hbar\omega_o \left(\hbar^2 p_z q/m_c - \hbar\omega - i\hbar\gamma_c\right)^{-1} \left(\delta_{q,q_o}\delta_{\omega,\omega_o} - \delta_{q,-q_o}\delta_{\omega,-\omega_o}\right), \\
b_o^c(p) &= (k_{\mathrm{B}} T m_c)^{1/2} \delta_{\omega,o} \left(\delta_{q,q_o} - \delta_{q,-q_o}\right), \\
b_1^c(p) &= \hbar p_z \delta_{\omega,o} \left(\delta_{q,q_o} - \delta_{q,-q_o}\right), \ldots
\end{aligned} \tag{26}$$

An *interpolation formula* for $\epsilon(q,\omega)$ can be given which reproduces the exact behaviour in the following limiting cases: $V = 0$ (collisionless plasma), $q \to 0$ (long wave length limit), $\omega \to 0$ (static limit):

$$\epsilon(q,\omega) = 1 + \frac{e^2}{\epsilon_o}\frac{n}{k_B T q^2}F(q,\omega)\left[1 - i\frac{\omega}{(qk_B T)^2}F(q,\omega)\frac{I}{N}\right]^{-1} \tag{27}$$

with

$$F(q,\omega) = \sum_c \left[1 + \left(\frac{m_c}{2\pi k_B T}\right)^{1/2}\frac{\omega}{q}D\left[\left(\frac{m_c}{2k_B T}\right)^{1/2}\frac{\omega}{q}\right]\right],$$

$$D(z) = \int_{-\infty}^{\infty} dx e^{-x^2}(x - z)^{-1},$$

$$I = Nn\frac{e^4}{\epsilon_o}(2\pi)^{-2}\left(\frac{\mu_{ei}}{k_B T}\right)^{1/2}\frac{\sqrt{2\pi}}{3}L; \tag{28}$$

L being the Coulomb logarithm (Born approximation of the collision term).

The dielectric function of strongly coupled plasmas has been considered by different authors [6]. The result given here allows also for the account of higher order approximations with respect to the interaction potential V improving the Born approximation (28) for the collision term. As an example, the dynamically screened interaction and the summation of ladder diagrams has been considered to derive the conductivity of the fully ioniyed plasma [4], [5], [7]. The approach given here can be generalized

- by inclusion of higher order distribution functions instead of only the single particle density (21),

- by describing also the response of the system with respect to other perturbations such as a temperature gradient.

5 Two–particle approximation for the entropy

Higher order distribution functions are included into the description of nonequilibrium if we consider also S_2 and higher order contributions to the entropy (12). Now, the determination of the Lagrange multipliers S_o, S_1, S_2 are more involved and must be done using approximation procedures. There are different reasons to consider higher order approximations for the entropy:

(i) The correct thermodynamic limit as well as the conservation of total energy is obtained if the two–particle correlations are taken into account.

(ii) A single–particle approximation for the entropy is possible only in the low density limit. Corrections to the transport coefficients etc. arise in higher orders of density by reason of nonequilibrium higher order distribution functions.

(iii) According to the chemical picture, two–particle bound states as well as higher clusters have to be included to describe strongly correlated systems.

The Lagrange multipliers are explicitly evaluated in the near-equilibrium regime using the theory of linear response. Then, the set of "induced currents" (21) contains also two–particle Wigner densities (2), and the response equations (20) contain also the response parameters $\varphi_\nu(t)$ of two–particle correlations.

By using this extension of the set of relevant observables including the lowest moment of the two–particle distribution, the *Debye–Onsager relaxation* term has been found [4] which is proportional to $n^{1/2}\ln n$;

$$\sigma^{-1}(T,n) = A(T)\ln n + B(T) + C(T)n^{1/2}\ln n + \dots \tag{29}$$

The formation of bound states is taken into account by considering the bound state part of the two–particle distribution function. We can introduce creation and anihilation operators for a bound cluster state [8]

$$a_{A\alpha}^{+}(P) = \sum_{1\dots A} \Psi_{A\alpha P}(1\dots A)a^{+}(1)\dots a^{+}(A) \tag{30}$$

where $\Psi_{A\alpha P}(1\dots A)$ is the wave function for the A–particle cluster state with internal quantum number α and total momentum P. The set of relevant observables can be extended including the Wigner transform of the *cluster density*

$$n_{A\alpha}(p,r) = \int \frac{d^3q}{(2\pi)^3} e^{iqr} a_{A\alpha,P-q/2}^{+} a_{A\alpha,P+q/2} \tag{31}$$

In the analogous way to the single–particle distribution, section 3, we obtain *kinetic equations* for the cluster distribution function $f_{A\alpha}(p,r,t) = \langle n_{A\alpha}(p,r)\rangle^t$:

$$\frac{d}{dt}f_{A\alpha}(p,r,t) = \mathrm{Tr}\left\{\rho(t)\frac{i}{\hbar}[H, n_{A\alpha}(p,r)]\right\} = -D_{A\alpha}^{\mathrm{Vlasov}} + I_{A\alpha}^{\mathrm{coll}} \tag{32}$$

The *Vlasov term* contains the mean field contributions due to single particles ($A = 1$) as well as due to clusters ($A > 1$) in the surroundings:

$$D_{A\alpha}^{\mathrm{Vlasov}}(p,r,t) = \frac{p}{m_A}\frac{\partial}{\partial r}f_{A\alpha}(p,r,t) - \left[\frac{\partial}{\partial r}\sum_{A'\alpha'}U_{A\alpha}^{A'\alpha'}(r)\right]\frac{\partial}{\partial p}f_{A\alpha}(p,r,t) \tag{33}$$

whereas the *collision term* $I_{A\alpha}^{\mathrm{coll}}$ contains the scattering as well as the reaction channels [3].

The inclusion of bound states into the linear response theory according to the chemical picture allows to treat also the response of a *partially ionized plasma*. We have to solve a coupled system of generalized linear Boltzmann equations for the single particle as well as for the cluster distribution function. The static conductivity for a partially ionized plasma has been considered, e.g., in [9]. Besides the determination of the Lagrange multipliers which yield the composition of the partially ionized plasma, we have also to treat different scattering processes between single particles and clusters in evaluating the time evolution operator. The equilibrium solution should give us the law of mass action.

The inclusion of bound state into the linear–response theory is of special interest considering the optical properties of the plasma. If bound states are present, we have transition between discrete energy levels and the formation of profiles of spectral lines.

The absorption of electromagnetic waves by a partially ionized plasma is described by the dielectric function $\epsilon(q,\omega)$ (25). The contribution of a bound state (atom), $c_{nP}^{+} = \sum_{1,2}\Psi_{nP}(1,2)a_1^+ a_2^+$, to the charge density operator is given by

$$\begin{aligned} n_q &= \sum_{Pnm}\frac{e}{\Omega_o}\sum_{p}\psi_n(p)\left[\psi_m(p) - \psi_m(p-q)\right]c_{nP}^+ c_{m,P-q} \\ &= \sum_{Pnm}M_{nm}(q)c_{nP}^+ c_{m,P-q}. \end{aligned} \tag{34}$$

The *cluster–RPA–approach* to the dielectric function and to the optical properties of nonideal plasmas is given in detail in [8]. It realizes the chemical picture. The collision term contains the contribution to the pressure broadening of spectral line shapes. The alternative approach by the perturbative expansion of the polarization function in form of the density–density correlation function (first fluctuation–dissipation theorem) is given, e.g., in [10].

6 Anomalous mean values

The most general form of the entropy (12) will contain also relevant observables which are not particle conserving:

$$S_1(t) = \sum_{k,s} \xi_s(k,t) a_{ks}^+ a_{ks} + \sum_{Pk} \Psi_P(k,t) a_{P-k\downarrow}^+ a_{P+k\uparrow}^+ + c.c. \tag{35}$$

A similar generalization is also possible for $S_2(t)$ etc. Such a generalization is of special interest to obtain a stationary solution in the case of superfluidity. The kinetic equations for the relevant observables have the form

$$\frac{d}{dt} \left\langle a_{ks}^+ a_{ks} \right\rangle^t = \frac{i}{\hbar} \mathrm{Tr}\{\rho(t) [H, a_{ks}^+ a_{ks}]\}$$

$$\frac{d}{dt} \left\langle a_{P-k\downarrow}^+ a_{P+k\uparrow}^+ \right\rangle^t = \frac{i}{\hbar} \mathrm{Tr}\{\rho(t) [H, a_{P-k\downarrow}^+ a_{P+k\uparrow}^+]\} \tag{36}$$

For $\Psi_P(k,t) \sim \delta_{P,P_o}$, a diagonalization of $S_1(t)$ (35) is possible (Bogolyubov transformation). The conditions of stationarity $\frac{d}{dt} \left\langle a_{ks}^+ a_{ks} \right\rangle^t = 0$, $\left\langle a_{P_o-k\downarrow}^+ a_{P_o+k\uparrow}^+ \right\rangle^t \sim e^{-i\omega t}$ allow to determine the Lagrange multipliers $\xi_s(k,t)$, $\Psi_{P_o}(k,t)$.

Neglecting higher order contributions to $S(t)$, the stationary solution is found from $\rho_{\mathrm{rel}}(t)$. We find $\hbar\omega = -2\mu$, and for the special case $P_o = 0$

$$\xi_s(k) = E_s + \sum_{k'} V(kk', kk') \frac{1}{2} \left[1 - \frac{\xi_k}{\sqrt{\xi_k^2 + \Psi_k^2}} \tanh\left(\frac{\beta}{2}\sqrt{\xi_k^2 + \Psi_k^2}\right) \right] \tag{37}$$

whereas Ψ_k is obtained from the Gorkov equation

$$\Psi_k = \frac{1}{2} \sum_{k'} V(k', -k'; k, -k) \frac{\Psi_{k'}}{\sqrt{\xi_{k'}^2 + \Psi_{k'}^2}} \tanh\left(\frac{\beta}{2}\sqrt{\xi_{k'}^2 + \Psi_{k'}^2}\right). \tag{38}$$

The collision term given by $[\rho(t) - \rho_{\mathrm{rel}}(t)]$ gives no contribution to the kinetic equation in the stationary state.

Including two–particle collelations, $S(t) \approx S_o(t) + S_1(t) + S_2(t)$, we can treat also the case of strong coupling where bound states are formed. In this case, the evaluation of the Lagrange parameters from the condition of stationarity becomes more involved. For instance, instead of the single–particle density which is related to the Fermi distribution function f_k yielding $\tanh(x) = \frac{1}{2}[1 - 2f(x)]$ in eqs. (37), (38), a contribution of the bound states according $\Sigma_{nq}^{(b)} g(E_{nq}) |\Psi_{nq}(k)|^2$ will arise in line of the single particle distribution f_k. A more detailed description of the case of superfluidity in correlated media will not presented here.

7 Summary

- A quantum statistical approach is needed to describe many particle effects and bound state formation in a dense plasma.

- Kinetic equations can be derived from the general approach to a nonequilibrium statistical operator, for single particle distribution function as well as for higher order distribution functions.

- Evaluation of the kinetic equations is possible within a sigle–particle approximation of the entropy $S(t)$. A Vlasov–Uehling–Uhlenbeck equation is obtained.

- An other approach which contains also two particle correlations is given within the linear response theory. The dielectric function, conductivity, but also different approximations for the collision term can be given by systematic many particle theory.

- Two particle correlations can be included into the linear response theory leading to such effects as the Debye–Onsager relaxation term and the formation of bound states, furthermore the shape of spectral lines in a plasma from the dielectric function.

- The use of higher order distribution functions to describe the nonequilibrium state gives the formation of clusters and has an influence in calculating macroscopic properties.

- The use of higher order correlation functions is also of importance in deriving kinetic equations for the anomal mean values in order to describe a strongly coupled system in the superfluid state.

References

[1] D.N. Zubarev, Nonequilibrium Statistical Thermodynamics (Plenum Press, New York, 1974)

[2] W. Kraeft, D. Kremp, W. Ebeling, G. Röpke, Quantum Statistics of Charged Particle Systems (Plenum Press, New York and Berlin, 1986)

[3] G. Röpke, H. Schulz, Nucl. Phys. A **477**, 472 (1988)

[4] G. Röpke, Phys. Rev. **A 38**, 3001 (1988)

[5] G. Röpke, R. Redmer, Phys. Rev. **A 39**, 907 (1989)

[6] V.B. Bobrov, S.A. Triger, Physica **A 151**, 482 (1988)
G. Kalman, K.I. Golden, Phys. Rev. **A 41**, 5516 (1990)
J.P. Hansen, I.R. McDonald, Phys. Rev. **A 23**, 2041 (1981)

[7] R. Redmer, G. Röpke, F. Morales, K. Kilimann, Phys. Fluids, B: Plasma Physics **2**, 390 (1990)

[8] G. Röpke, Contrib. Plasma Phys. **29**, 469 (1989)

[9] R. Redmer, G. Röpke, Contrib. Plasma Phys. **29**, 343 (1989)

[10] S. Günter, L. Hitzschke, G. Röpke, Phys. Rev. A **44**, 6834 (1991)

KINETIC THEORY OF IONIZATION PROCESSES IN DENSE PLASMAS

M. Schlanges, Th. Bornath and D. Kremp

Fachbereich Physik, Universität Rostock, O-2500 Rostock, Germany

Abstract

The impact ionization and the three–body recombination coefficients are calculated for a nonideal hydrogen plasma. Statistical expressions are derived for these rate coefficients on the basis of generalized kinetic equations in order to account for many particle effects as self energy, screening and Pauli–blocking. The ionization cross section was calculated numerically in first Born approximation. Quasiparticle energies and effective interaction potentials are considered in the static limit. The influence of the medium effects on the cross section as well as on the rate coefficients is discussed.

1 Introduction

The determination of reaction rates is one of the basic problems in order to study the nonequilibrium properties of chemical systems as well as the ionization kinetics in plasmas. At present the progress in the field of plasma–laser interaction leads to a growing interest in the population kinetics of dense plasmas and therefore in the calculation of the corresponding rate coefficients.

The standard approach to the ionization kinetics of plasmas is to assume density independent rate coefficients which correspond to ideal plasma physics [1]–[4]. This must be changed for dense nonideal plasmas, where the plasma properties are essentially determined by many particle effects as

- dynamical screening

- self energy

- Pauli–blocking

- lowering of the ionization energy of bound states.

In order to derive rate equations for nonideal plasmas these many particle effects must be taken into account on the basis of quantum statistical theory.

First generalized expressions for the impact ionization and recombination coefficients were given in [5, 6]. The results can be considered as a first step because nonideality was included in an approximate way by energy shifts only. The nonideality effects lead to a strong density dependence of the ionization coefficient whereas the recombination coefficient remains density independent in this approximation.

In the following we will consider the influence of many particle effects on the rate coefficients more in detail. Especially we are interested in the density dependence of the recombination coefficient. The aim of this paper is to present improved results for the impact ionization and three–body recombination coefficients of dense hydrogen plasmas. Furthermore we will discuss the effects of improved energy shifts and Pauli–blocking.

2 Kinetic Equations for Dense Many Particle Systems with Two–Particle Bound States

In order to solve the problem we start from generalized kinetic equations which we have derived using the method of nonequilibrium Green's functions [7, 8]. In this paper we will restrict us to the description of the results.

We consider a spatially homogeneous quantum systems with the possibility of two–particle bound states. The kinetic equation for the bound particles in the state $|j>$ can be written then in the following form

$$\frac{\partial}{\partial t} F_j(P, t) = I_j^{sc}(P, t) + I_j^{react}(P, t) + I_j^{rearr}(P, t) \tag{1}$$

On the r.h.s. of (1) we have the collision terms which account for the different scattering processes of the bound states with the free particles. Let us consider the different collision terms more in detail.

i) elastic and inelastic scattering

$$(1 + 2) + 3 \leftrightarrow (\bar{1} + \bar{2}) + \bar{3}$$

$$
\begin{aligned}
I_j^{sc} &= \frac{2\pi}{V} \int \frac{dp_3}{(2\pi)^3} \sum_{\bar{j}} \int \frac{d\bar{P}}{(2\pi)^3} \frac{d\bar{p}_3}{(2\pi)^3} |< jPp_3| T_{123}^{33} |\bar{p}_3 \bar{P}\bar{j} >|^2 \\
&\times \delta\left(E_{123}^3 - \bar{E}_{123}^3\right) \left\{ \bar{F}_{\bar{j}} \bar{f}_3 (1 \pm f_3)(1 + F_j) - F_j f_3 (1 \pm \bar{f}_3)(1 + \bar{F}_{\bar{j}}) \right\}
\end{aligned} \tag{2}
$$

ii) reactions

$$(1 + 2) + 3 \leftrightarrow \bar{1} + \bar{2} + \bar{3}$$

$$
\begin{aligned}
I_j^{react} &= \frac{2\pi}{V} \int \frac{dp_3}{(2\pi)^3} \int \frac{d\bar{p}_1}{(2\pi)^3} \frac{d\bar{p}_2}{(2\pi)^3} \frac{d\bar{p}_3}{(2\pi)^3} |< jPp_3| T_{123}^{30} |\bar{p}_3 \bar{p}_2 \bar{p}_1 >|^2 \\
&\times \delta\left(E_{123}^3 - \bar{E}_{123}^0\right) \left\{ \bar{f}_1 \bar{f}_2 \bar{f}_3 (1 \pm f_3)(1 + F_j) - F_j f_3 (1 \pm \bar{f}_1)(1 \pm \bar{f}_2)(1 \pm \bar{f}_3) \right\}
\end{aligned} \tag{3}
$$

Similar expressions follow for the third term which accounts for rearrangements

$$(1 + 2) + 3 \leftrightarrow (\bar{1} + \bar{3}) + \bar{2} \qquad (\bar{1} \leftrightarrow \bar{2})$$

In order to classify the different processes in the collision integrals we have used the notation of multi–channel scattering theory [9].

As already mentioned the derived kinetic equations are generalized equations in the sense that many particle effects are taken into account as self energies, Pauli–blocking and medium dependent scattering. In our case the scattering energies are given by quasiparticle energies for the free and bound particles which are

$$
\begin{aligned}
E(p, t) &= \frac{p^2}{2m} + \Delta(p, t) \\
E_j(P, t) &= \frac{P^2}{2M} + E_j + \Delta_j(P, t)
\end{aligned} \tag{4}
$$

Here the influence of the surrounding medium is condensed in the energy shifts which are determined by the real part of the corresponding retarded self energy function. For the one–particle energies we have

$$E(p, t) = \frac{p^2}{2m} + Re\Sigma^R(p, \omega, t)\Big|_{\omega = E(p, t)} \tag{5}$$

The retarded self energy Σ^R in (5) can be calculated from the causal one using the following cluster expansion [10]

$$\left\langle \Sigma \right\rangle \;=\; \pm\; i\; \boxed{T_{12}} \;-\; \frac{1}{2}\; \boxed{T^{11}_{123}}\;\Bigg|_{\text{connected}}$$

The first diagram represents the two–particle ladder contribution and the second term accounts for the three–particle processes.

The inclusion of energy shifts in an exact manner is a difficult task. Therefore we look for possible approximations. One possibility is to work with thermally averaged shifts [11] which are given by the following expression

$$\Delta(t) = \frac{\int \frac{dp}{(2\pi)^3} \Delta(p,t) \frac{\partial}{\partial \mu^{id}} f^0(p,t)}{\int \frac{dp}{(2\pi)^3} \frac{\partial}{\partial \mu^{id}} f^0(p,t)} \tag{6}$$

Here we have assumed local equilibrium distribution functions. One can show that the averaged shift (6) is equal to the correlation part of the chemical potential, that means

$$\mu(t) = \mu^{id}(t) + \Delta(t)$$

A consequence of the inclusion of many particle effects is that the two– and three–particle T–matrices describe collision processes in the nonequilibrium medium. The T–matrices are defined by generalized Lippmann–Schwinger equations where the hamiltonians are given by effective ones modified by quasiparticle energies and Pauli–Blocking.

We cannot give here the details of the relations of medium dependent scattering theory. Let us consider only the equation for the T–operator T^{33} which is of importance in our further description of three–body reaction processes. We have

$$T^{33}_{123}(z) = V^3_{123} + i^2 V^3_{123} \frac{1 \pm f_1 \pm f_2 \pm f_3}{z - H^{eff}_{123}} V^3_{123} \tag{7}$$

with the scattering potential

$$V^3_{123} = (1 \mp f_2)\, V_{13} + (1 \mp f_1)\, V_{23}$$

In (7) H^{eff}_{123} is the effective hamiltonian for the considered three–particle system in the medium.

3   Rate Equations for a Dense Hydrogen Plasma

In the following we will apply the theory presented in the previous section to a hydrogen plasma consisting of electrons and protons with the number densities n_e, n_p and hydrogen atoms in the states $|j>$ with the densities n_H^j. If we integrate the kinetic equation (1) with respect to the atomic momentum P it is easy to obtain the following rate equation

$$\frac{d}{dt} n_H^j = \sum_{\bar{j}} \left(n_e n_H^{\bar{j}} K_{\bar{j}j} - n_e n_H^j K_{j\bar{j}} \right) + \left(n_e^3 \beta_j - n_e n_H^j \alpha_j \right) \tag{8}$$

Here $K_{j\bar{j}}$ are the coefficients of impact excitation, α_j is the impact ionization and β_j the three-body recombination coefficient. Because we start from a quantum kinetic equation explicit expessions can be given for these rate coefficients. We will consider the reaction terms only. For the ionization coefficients then follows

$$\alpha_j = \frac{2\pi}{\hbar V} \sum_{a=e,p} \int \frac{dP}{(2\pi\hbar)^3} \frac{dp_a}{(2\pi\hbar)^3} \frac{d\bar{P}}{(2\pi\hbar)^3} \frac{d\bar{p}}{(2\pi\hbar)^3} \frac{d\bar{p}_a}{(2\pi\hbar)^3} \delta\left(E_{123}^3 - \bar{E}_{123}^0\right)$$

$$\times \; |<jP|<p_a\,|T_{123}^{33}|\,\bar{p}_a> |\bar{P}\bar{p}+>|^2 \frac{f_a^0}{n_a} \frac{F_j^0}{n_H^j}\left(1 - \bar{f}_a^0\right)\left(1 - \bar{f}_e^0\right)\left(1 - \bar{f}_p^0\right) \tag{9}$$

The corresponding expression for the three-body recombination coefficients reads

$$\beta_j = \frac{2\pi}{\hbar V} \sum_{a=e,p} \int \frac{dP}{(2\pi\hbar)^3} \frac{dp_a}{(2\pi\hbar)^3} \frac{d\bar{P}}{(2\pi\hbar)^3} \frac{d\bar{p}}{2\pi\hbar)^3} \frac{d\bar{p}_a}{(2\pi\hbar)^3} \delta\left(E_{123}^3 - \bar{E}_{123}^0\right)$$

$$\times \; |<jP|<p_a\,|T_{123}^{33}|\,\bar{p}_a> |\bar{P}\bar{p}+>|^2 \frac{\bar{f}_a}{n_a} \frac{\bar{f}_e}{n_e} \frac{\bar{f}_p}{n_p}\left(1 - f_a\right)\left(1 + F_j\right) \tag{10}$$

The calculation of α_j and β_j requires the knowledge of the distribution functions and the solution of the scattering problem. A considerable simplification is possible if we take into account that there are different time scales during the approach to thermodynamic equilibrium. In the following we consider the stage where local equilibrium distribution functions for quasiparticles can be used, that means we have Fermi– and Bose–distribution functions. With this approximation and using thermally averaged shifts (6) one can derive the following relations between α_j and β_j.

$$\beta_j = \frac{n_H^j}{n_e^2} \alpha_j e^{\left(\mu_e + \mu_p - \mu_H^j\right)/k_B T} \tag{11}$$

Therefore only one of these coefficients must be calculated explicitly. In comparison to the ideal case the relation (11) represents a generalization to a degenerate nonideal plasma.

4 The Coefficients of Impact Ionization and Three–body Recombination

As one can see from the expression (9) the dynamics of the impact ionization process is described by the square of the corresponding three–particle T–matrix. In first Born approximation and if we consider electron impact ionization only we arrive at

$$\alpha_j = \frac{(2\pi\hbar)^3}{\hbar} \int \frac{dp_e}{(2\pi\hbar)^3} \frac{d\bar{p}}{(2\pi\hbar)^3} \frac{d\bar{p}_e}{(2\pi\hbar)^3} 2\pi\delta\left(E_j + \frac{p_e^2}{2m_e} - \frac{\bar{p}^2}{2m_e} - \frac{\bar{p}_e^2}{2m_e} + \Delta_j - \Delta_e - \Delta_p\right)$$

$$\times \; \left|V_{ee}^{eff}(q)P_{j\bar{p}}(q)\right|^2 f_e^0\left(1 - \bar{f}_e^0\right)\left(1 - \bar{f}_e^{\prime 0}\right) \tag{12}$$

where $\bar{f}_e^{\prime 0}$ denotes the distribution function of the ejected electron. Because of the great mass difference between electrons and protons the adiabatic approximation was applied.

In order to study the influence of many particle effects on the scattering quantities let us introduce the impact ionization cross section which is in first Born approximation for electron impact

$$\sigma_j^{ion} = \frac{8\pi\hbar^2}{p_e^2 a_B^2} \int\limits_0^{\bar{p}_{max}} d\bar{p}\bar{p}^2 d\Omega_{\bar{p}} \int\limits_{q_{min}}^{q_{max}} qdq \left|V_{ee}^{eff}(q)P_{j\bar{p}}(q)\right|^2 \tag{13}$$

Here V_{ee}^{eff} is the electron–electron effective interaction potential and q describes the momentum transfer. The limits of integration follow from energy conservation in the three–body reaction process. With $P_{j\bar{p}}$ we denote the atomic form factor

$$P_{j\bar{p}}(q) = \int dr \Psi_j^*(r)\Psi_{\bar{p}}^+(r)e^{\frac{i}{\hbar}q\cdot r} \tag{14}$$

which is given in terms of the two–particle states of the atomic subsystem. It describes the transition probability from the discrete state Ψ_j to the continuum states $\Psi_{\vec{p}}^+$ which correspond to the ionized atom. The two–particle wave functions are detemined by an effective wave equation which accounts for the influence of the surrounding plasma.

Such a wave equation was derived on the basis of Green's function techniques [12, 13] and can be written in the following form

$$\left(\frac{p_e^2}{2m_e} + \frac{p_p^2}{2m_p} + \Delta_{ep}^{eff}\left(p_e p_p z\right) - z \right) \Psi\left(p_e p_p z\right)$$
$$+ \ \left(1 - f_e - f_p\right) \int dq V_{ep}^{eff}\left(p_e p_p q z\right) \Psi\left(p_e + q, p_p - q, z\right) = 0 \tag{15}$$

In this wave equation the many particle effects as self energy, Pauli–blocking and dynamical screening are taken into account in the energy shifts Δ_{ep}^{eff} and in the dynamical screened potential V_{ep}^{eff}.

The solution of eq(15) especially for the dynamical case is a difficult task. Approximate results are possible in the static limit. Therefore we will consider the effective wave equation in the static approximation with the energy shifts

$$\Delta_{ep}^{eff} \ = \ \Delta_e + \Delta_p$$
$$\Delta_e \ = \ \Delta_p = -\kappa e^2 / 2 \qquad \kappa^2 = 4\pi \sum_{a=e,p} \frac{d}{d\mu_a} n_a(\mu_a) \tag{16}$$

and the effective potential

$$V_{ab}^{eff}(q) = \frac{V_{ab}(q)}{\varepsilon(q,0)} \qquad \varepsilon(q,0) = \left(\frac{q^2}{q^2 + \kappa^2} \right)^{-1} \tag{17}$$

Here $V_{ab}(q)$ is the Coulomb potential and $\varepsilon(q,0)$ is the static dielectric function.

Now we assume (16) and (17) in order to apply the theory presented in section 2 and 3 for the description of the ionization kinetics in a dense hydrogen plasma. We will restrict us to ionization from the atomic ground state.

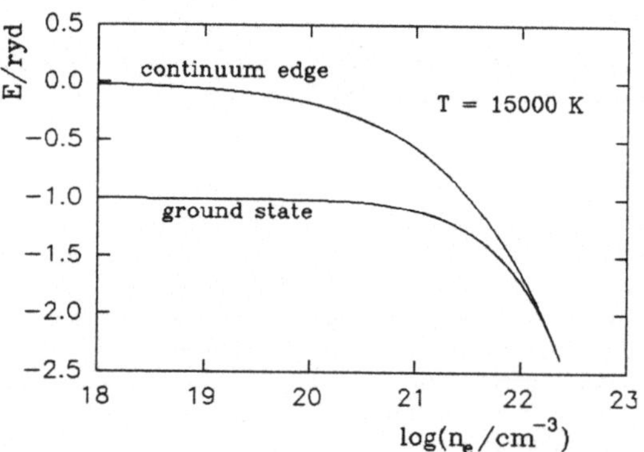

In Fig.1 the numerical results for the binding energy and the continuum edge are shown which follow from our statically screened effective wave equation. We observe a lowering of the effective ionization energy with growing plasma density. This lowering is mainly determined by the lowering of the continuum edge with the energy shifts (16).

Fig.1: Effective ionization energy for hydrogen atoms in the ground state

For the calculation of the cross section (13) we start from a partial wave expansion of the form factor. Introducing the auxiliary quantity $F_{1k} = \int d\Omega_k \left| P_{1k} \right|^2$ we get:

$$F_{1k} = \sum_{\ell} (2\ell + 1) \left[\frac{1}{k} \int_0^\infty dr\, r^2 R_{10}(r) R_{k\ell}(r) j_\ell(qr) \right]^2 \tag{18}$$

where $j_\ell(qr)$ are the spherical Bessel functions. The wave functions in (18) are obtained from radial Schrödinger equations with (16) and (17) which we have solved numerically for different screening parameters $\lambda_D = \kappa a_B$ (a_B– Bohrradius).

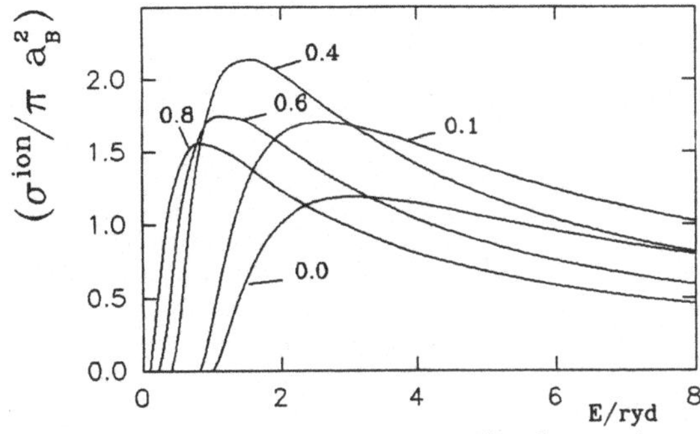

Fig.2: Impact ionization cross section for the atomic ground state

The results for the ionization cross section as a function of the electron impact energy for different screening parameters are shown in Fig.2. One can see that with increasing plasma screening the threshold energy moves to zero energy which corresponds to the effective ionization energy. Near the threshold the cross section shows a change of an increasing and a decreasing behaviour. This corresponds to a competition between the screening effects in the three–body scattering potential and in the atomic form factor.

Now let us discuss the results for the ionization and recombination coefficients which follows from (12) and (11). First we consider the nondegenerate case with screening parameters up to values $\kappa a_B = 0.8$.

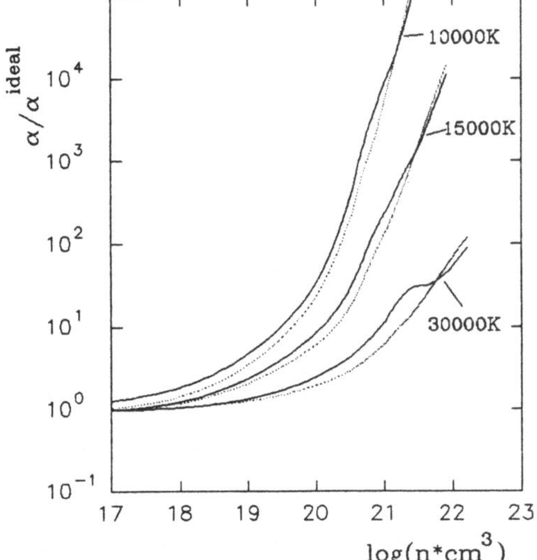

Fig.3: Ionization coefficient for a dense hydrogen plasma.

In fig.3 we see the ionization coefficient as a function of the free electron density for different temperatures. Here α_1^{id} denotes the coefficient for the ideal case with $\kappa a_B = 0$.
With the solid lines the results are given which follow from the numerical calculation presented in this paper. The dotted lines show the curves of a generalized Bethe–type formulae which we have given in an earlier paper [5, 6]

$$\alpha_1 = \alpha_1^{id} e^{(\Delta_1 - \Delta_e - \Delta_p)/k_B T}$$

$$\alpha_1^{id} = \frac{10\pi a_B^2 E_1}{(2\pi m_e k_B T)^{3/2}}$$

$$\times Ei\left(-\frac{|E_1|}{k_B T}\right) \qquad (19)$$

where $Ei(x)$ is the exponential integral function $Ei(x) = \int\limits_{-\infty}^{x} \frac{e^t}{t}\, dt$

In (19) many particle effects were included in an approximate way by energy shifts only. One can see that the numerical results show relatively small deviations from the exponential behaviour given by the simple formula (19).

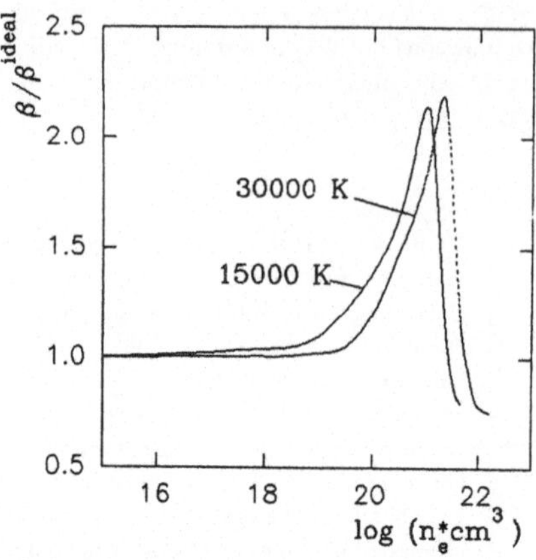

Fig.4: The three–body recombination coefficient for a dense hydrogen plasma for different temperatures

In Fig.4 the three–body recombination coefficient is given which can be calculated from relation (11). As already mentioned β_1 remains density independent $\left(\beta_1/\beta_1^{id} = 1\right)$ if we use the simple formulae (19) in (11). As expected, our numerical results show that the influence of plasma effects is small in β_1 in comparison to the ionization coefficient α_1. This follows from the fact that we have an effective ionization energy which affects essentially the ionization coefficient only. The density dependence of the recombination coefficient first is given by an increasing behaviour, then a strong deacreasing follows up to densities corresponding to $\kappa a_B = 0.8$. A similar increasing behaviour for lower densities was presented by Ebeling et al. [15].

At the end of this paper we want to show some results for the rate coefficients concerning the effects of improved energy shifts and Pauli–blocking.

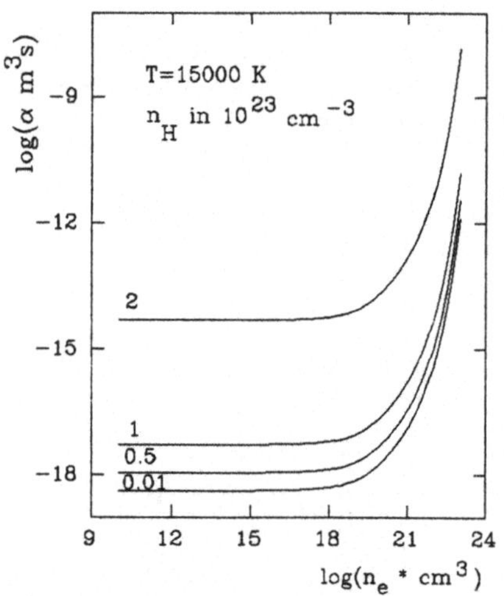

Fig.5: Ionization coefficient (19) with improved energy shifts

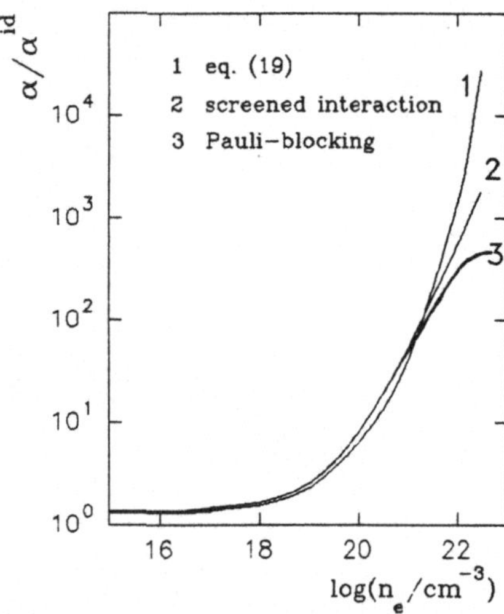

Fig.6: Degeneration effects in the ionization coefficient ($T = 15000K$, $n_H = 10^{-23} cm^{-3}$)

In the shifts of the charged particles we included the Hartree–Fock and Montroll–Ward contributions and we have used the Padé formulaes proposed by Ebeling et al. [16]. For the atoms the hard sphere interaction model was applied using the Carnahan–Starling formulae [17] for the energy shift. In Fig.5 the ionization coefficent is shown which was calculated from (19) with the shifts explained above.

160

In the last figure we show the influence of degeneration effects on the ionization coefficient. Starting from (12) the degeneration effects are taken into account by Fermi–distribution functions and Pauli–blocking for the electrons. In the calculation of the results shown in Fig.6 screening was included only in the scattering potential. As expected, Pauli–blocking effects reduce the ionization coefficients at higher plasma densities. A similar behaviour follows for the recombination coefficient [14].

References

[1]   D.R. Bates, A.E. Kingston, R.W.P. McWhirter, Proc.Roy.Soc. London **A270**, 155 (1962)

[2]   H.W. Drawin, F. Emard, Physica **85C**, 333 (1977)

[3]   L.M. Biberman, V.S. Vorobev, I.T. Yakubov, Kinetics of the nonequilibrium Low–Temperature Plasma (Nauka, Moskow, 1982)

[4]   Yu.L. Klimontovich, Kinetic Theory of Nonideal Gases and Nonideal Plasmas (Pergamon, New York, 1982)

[5]   M. Schlanges, Th. Bornath, D. Kremp, Phys.Rev. **A38**, 2174 (1988)

[6]   Th. Bornath, M. Schlanges, D. Kremp, Contrib. Plasma Phys. **28**, 57 (1988)

[7]   M. Schlanges, Ph.D.thesis B (Rostock University, 1985)
      Th. Bornath, Ph.D.thesis A (Rostock University, 1987)

[8]   D. Kremp, M. Schlanges, Th. Bornath, in: The Dynamics of Systems with Chemical Reactions, J.Popielawski,Ed. (World Scientific, Singapore, 1989)

[9]   J.R. Taylor, Scattering Theory (John Wiley and Sons., Inc., New York, 1972)

[10] D. Kremp, M. Schlanges, Th. Bornath, The Method of Green's Functions in Statistical Mechanics of Nonequilibrium Systems (in: Zentralinstitut für Elektronenphysik, preprint 86-3, 1986)

[11] R. Zimmermann, Many Particle Theory of Higly Excited Semiconductors (Teubner, Leipzig, 1984)

[12] R. Zimmermann, K. Kilimann, W.-D. Kraeft, D. Kremp, G. Röpke, Phys.stat.sol.(b) **90**,175 (1978)

[13] W.-D. Kraeft, D. Kremp, W. Ebeling, G. Röpke, Quantum Statistics of Charged Particle Systems (Plenum, New York, 1986)

[14] M. Schlanges, Th. Bornath, to appear
      D. Kremp, M. Schlanges, 8th Am.Phys.Soc.Conf. Atomic Processes in Plasmas (Portland, Maine, 1991)

[15] W. Ebeling, L. Leike, U. Leonhardt, 8th Am.Phys.Soc.Conf. Atomic Processes in Plasmas (Portland, Maine, 1991)

[16] W. Ebeling, W. Richert, Ann.Phys.(Leipzig) **39**, 227 (1982)

[17] N.E. Carnahan, K.E. Starling, J.Chem.Phys. **51**, 635 (1969)

IONIZATION AND RECOMBINATION COEFFICIENTS OF EXCITED STATES IN NONIDEAL HYDROGEN PLASMAS

Ulf Leonhardt,
Humboldt University, Institute of Theoretical Physics,
Invalidenstaße 42, O-1040 Berlin, Germany

Abstract

Impact ionization and three-body recombination was studied for a nonideal hydrogen plasma. Arrhenius-type expressions depending on the effective ionization energies were derived for the ionization coefficients. The deviations of the exponential Arrhenius law were calculated numerically in Born approximation taking into account thermally averaged shift approximation for the effective ionization energies and static screening in the electron-electron interaction. They correspond to density dependent recombination coefficients.

1 Introduction

The coefficients of the elementary ionization and recombination processes in plasmas depend on atomic parameters like the ionization energy, the shape of the wave functions and the type of interaction with ionizing particles. It is well known that many particle effects modify atomic parameters, for example they decrease the effective ionization energies of atoms in dense hydrogen plasmas. Thus, atomic constants like ionization and recombination coefficients become density dependent in the nonideal case.

In this paper we consider electron impact ionization of the hydrogen ground state and excited states up to the 5th level and the corresponding three-body recombination. First, we give Arrhenius-type expressions for the ionization coefficients and a general formula for the recombination coefficients. Second, we calculate the deviations of a simple exponential Arrhenius law in Born approximation taking into account thermally averaged shift approximation for the effective ionization energy and static screening in the electron-electron interaction. They correspond to density dependent recombination coefficients.

2 The Arrhenius law

In the following we are interested in bound states characterized by the main quantum number n. All atomic quantities have to be considered as averaged over the $l, m-$ quantum numbers.
The effective ionization energies of the bound states are expressed by

$$I_{eff}^{(n)} = Ry/n^2 - \Delta_n \, . \tag{1}$$

Here Δ_n denotes the effective shifts of the ionization energies due to many particle effects, which must not be specified yet.
The electron impact ionization coefficients of the bound states are expressed by

$$\alpha_n = \int \frac{d^3 p}{(2\pi\hbar)^3} \, \sigma_n \, \frac{p}{m} \, \lambda^3 \exp(-\frac{p^2}{2mkT}) \, , \quad \lambda = \frac{2\pi\hbar}{\sqrt{2\pi mkT}} \, , \tag{2}$$

where σ_n denotes the ionization cross sections. Applying the method of detailed balancing the recombination coefficients β_n are closely related to α_n :

$$\beta_n = \alpha_n \, n^2 \, \lambda^3 \exp(I_{eff}^{(n)}/kT) \, . \tag{3}$$

The ionization cross sections can be expressed as

$$\sigma_n \frac{p}{m} = \sum_{lm} \frac{1}{n^2} \int \frac{d^3 p'}{(2\pi\hbar)^3} \frac{d^3 \bar{p}}{(2\pi\hbar)^3} \frac{2\pi}{\hbar} \, \delta\left(\frac{p^2}{2m} - I_{eff}^{(n)} - \frac{p'^2}{2m} - \frac{\bar{p}^2}{2m}\right) | <p',\bar{p}\,|\,\mathrm{T}\,|\,p,nlm> |^2 . \tag{4}$$

Here T denotes the T-matrix of the quantum transition from the bound state $|\,nlm>$ to a scattering state $|\,\bar{p}>$ due to electron impact. We see that σ_n vanishes at impact energies lower than the effective ionization energy. Performing a simple transformation of Eq.(2) and introducing a collision strength s_n as follows

$$\sigma_n =: \frac{a_B^2 \, Ry/n^2}{p^2/2m} \, s_n \left(\frac{p^2}{2m}, I_{eff}^{(n)}\right) , \tag{5}$$

we obtain an Arrhenius-type expression for α_n and following Eq.(3) a general formula for β_n :

$$\alpha_n = \frac{4 a_B^3 n^{-2} Ry/\hbar}{\sqrt{\pi kT/Ry}} \, e^{-I_{eff}^{(n)}/kT} \int_0^\infty dx \, e^{-x} s_n \left(I_{eff}^{(n)} + kTx, I_{eff}^{(n)}\right) , \tag{6}$$

$$\beta_n = \frac{32\pi^3 a_B^6 Ry/\hbar}{(\pi kT/Ry)^2} \int_0^\infty dx \, e^{-x} s_n \left(I_{eff}^{(n)} + kTx, I_{eff}^{(n)}\right) . \tag{7}$$

There are two conditions in order to ionize an atom by electron impact. First, the kinetic energy of the electron has to exceed the threshold $I_{eff}^{(n)}$. Second, if this is fulfilled there is only a certain success probability of the elementary ionization process due to its quantum nature. The collision strength is a measure of this probability. The first condition leads to an Arrhenius factor $\exp(-I_{eff}^{(n)}/kT)$ in the expression of the ionization coefficient α_n. Recombination takes place without an energy threshold and therefore the Arrhenius factor is missing in the expression of β_n. Due to the second point both α_n and β_n are proportional to a factor containing the collision strength s_n. This factor modifies the exponential Arrhenius dependence of α_n and gives rise to a dependence of β_n on the effective ionization energy. As a consequence β_n becomes density dependent.

3 Deviation from the exponential Arrhenius law

In order to calculate the deviation from the exponential Arrhenius dependence of α_n and the corresponding density dependence of β_n we consider the effective ionization energies of the bound states in the thermally averaged shift approximation [1], giving

$$I_{eff}^{(n)} = Ry/n^2 - \Delta , \quad \Delta = \frac{e^2}{4\pi\epsilon_0} \kappa , \quad \kappa = \sqrt{\frac{2e^2 n_e}{\epsilon_0 kT}} . \tag{8}$$

In Born approximation the collision strength is given by

$$s_n \left(I_{eff}^{(n)} + kTx, I_{eff}^{(n)}\right) = \int_0^{\bar{k}_{max}} d\bar{k} \int_{q_{min}}^{q_{max}} dq \, \mathcal{P}_n (q, \bar{k}) , \tag{9}$$

$$q_{min} = \sqrt{(I_{eff}^{(n)} + kTx)/Ry} - \sqrt{\bar{k}_{max}^2 - \bar{k}^2} , \quad q_{max} = \sqrt{(I_{eff}^{(n)} + kTx)/Ry} + \sqrt{\bar{k}_{max}^2 - \bar{k}^2} , \tag{10}$$

$$\bar{k}_{max} = \sqrt{(kT/Ry) x} , \tag{11}$$

The quantity $\mathcal{P}_n$ is a measure of the microscopic transition probability. It depends on the momentum transition q and the momentum $\bar{k}$ of the ejected electron. The region of q and $\bar{k}$ is constrained by energy conservation. It is the upper half of a circle area in the $(q, \bar{k})-$plain, the centre being

$$(q_0, \bar{k}_0) = (\sqrt{(I_{eff}^{(n)} + kTx)/Ry} , \, 0) , \tag{12}$$

and the radius is given by $\bar{k}_{max}$. P_n was derived in [2]. It can be simplified by means of computer formula manipulation and has the structure :

$$\mathcal{P}_n(q,\bar{k}) = \frac{2^8 \pi \bar{k}}{1 - e^{-2\pi/\bar{k}}} \frac{1}{q} \frac{P_n(q^2,\bar{k}^2)}{Q_n(q^2,\bar{k}^2)} e^{-2\phi/\bar{k}} \; , \quad \phi = \arctan \frac{2n\bar{k}}{1 + n^2 q^2 - n^2 \bar{k}^2} \; , \quad 0 \le \phi < \pi \; , \qquad (13)$$

$$Q_n = \left((1 + n^2 q^2 + n^2 \bar{k}^2)^2 - 4n^4 \bar{k}^2 q^2 \right)^{n+2} \; , \qquad (14)$$

with P_n being certain polynomials in q^2 and $\bar{k}^2$. As suggested in [3] screening has to be taken into account of the interaction potential of the impact electron and the bound target electron. Then we have to replace $\mathcal{P}_n$ by

$$\mathcal{P}_n^{sc}(q,\bar{k}) := q^4/(q^2 + \kappa^2 a_B^2)^2 \, \mathcal{P}_n(q,\bar{k}) \; . \qquad (15)$$

Now we discuss the qualitative picture of s_n and β_n. With increasing electron density the effective ionization energy is decreasing. According to Eq.(12) the region of allowed q and $\bar{k}$ is moving to the left in the $(q,\bar{k})$-plain (see Fig. 1). If we exclude screening in the electron-electron interaction potential the function $\mathcal{P}_n(q,\bar{k})$ does not change, but the circles of allowed q and $\bar{k}$ are moving into regions of higher microscopic transition probability $\mathcal{P}_n$, with x and kT being fixed (see Fig. 2). Hence s_n and consequently β_n are increasing. If screening is included it can compensate the growing of β_n because of $\mathcal{P}_n^{sc} \le \mathcal{P}_n$ (see Fig. 3).

Fig. 1 Regions of q and $\bar{k}$,   (a) $I_{eff} = 1\,Ry$, (b) $I_{eff} = 0.25\,Ry$;   at $kT = 3eV$, $x = 1,2,3,4$

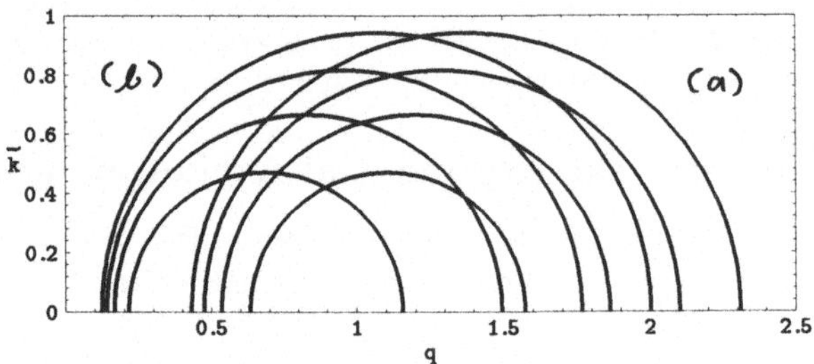

Fig. 2 Transition probability $\mathcal{P}_1(q,\bar{k})$

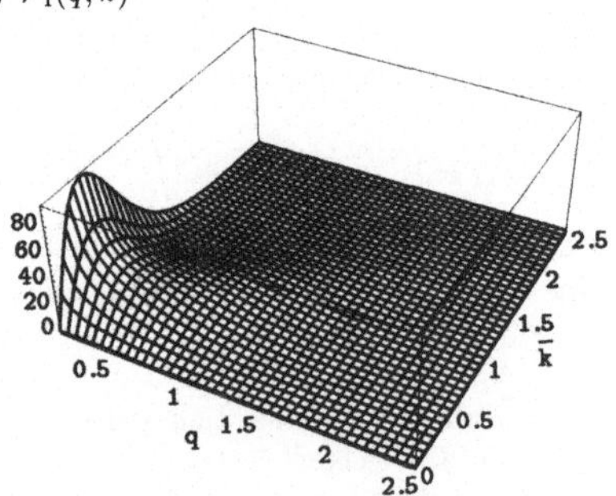

Fig. 3 Density dependent recombination coefficients $\beta_n(n_e)/\beta_n(0)$

(1) $kT = 1eV$, **(2)** $kT = 2eV$, **(3)** $kT = 3eV$

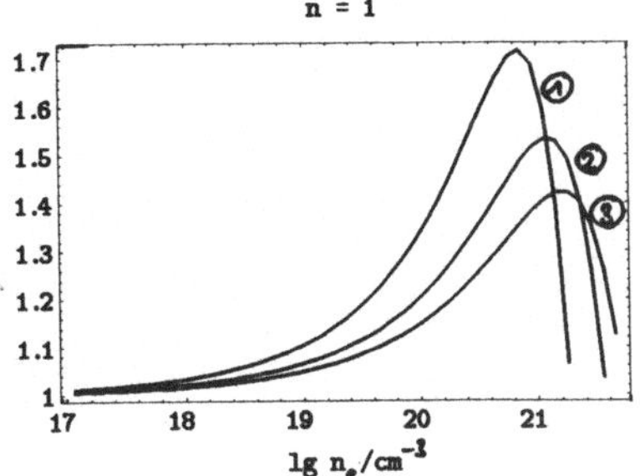

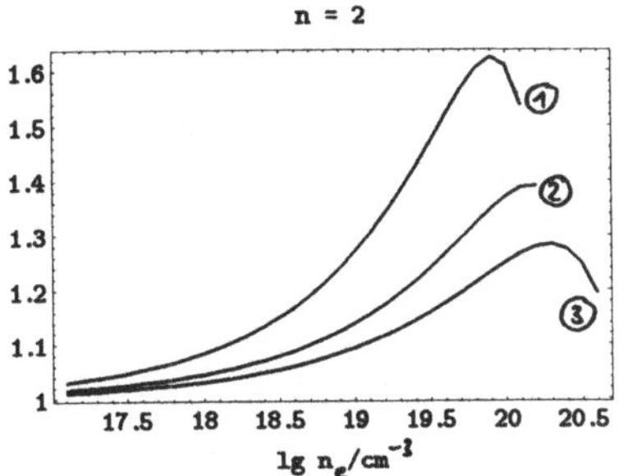

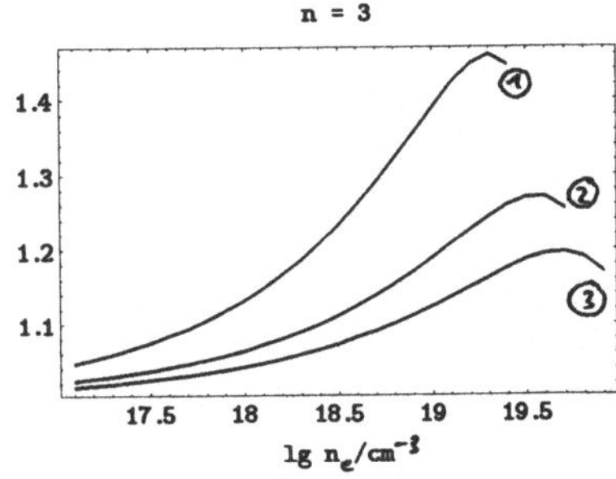

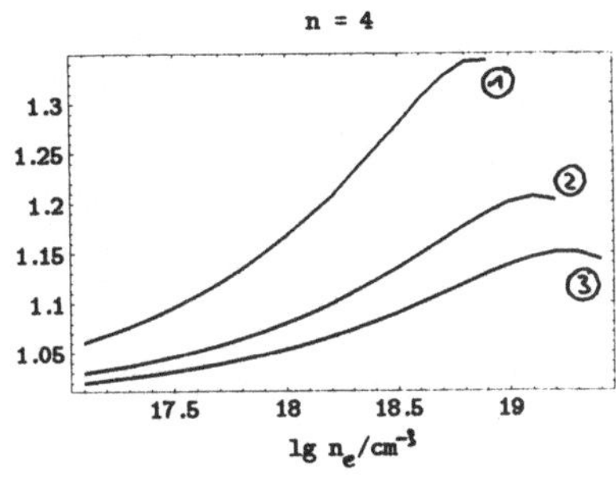

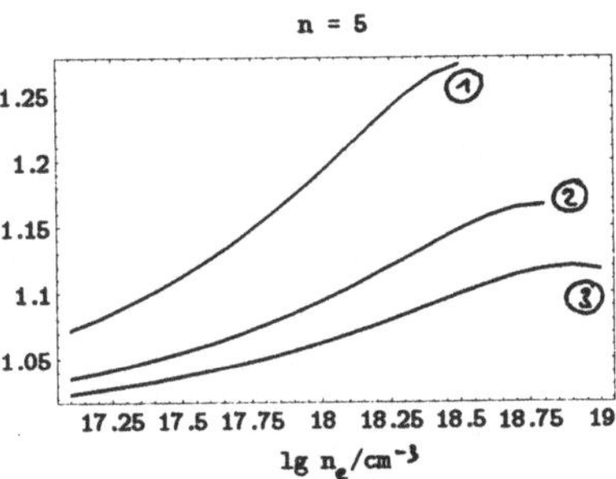

The density dependence of β_n or the corresponding deviation of the Arrhenius dependence of α_n is weak with respect to the Arrhenius factor (see also [4] , [5] and [6]) and can be neglected if

$$kT \ll I_{eff}^{(n)} \ .$$ (16)

Thus, as assumed in [7],

$$\alpha_n = \alpha_n^{id} \, e^{\Delta_n/kT} \ , \quad \beta_n = \beta_n^{id}$$ (17)

is a good approximation for practical calculations. α_n^{id} and β_n^{id} denote the ideal ionization and recombination coefficients and can be taken from semiempirical formulas.

Acknowledgements

I would like to thank Yu. L. Klimontovich, I. T. Yakubov and the group of D. Kremp for useful discussions.

References

[1] R. Zimmermann,
Many Particle Theory of Highly Excited Semiconductors (Teubner, Leipzig, 1984)

[2] K. Omidvar, Phys. Rev. **140** (1965) A 26

[3] D. Kremp, M. Schlanges, T. Bornath, phys. stat. sol. (b) **147** (1988) 747

[4] M. Schlanges, Th. Bornath, D. Kremp, Phys. Rev. **A 88** (1988) 2174

[5] W. Ebeling, I. Leike, U. Leonhardt,
8th Am. Phys. Soc. Conf. Atomic Processes in Plasmas (Portland, Maine, 1991)

[6] M. Schlanges, Th. Bornath, D. Kremp,
8th Am. Phys. Soc. Conf. Atomic Processes in Plasmas (Portland, Maine, 1991)

[7] W. Ebeling, I. Leike, Physica **A 170** (1991) 682

C O L L E C T I V E   M O D E S   I N
S T R O N G L Y   C O U P L E D   P L A S M A S

G. Kalman
Department of Physics, Boston College
Chestnut Hill, MA 02167, USA

I. INTRODUCTION

In this paper we give a rather incomplete discussion of the collective excitation spectra of strongly coupled plasmas. The discussion is incomplete because our understanding of these phenomena is still rudimantary. While there is vast literature on the collective motions of weakly coupled plasmas – both classical and quantum – the study of the corresponding processes in strongly coupled systems has started relatively recently.

The measure of the correlations or coupling is the ratio of potential and kinetic energies in the system. As long as the system can be regarded classical the latter can be characterized by the temperature, and if only one species of particles is of interest the coupling parameter is

$$\Gamma = Z^2 e^2/akT$$

with a being the Wigner-Seitz radius, $\frac{4\Pi}{3} a^3 n = 1$. In the other limit, for a zero temperature electron gas the kinetic energy is given by the Fermi energy, and the coupling parameter is

$$r_s = a/a_B$$

where $a_B = \dfrac{h^2}{e^2 m}$ is the Bohr radius. Even though classical and quantum systemsare in many ways different, inference can be drawn for one system from information pertaining to the other: in order to do this the useful correspondence relation (note that Z = 1)

$$\Gamma \rightarrow 1.36 \ r_s$$

based on the comparison of the potential energy/kinetic energy ratios, can be established. The strong coupling domain is now characterized by

$$\Gamma \geq 1 \quad \text{or} \quad r_s \geq 1$$

The simplest model for a Coulomb system is the one component plasma (OCP), consisting of one single species of particles dispersed in a neutralizing background. In addition to the normal three dimensional (3-d) OCP, a two-dimensional (2-d) OCP can serve as a physical model for 2-d electron films and layers. With two different ion species dispersed in neutralizing background, one of faced with more involved model of a binary ionic mixture (BIM). These are the systems we consider in the present paper. The more realistic models with two oppositely charged species present additional problems (bound states, etc.) in the strong coupling domain, which are outside the scope of this description.

When correlations are negligible (formally $\Gamma = 0$), the Vlasov or Random Phase Approximation (RPA) is appropriate. In this approximation the longitudal dielectric response functions $\varepsilon_L(k\omega)$ and $\varepsilon_T(k\omega)$ are given by

$$\varepsilon_L(k\omega) = 1 - \varphi(k)\chi_{LO}(k\omega)$$

$$\varepsilon_T(k\omega) = 1 + \varphi(k)\,\frac{k^2 c^2}{\omega^2}\,\chi_{TO}(k\omega)$$

where

$$\varphi(k) = \frac{4\Pi e^2 Z^2}{k^2}$$

represents the Coulomb potential and $\chi_{LO}(k\omega)$ and $\chi_{TO}(k\omega)$ are the density and current response functions of non-interacting (ideal) gas. The salient features of the collective mode spectrum in this approximation are the following:

(i) The longitudinal plasmon mode $\omega(k)$ for 3-d systems starts, even for multicomponent plasmas, at the plasma frequency

$$\omega_p = \left\{ \sum_A \frac{4\Pi Z_A^2 e^2 n_A}{m_A} \right\}^{1/2} \qquad \text{and for finite k-values develops a positive}$$

dispersion $\left(\dfrac{d\omega}{dk} > 0 \right)$; for 2-d systems $\omega_L(k) \sim \sqrt{k}$.

(ii) The only damping mechanism is the collisionless Landau damping, which vanishes exponentially for $k \to 0$.

(iii) There is no genuine shear generated transverse mode (shear mode) in the system: this is manifested by the fact that in the $c \to \infty$ limit the existing transverse photon mode doesn´t survive.

These features may be contrasted with the behavior in the very strong coupling ($\Gamma \gg 1$) limit. It is known that the 3-d OCP crystallizes into a bcc lattice around $\Gamma = \Gamma_m = 178$ and the 2-d OCP into a hexagonal lattic around $\Gamma = \Gamma_m = 137$. In the lattic the collective excitations are the optical and acoustic plasmons, whose relevant features now can be listed as follows [1, 2]:

(i) The equivalent of the plasmon, the longitudinal optical phonon $\omega(k)$ for the 3-d OCP starts at the plasma frequency and for finite k develops a negative dispersion; for the 2-d OCP $\omega(k) \sim \sqrt{k}$ only for small k, for higher k-values a maximum develops.

(ii) There exist two transverse shear modes $\omega(k), \omega(k)$ which for small k-values have acoustic type $\omega(k) \sim k$ dispersion both in the 3-d and in the 2-d system; in addition in the 3-d lattic the phonons satisfy the Kohn sum rule: $\sum_i \omega_i^2(k) = \omega_p^2$.

(iii) In the linear approximation the phonon modes are undamped.

The theoretical challenge is to understand the development of the dynamical properties of Coulomb systems as correlations become more important from $\Gamma = 0$ through $\Gamma \gg 1$.

II. APPROXIMATION METHODS

The general correlation problem can be approached from different angles. Starting from the RPA, a systematic perturbation technique of the IBBGKY hierarchy can be worked out [3, 4, 5]. This technique has been around for a long time, but by its very nature, perturbation technique is limited to the $\Gamma < 1$ domain. From the strong coupling side one can generate an approximation [6] which exploits the main feature of the strongly coupled systems, namely the quasi-localization of the particles (Quasilocalized Charges-QLC-model). The principal idea of the QLC model is that for strong coupling ($\Gamma \geq 10$) the particles are trapped in local potential minima and oscillate around their quasi-equilibrium positions. Averaging over these quasi-sites the Hamiltonian for the particles can be written down in terms of a dynamical matrix $D_{\alpha\beta}^{AB}(k)$ (A, B, etc. are species indices, α, β, etc. designate Cartesian components; $\omega^A(q)$ is the plasma frequency of species A; $g_{AB}(k)$ is the pair correlation function between species A and B).

$$D_{\alpha\beta}^{AB}(k) = \frac{1}{V} \sum_q \frac{q_\alpha q_\beta}{q^2} \omega^A(q)\omega^\beta(q)$$

$$\times \left[g^{AB}(k-q) - \delta^{AB} \frac{N_{\bar C} Z^{\bar C}}{N_A Z^A} g^{A\bar C}(q) \right] \tag{1a}$$

or for the 3-d OCP

Table I. Overview of approximation methods used for the physical systems listed. The entries refer to the plasmon (P) and shear (S) modes.

		QLC	MFT	DMFT
3-D	OCP	P S	S	P
2-D	OCP	P S	P S	
B I M		P (k = 0) P_(k = 0)		

$$D_{\alpha\beta}(k) = \frac{1}{V} \sum_q \frac{q_\alpha q_\beta}{q^2} \omega_p^2 [g(k-q) - g(q)] \tag{1b}$$

which then provides the Hamiltonian

$$H = \frac{1}{2} \sum_k \Pi_{k,\alpha}^{\bar A} \Pi_{-k,\alpha}^{\bar A} + \frac{1}{2} \sum_k \left[D_{\alpha\beta}^{\bar A \bar B}(k) + \frac{k_\alpha k_\beta}{k^2} \omega^{\bar A}(k)\omega^{\bar B}(k) \right] \xi_{k,\alpha}^{\bar A} \xi_{-k,\beta}^{\bar B}$$

$$- \frac{i}{V} \left[\frac{N_{\bar A}}{m_{\bar A}} \right]^{1/2} \sum_k k_\alpha \hat{\Phi}^{\bar A}(-k,t) \xi_{k\alpha}^{\bar A} , \tag{2}$$

(Barred indices are summed over; $\hat{\Phi}$ is external perturbing potential, $\xi_{k,\alpha}$ and $\Pi_{k,\alpha}$ are the dynamical coordinates and momenta).

From the above Hamiltonian a fairly straightforward calculation of the equation of motion provides the density and current responses to an external perturbation and the dielectic response tensor.

A quite general method, valid in principle for arbitrary values of the coupling, is the Mean Field Theory [7], (MFT) which represents the dielectric function as

$$\varepsilon(k\omega) = 1 - \varphi(k) \frac{X_{LO}(k\omega)}{1+\varphi(k)G(k\omega)X_{LO}(k\omega)}$$

$$\varepsilon_T(k\omega) = 1 + \varphi(k)\frac{k^2c^2}{\omega^2}\frac{X_{LO}(k\omega)}{1-\varphi(k)\frac{k^2c^2}{\omega^2}H(k\omega)X_{TO}(k\omega)} \tag{3}$$

and collects [8] all correlational effects in the frequency and wavenumbers dependent mean fields $G(k\omega)$ and $H(k\omega)$. We have detailed experience with the longitudinal $G(k\omega)$ only: the frequency dependence of this latter is crucial for the correct treatment of dynamical effects and its proper inclusion distinguishes the dynamical mean field theory (DMFT) from static mean field theories (MFT) with frequency independent $G(k)$-s.

In this paper we discuss recently obtained results [6, 8, 9, 10] through the application of QLC to various physical systems, such as the 3-d and 2-d OCP-s and the BIM. We show how the QLC description accounts for the profound modification of the longitudinal plasmon mode in the strong coupling domain and how the transverse shear mode appears as an immediate consequence of the quasi-localization. Next we combine it with the DMFT proposed by Golden and Kalman (11) and we show how a reliable $\varepsilon_L(k\omega)$ can be obtained for the 3-d OCP over a broad range of Γ-values. For the transverse $\varepsilon_T(k\omega)$ and for the properties of the 2-d system we have only the less accurate, but from the point of view of the dispersion still quite satisfactory static MFT.

III. QLC DESCRIPTION

A. 3-D one Component Plasma

From $E_q(3)$ one can derive the logithdinal dielectric function as

$$\varepsilon_L(k\omega) = 1 \frac{\omega_p^2}{\omega^2 - D_L(k)} \tag{4a}$$

with

$$D_L(k) = \frac{1}{k^2} k_\alpha D_{\alpha\beta}(k)k_\beta \tag{4b}$$

170

$$\omega_L(k) = \sqrt{\omega_p^2 + D_L(k)} \tag{5}$$

The equilibrium pair correlation function is known from Monte Carlo (MC) simulations and Hypernetted Chain (HNC) calculations [12]. The calculated $\omega(k)$ curves for different Γ-values are given in Fig.1.

For high Γ values one can observe the oscillating behavior, which reflects the short range order in the liquid phase. For $\Gamma > 100$ the oscillation period fairly well matches the Brillouin zone structure of the Wigner crystal; the first minimum of $\omega_L(k)$ (at ka = 4.2) almost coincides with the boundary of the first Brillouin zone in the (1,1, 0] direction (ka = 4.44). In addition, it can be shown that for small ka and high Γ Eq. (5) becomes identical with the angle-averaged optical phonon dispersion of the Wigner crystal, as calculated by Caldwell-Horsefall and Maradudin [1].

In addition to the longitudinal plasmon mode the strongly coupled Coulomb liquid possesses transverse shear excitations [10]. These are analogous to the acoustic phonons in the Wigner crystal. The dispersion relation can be derived from the transverse dielectric function, which in the QLC becomes

$$\varepsilon_T(k\omega) = 1 - \frac{\omega_p^2}{\omega^2 - D_T(k)} \tag{6}$$

where, from Eq. (2)

$$D_T(k\omega) = -\frac{1}{2} D_L(k\omega) \tag{7}$$

The transverse dispersion relation is

$$\frac{1}{\varepsilon_T(k\omega)} = 0 \tag{8}$$

One obtains a doubly degenerate shear-mode

$$\omega_T(k) = \sqrt{D_T(k)} \tag{9}$$

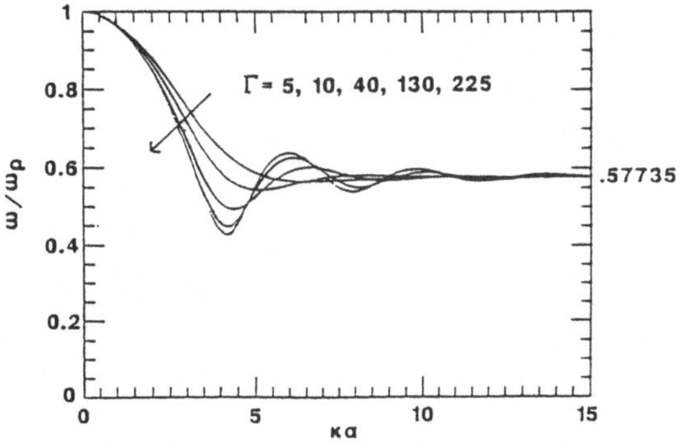

Fig. 1 Plasmon dispersion in the 3-d OCP, as calculated in the QLC approximation taken from Ref. [8]

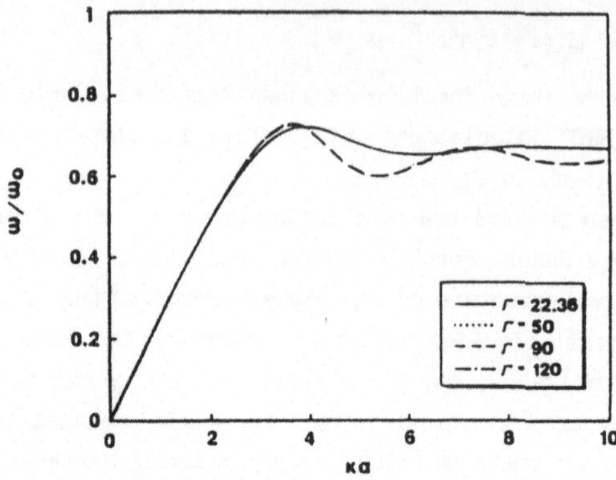

Fig. 2 Transverse shear mode dispersion in the 3-d OCP as calculated in the
        QLC approximation; taken from Ref. [11b]

We note that as $k \rightarrow \infty$, both $\omega_L(k)$ and $\omega_T(k)$ go to the value $\dfrac{\omega_p}{\sqrt{3}} = 0.577\ \omega_p$. This can
be interpreted as the frequency of individual particle motion in the background of a
uniform distribution of opposite charges which has, for small k an acoustic type
dispersion relation with the phase velocity

$$V_O = \sqrt{\frac{|E_{corr}(\Gamma}{\frac{15}{2}\,m}}$$
(10)

(10)For larger __k__ and high Γ values the dispersion curve emulates the behavior of
the angle averaged acoustic photon of the Winner crystal (See Fig. 2). The mode
frequencies satisfy the John sum rule in the form

$$\omega_L^2(k) + 2\omega_T^2(k) = \omega_p^2$$
(11)

B. 2-D ONE COMPONENT PLASMA

Turning now to the 2-d system, it is well-known that in the PA description the
longitudinal plasmon mode has an $\omega \sim \sqrt{k}$ dispersion. In the QLC description this
feature is maintained for $k \rightarrow 0$, but strongly modified for higher k-values. The
dispersion relation is similar to (3):

$$\omega_L(k) = \sqrt{\omega_p^2(k) + D_L(k)}$$
(12)

with the difference that $\omega(k)$ is the k-dependent RPA plasma frequency and$D(k)$ has
its 2-d form [9]:

$$\omega_p(k) = \sqrt{2\Pi e^2 nk/m}$$

$$D_L(k) = \omega_p^2(k)\,\frac{1}{A}\sum_q \frac{(k \cdot q)^2}{k^3 q}\left\{g(k-q)-g(q)\right\}$$
(13)

The correlation function g(k) can be obtained from Monte Carlo or HNC [13] data. The resulting dispersion curve is plotted in Fig.3. Again, for $\Gamma > 100$ the dispersion curve matches the optical photon branch of the 2-d Wigner lattice [2]. As to the transverse shear mode, its behavior is also very similar to that of its 3-d counterpart (See Fig.4). The important difference is, however, that in 2-d the Kohn sum rule doesn't hold, and therefore $D_T(k)$ is independent of $D_L(k)$:

$$D_T(k) = \frac{1}{A} \sum_q \left\{ \frac{q}{k} - \frac{(k \cdot q)^2}{k^3 q} \right\} \{g(k-q) - g(q)\} \tag{14}$$

The sher velocity is now

$$V_o = \sqrt{\frac{|E_{corr}(\Gamma)|}{8m}} \tag{15}$$

very close to the 3-d value. In the $k \to \infty$ limit both $\omega_L(k)$ approach the individual particle oscillation frequency which now becomes $0.641 \; x \left[\dfrac{2\Pi e^2 n}{ma} \right]^{1/2}$ (a is 2-d Wigner-Seitz radius, $\Pi na^2 = 1$

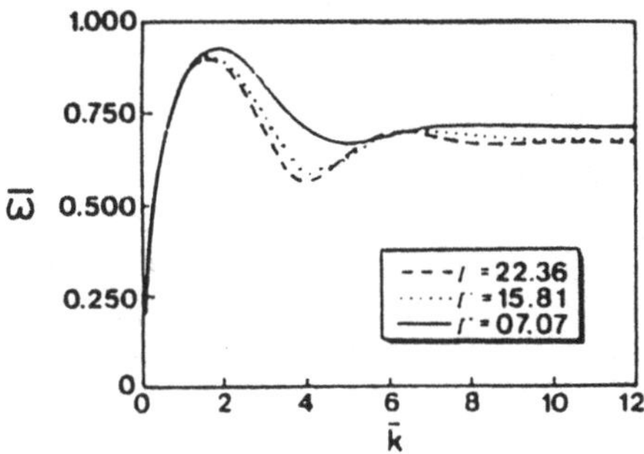

Fig. 3 Plasmon dispersion in the 2-d OCP, as calculated in the QLC approximation; taken from Ref. [9]

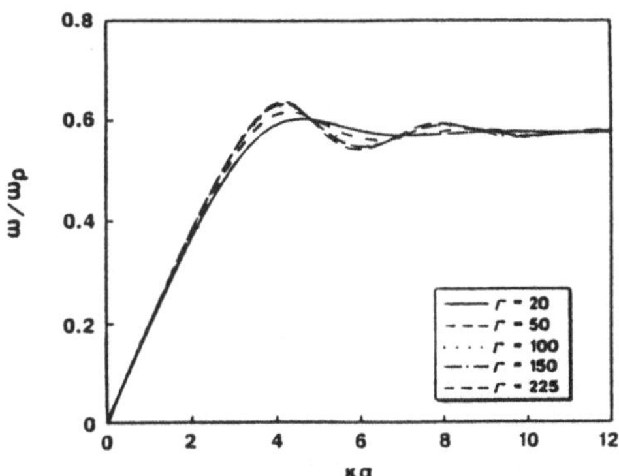

Fig. 4 Transverse sher mode dispersion in the 2-d OCP, as calculated in the QLC approximation; taken from Ref. [10]

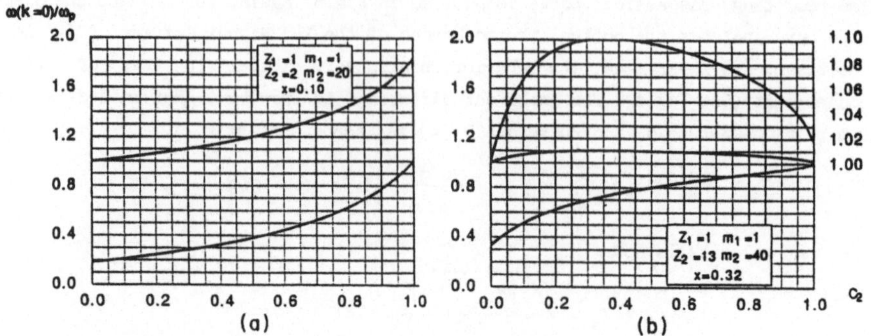

Fig. 5   Upper curve (to be read by the right-hand side scale): variation of the
plasma frequency at k=0 in terms of the RPA plasma frequency of the mixture,
$\omega_p=(\omega_1^2+\omega_2^2)^{1/2}$ vs. the heavier species in hydrogen. The lower curve (to be
read by the left-hand side scale) represents a   second, low frequency
longitudinal mode; taken from Ref. [6]

C. B I N A R Y   I O N I C   M I X T U R E

In the OCP momentum conservation precludes any deviation from the RPA behavior at
k=0. In contrast, in the case of two or more components, both a collisional damping
and a shift of the plasma frequency develop at k = 0. The latter has been measured
in Molecular Dynamics experiments and is easily calculable from the QLC formalism
[6]. The result for the shifted plasma frequency in the case of the BIM is

$$\omega^2(k{=}0) = \frac{\omega_p^2}{2}\left\{1 + p \pm \left[(1 - p)^2 + 4d^2\right]^{1/2}\right\},\qquad(16)$$

with

$$p = \frac{1}{3}\,\frac{x+y}{1+xy}$$

and

$$d = \sqrt{y/3}\,\frac{1-x}{1+xy}$$

$$x = \frac{Z_2 m_1}{Z_1 m_2},\qquad y = \frac{Z_2 c_2}{Z_1 c_1}\,.$$

Fig. 5   Shows the calculated variation of $\omega(k{=}o)$ as a function of concentration, for
a few typical mixtures

The calculated frequency shifts can be compared with the MD results of Hansen and
his collaborators [14] for the 50% - 50% H^+ - He^{2+} BIM system. Result of the latter
for Γ = 40 show a 3.9% upward shift of the plasma frequency, which is higher   than
the 1.9 ~ 2% shift predicted by naive sum rule approximation [14, 15]. Our calculated value at this point is 3.2%, quite close to the MD result.

The QLC gives a good physical description of the collective excitations in the high

Γ limit, but it has a number of deficiencies. Since both mode-particle and mode-mode interaction are absent in the model, it is incapable to describe damping; this is especially manifest with regard to the sher mode. Contrary to the QLC predicted propagation in the $k \to 0$ limit, the liquid phase doesn't have the ability to maintain a shear mode. The merits and the deficiencies of the QLC approach can now be summarized as follows.

The good features are:

(i)   the smooth approach from the liquid state to the Wigner crystal behavior of the dispersion relation;

(ii)   the prediction of the shear mode;

(iii)   the satisfaction of the Kohn and ω^{-4} third moment sum rules;

(iv)   the prediction of the shift of the plasma frequency for multi-component systems.

On the other hand the unsatisfactory features are:

(i)   the inability of the model to describe damping;

(ii)   the poor $\omega \to 0$ properties;

(iii)   no transition to weak coupling regime.

In order to address these problems one has to turn to the mean field theory description. This is done in the next Section.

IV. DYNAMICAL MEAN FIELD DESCRIPTION OF THE OCP

As discussed in the Introduction, in the DMFT approach one concentrates on the determination of the dynamical mean field $G(k\omega)$. There is no simple way to do this. A method that has proved to be quite successful [16] has been worked out by Golden and Kalman [11]. The general idea is to achieve self-consistency by simultaneously using kinetic equations and fluctuation-dissipation relations.

The central approximation used in the scheme is velocity average approximation (VAA): this consists of replacing the perturbed (symbol$^{(1)}$ two.body function $G^{(1)}(x_1,v_1;\ x_2,v_2)$ by its average over the velocities v_1 and v_2, $G^{(1)}(x_1,\ x_2)$. As a result $\chi(k\omega)$ becomes a functional of the perturbed two point functior $\langle n(x_1t_1)\ n(x_2t_2)\rangle^{(1)}$. At this point the quadratic Fluctuation-Dissipation theorem (QFDT) derived by Golden and Kalman [17] and by Kalman and Gu [18] comes to one's help; it relates $\langle n\ n\rangle^{(1)}$ to its three-point equilibrium (symbol$^{(0)}$) counterpart; $\langle n\ n\rangle^{(1)} \leftrightarrow \langle n\ n\ n\rangle^{(0)}$. Further application of the QFDT allows one to express $\langle n\ n\ n\rangle^{(0)}$ in terms of the quadratic response function $\chi(k_1\omega_1;\ k_2\omega_2)$. Finally, an additional approximation helps to express $\chi(k_1\omega_1;\ k_2\omega_2)$ in terms of the linear χ-s, $\chi(k_1\omega_1)$ and $\chi(k_2\omega_2)$. The final outcome of this procedure is a self-consistency relation (integral equation) for $\chi(k\omega)$ or for $G(k\omega)$

$$G(k) = -\{D_L(k) + F(k\omega)\} \tag{17}$$

$D_L(k)$ is given by Eq. (4b); for small k, $F(k\omega)$ is determined by the equation

$$F(k\omega) = -(ka)^2 \frac{1}{N} \sum_q (1-6\lambda^2 + 8\lambda^4)[\varphi(q)]^2 \int_0^\infty d\mu \; \delta-(\mu) \frac{\chi(q\mu)}{\varepsilon(q\mu)} \frac{\chi(q\omega-\mu)}{\varepsilon(q\omega-\mu)} \tag{18}$$

where
$$\lambda = \frac{k \cdot q}{kq}$$

Based on our experience with the QLC model, it is reasonable – although certainly not rigorously justifiable – to extrapolate the k-dependence so that

$$G(k) = -D(k) \, X(\omega) \tag{19}$$

where $X(\omega)$ is evaluated from (18). The calculation is made possible by an approximation on the integrand in (18): this is the "two-pole" approximation [16]. This approximation picks up the contribution from $\varepsilon^{-1}(\omega)$ in the vicinity of the pole; the position of the pole is taken as $\omega = \omega_p + A(ka)^2$, with floating $A = A(\Gamma)$ and $B = B(\Gamma)$ coefficients: the coefficients are ultimately determined by requiring the satisfaction of (18). (For details of the latter involved procedure the reader is referred to Ref. [8]). The resulting $G(k\omega)$ is depicted in Fig. 6. The main characteristics of this function are as follows:

(i)   The relatively large positive Re $G(k\omega=0)$ has an important effect on the static isothermal compressibility.

(ii)   Re $G(k\omega)$ develops a cusp-like singularity around $\omega = 2\omega_p$; the Im $G(k\omega)$ develops a bump in the same region: these features are the consequences of the enhanced plasmon-plasmon interaction for $\Gamma \simeq \Gamma_{crit}$ and $\Gamma > \Gamma_{crit}$

(iii) As $\omega \to \infty$, Im $G(k\omega)$ decays slowly though a power-law behavior ($\sim \omega^{-5/2}$). The formation of this "algebraic tail" is a well established feature [19, 20, 21] of Coulomb systems.

(iv)   For $\omega < 2\omega_p$ Im $G(k\omega)$ becomes significant only if $\Gamma > \Gamma_{crit}$.

With the aid of $G(k\omega)$ one can now construct $\varepsilon(k\omega)$. To, examine the static $\varepsilon(k0)$ we observe the isothermal compressibility, K_T which can be derived from $\varepsilon(k0)$ (K_T^o is the compressibility of the ideal gas) is now

$$K_T = K_T^o(1 + \frac{4}{15} \beta E_{corr} X(\omega=0)) \tag{20}$$

The appearance of $X(\omega=0)$ is responsible for a dramatic improvement in $\varepsilon(k0)$ and in the compressibility over static MFT results.

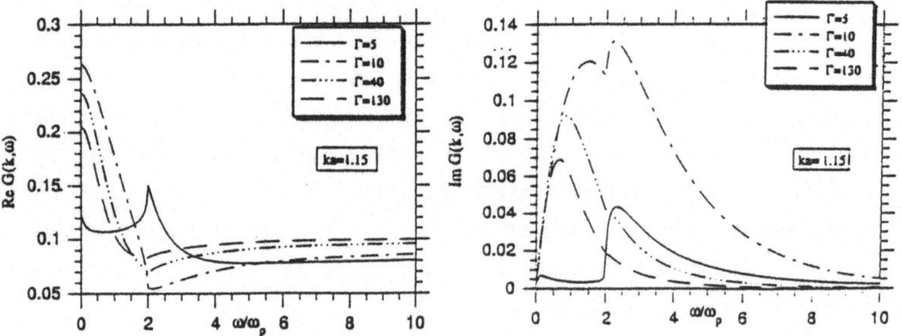

Fig. 6 Real (a) and imaginary (b) parts of the dynamical mean field, $G(k\omega)$. Note the singular behavior around $\omega \cong 2\omega_p$; taken from Ref. [8].

We have calculated the dynamical structure function S(kω) from ε(kω) via the fluctuation-dissipation theorem and compared it with the S(kω) values obtained by Hansen, Pollock and McDonald [22] through MD calculations. Fig 7 shows the results for a broad range of Γ values. The agreements is very satisfactory, especially in view of the fact that the derivation has been effected from first principles, without the introduction of any adjustable parameter.

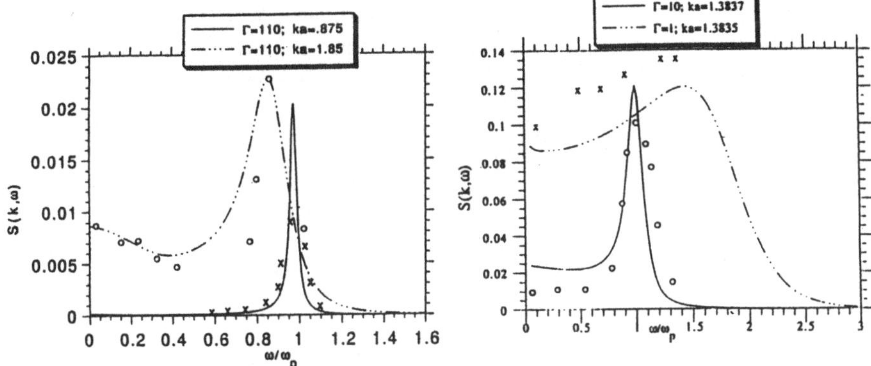

Fig. 7 Comparison of the calculated results for the dynamical structure function S(kω) given in this paper, with the MD results of Ref. [22] ((a) Γ = 110.4, ka = 0.875 (x) and ka = 1.85 (o); damping (ν = Imω/ωp) as (b) Γ = 0.99 (x)); taken from Ref. [8].

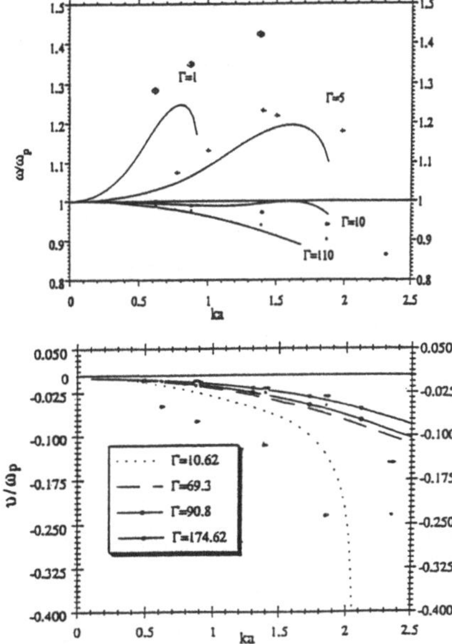

Fig. 8 Comparison of the calculated results for (a) plasmon dispersion and(b) damping (ν = Imω/ωp) given in this paper, with the MD results of Ref. [22] Γ = 0.99 (Φ), Γ 110.4(*) Γ = 152.4 ( ) taken from Ref. [8].

Finally, Fig. 8 shows the dispersion and damping of plasma oscillations compared
with MD data [22]. The agreement for the dispersion is very good and for the damping
quite good. Further corroboration for the correctness of the theory is provided by
comparing [23] its predictions for plasmon dispersion with recent experimental
results [24] obtained on alkali metals. The coefficient $A(\Gamma)$ (or $A(r_s)$) has been
extracted from the measured data and is compared with the calculated data (Fig. 12).
It is evident that the present theory is much superior to previous static [25, 26]
and semi-dynamic [27] theories. The mechanism of the abrupt change in the vicinity
of $r_{s,crit}$ is related to the central role played by the plasmon-plasmon interaction
in the theory [23].

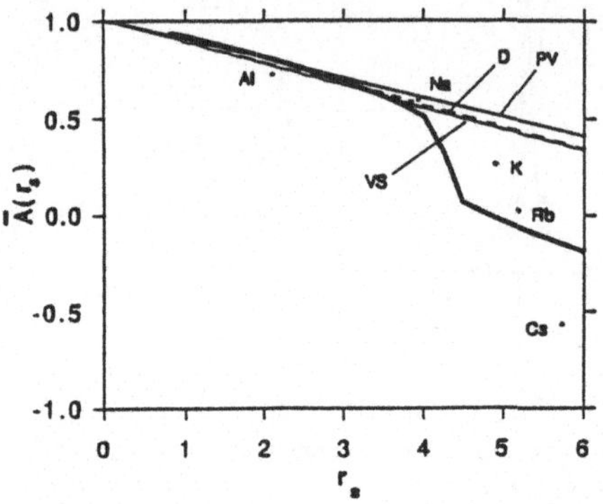

Fig. 9   Comparison of experimental results on plasmon group velocity with various
         theories. Experiments on Al from Ref. [24], on Na, K, Rb and Csfrom Ref.
         [24b]. Thin solid lines refer to different approaches   neglecting genuine
         dynamical correlations: Ref. [26a] (VS), Ref. [26b](PV), Ref. [27] (D);
         heavy line refers to calculated values given in this paper (Ref. [8] and
         [16]; taken from Ref. [23].

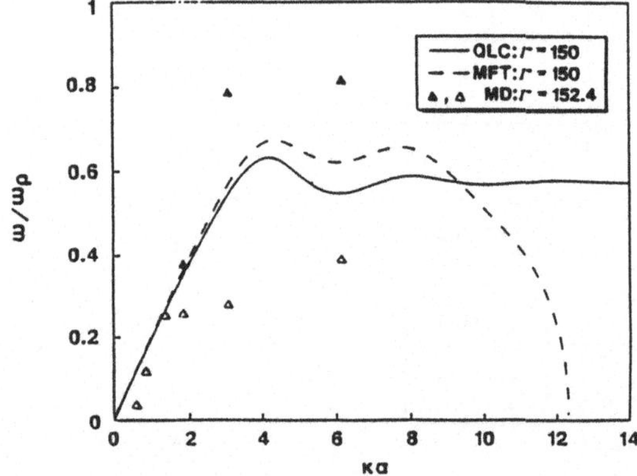

Fig. 10   Comparison of the calculated results for shear mode dispersion in the 3-d
          OCP as given in this paper, with the MD data of Ref. [14] (triangle
          points); the graph also shows the comparison between the QLC and MFT data;
          taken from Ref. [10].

V. STATIC MEAN FIELD DESCRIPTION OF THE 3-D AND 2-D ONE COMPONENT PLASMA

For the cases where the dynamical mean field theory is not available, a static mean field theory extension of the QLC can be generated [9]. For the longitudinal $\varepsilon_L(k\omega)$ this is done simply by omitting $F(k\omega)$ in Eq.(17). For the transverse $\varepsilon_T(k\omega)$ one replaces $D_L(k\omega)$ by $D_T(k)$ in Eq. (1). The MFT generalization of the QLC is expected to account for most of the "indirect" thermal effects, ignored in the original QLC. These originate from the slow thermal migration and diffusion of the quasi-sites. To be more exact, one expects the MFT to work reasonably well for frequencies $\omega > t_D^{-1}$ where t_D is the (k-dependent) diffusion time. In particular, the MFT is not reliable for the acoustic shear mode below a k_{min} determined by t_D. The main effect of the introduction of the MFT description emerges in the higher $\underline{k}$ domain: here the onset of Landau damping and the disappearance of the modes beyond a k_{max} are the principal new features. In this Section we discuss

(i)   the transverse shear mode in the 3-d OCP;

(ii)   the longitudinal plasmon mode in the 2-d OCP;

(iii) and the transverse shear mode in the 2-d OCP.

A. 3-D One Component Plasma

The modification of the 3-d shear mode brought about by the MFT as compared to the QLC, is illustrated in Fig. 10. The main feature, namely the acoustic behavior up to about ka $\simeq$ 4, and the shear velocity given by Eq. (10) are un-changed. The principal new feature is that the shear mode terminates at a maximum $\underline{k}$ value around k_{max} a $\simeq$ 12.5. Landau damping (not shown in the graph) increases with increasing k, and by time k_{max} is reached extinguishes the shear wave.

Fig. 10 also shows the comparison with the MD data of Hansen and collaborators [22]. The agreement for the shear velocity in the acoustic domain is very good. However, for higher k values the MD data indicate a splitting of the mode into a low-frequency and a high-frequency branch. No such effect is suggested by the calculated behavior: we believe that nonlinear shear-plasmon interaction (which is not a part of the theoretical model) may be responsible for this effect. For lower Γ values one finds that k_{max} decreases, roughly k_{max} a $\simeq \sqrt{\Gamma}$, and the shear mode completely ceases to exist for $\Gamma < 9.41$.

B. 2-D One Component Plasma

Turning now to the analysis of the 2-d system, one finds that the introduction of the direct thermal effects through the MFT in the QLC has qualitatively the same effect as in the case of the 3-d shear mode. Fig 11 shows the behavior of the plasmon mode [11]. One can observe the disappearance of the mode for $k > k_{max}(\Gamma)$ where k_{max} a increases with Γ towards the value k_{max} a $\simeq$ 3.65. In the vicinity of k_{max} the mode heavily Landau damped. The comparison of our dispersion curve with the

MD data of Totsuji and Kakeya [28]. The agreement, including the cut-off near k_{max}, is very good.

Fig. 12 shows the shear mode in the 2-d system for $\Gamma = 50$. Most of the comments made for the 3-d shear wave apply in this situation as well. The shear mode exists now down to $\Gamma \simeq 1.82$, a value substantially lower than its 3-d equivalent. The high-k cutoff is now given by $k_{max} a = 0.9 \sqrt{\Gamma}$, but heavy Landau damping extinguishes propagation for somewhat lower values.

Fig. 12 also shows the comparison with the MD data of Totsuji and Kakeya, [28]. Overall qualitative agreement with the theory is good. In contrast to the 3-d case, the MD data do not indicate any splitting of the shear mode. The observed k_{max} values are also in good agreement with theory, once allowance for the reduction of k_{max} because of Landau damping is made. On the other hand, the MD data clearly indicate the existence of a k_{min}, below which the shear mode doesn't propagate. This feature is not predicted by the theory: this is, however not surprising in view of the limitations of the QLC in the small-k domain.

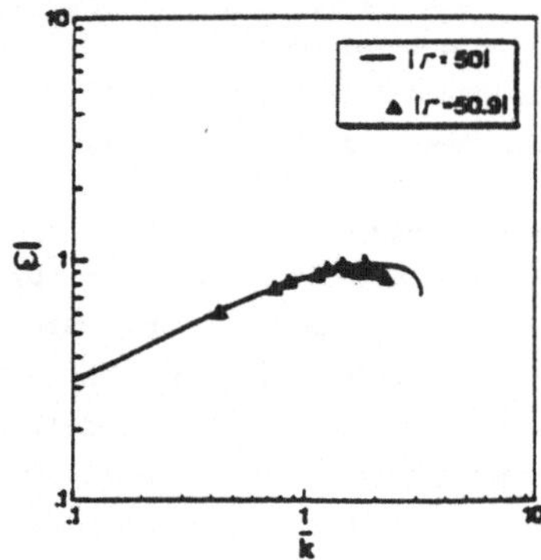

Fig. 11 Comparison of the plasmon dispersion in the 2-d OCP as given by the MFT theory calculations and by the MD data of Ref. [28] (triangle points); taken from Ref. [9].

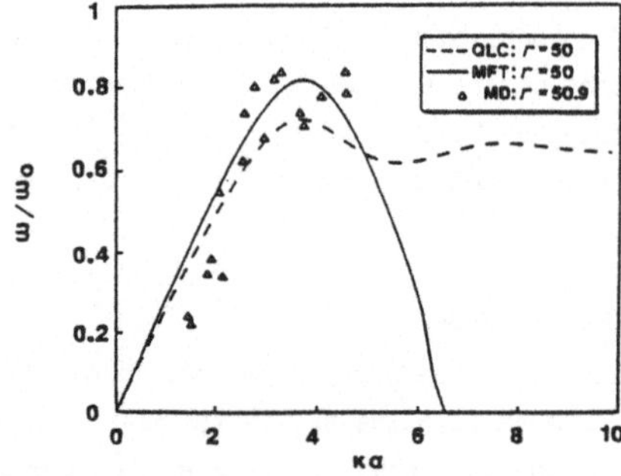

Fig. 12 Comparison of the calculated shear results for the shear mode dispersion in the 2-d OCP as given in this paper with the MD data of Ref. [28] (triangle points), the graph also shows the comparison between the QLC (dashed lines) and MFT data (full line); taken from Ref. [10].

VI. C O N C L U S I O N S

In this paper we have shown that a theoretical framework based on the quasi-localization of the particles and combined with a mean field formalism gives a satisfactory description of most of the MD analyzed properties both of 3-d and 2-d plasmas. A rather sophisticated dynamical mean field theory with a frequency dependent mean field, based on the relationship between the linear and quadratic response function through the quadratic fluctuation-dissipation theorem, is capable to generate a detailed formulation of $\varepsilon_L(k\omega)$ and $S(k\omega)$ for the 3-d OCP and of the Γ-dependence of the plasmon dispersion and damping. The more modest static mean field theory with a frequency independent mean field provides a satisfactory description of the 2-d plasmon dispersion and of the transverse sher mode both in 3 and in 2 dimensions.

A C K N O W L E D G E M E N T S

The works reported in the paper have been the result of a longstanding collaboration with Kenneth Golden and of outstanding contributions by number of former graduate students: Paul Carini, Massimo Minella, Xiaoyue Gu, Zichi Tao, Hong Zhang and Philippe Wyns. This work has been partially supported by the National Science Foundations through Grant ECS-87-13337 and PHY-91- 15714.

R E F E R E N C E S

[1] R.A. Caldwell-Horsfall and A.A. Maradudin, J. Math.Phys. 1, 395 (1960)

[2] L. Bonsall and A.A. Maradudin, Phys. Rev. B15, 1959 (1977)

[3] (a) V.I. Perel and G.M. Eliashberg, Zh. Eksperim. i Teor. Fiz. 41, 886 (1961) [translation: Sov. Phys. JETP 14, 633 (1962)] (b) J. Dawson and C. Oberman, Phys. Fluids 5, 517 (1962); Phys. Fluids 6, 394 (1962); C. Oberman, A. Ron and J. Dawson, Phys. Fluids 5, 1514 (1962)

[4] (a) J. Coste, Nuclear Fusion 5, 284, 293 (1965) (b) M.G. Kivelson and D.F. DuBois, Phys. Fluids 7, 1578 (1964)

[5] P. Carini, G. Kalman and K. Golden, Phys. Rev. A26, 1686 (1982)

[6] G. Kalman and G. Golden, Phys. Rev. A41, 5516 (1990)

[7] e.g. V.D. Gorbchenko and E.G. Maksimov, Usp. Fiz Nauk 130, 65 (1980)

[8] M. Minella, G. Kalman, K.I. Golden, and P. Carini, submitted to Phys. Rev. A

[9] K.I. Golden, G. Kalman and P. Wyns, Phys. Rev. A41, 6940 (1990)

[10] K.I. Golden, G. Kalman and Ph. Wyns, submitted to Phys. Rev. A

[11] K.I. Golden and G. Kalman, Phys. Rev. A19, 2112 (1979)

[12] (a) G.S. Stringfellow, H. DeWitt and W. Slattery, Phys. Rev. A41, 1105 (1989) (b) F.J. Rogers, D.A. Young, H.E. DeWitt and M. Ross, Phys.Rev. A28, 2990 (1983)

[13] (a) H. Totsuji and N. Kakeya, Phys. Rev. A22, 1220 (1980) (b) R.C. Gann, S. Chakravarty, and G.V. Chester, Phys. Rev. B20, 326 (1979)

[14] I.R. McDonald, P. Vieillefosse and J.P. Hansen, Phys. Rev. Lett 39, 271 (1977); J.P. Hansen, I.R. McDonald, and P. Vieillefosse, Phys. Rev. A20, 2590 (1979); J.P. Hansen, F. Joly, and I.R. McDonald, Physica 132A, 4721 (1985)

[15] K.I. Golden, F. Green, and D. Neilson, Phys. Rev. A31, 3229 (1985); 32, 1969 (1985)

[16] P. Carini, G. Kalman, and K.I. Golden, Phys. Lett. 78A, 450 (1980); P. Carini and G. Kalman, ibid. 105A, 229 (1984)

[17] K.I. Golden, G. Kalman and M.B. Silevitch, J. Stat. Phys. 6, 87 (1972)

[18] G. Kalman and Xiao-Yue Gu, Phys. Rev. A36, 3399 (1987)

[19] A.J. Glick and W.F. Long, Phys. Rev. B4, 3455 (1971)

[20] F. Family, Phys. Rev. Lett. 34, 1375 (1975)

[21] Z.C. Tao and G. Kalman, Phys. Rev. A43, 973 (1991); G. Kalman and Z.C. Tao, Phys. Rev. A43, 7073 (1991)

[22] J.P. Hansen, E.L. Pollock, and I.R. McDonald, Phys. Rev. Lett. 32, 277 (1974); J.P. Hansen, I.R. McDonald, and E.L. Pollock, Phys. Rev. A11, 1025 (1975)

[23] G. Kalman, K. Kempa, and M. Minella, Phys. Rev. B43, 14238 (1991)

[24] (a) A. vom Felde, J. Sprosser-Prou, and J. Fink, Phys. Rev. B40, 10181 (1989)
(b) J. Sprosser-Prou, A. vom Felde, and J. Fink, Phys. Rev. B40, 5799 (1989)

[25] P. Vashishta and K.S. Singwi, Phys. Rev. B6, 875 (1972)

[26] K.N. Pathak and P. Vashishta, Phys. Rev. B7, 3649 (1973)

[27] B. Dabrowski, Phys. Rev. B34, 4989 (1986)

VISCOUSITY AND ION SOUND IN STRONGLY COUPLED PLASMA

I.T. Iakubov

Institute for High Temperatures, USSR Academy of Sciences,
Izhorskaya 13/19, Moscow 127412, USSR

Abstract

The diapason of plasma parameters is considered which
includes the domains of liquid and crystal states. In the vicinity
of melting the time-flight mechanism of viscosity is changed to the
vacansion mechanism. The increase of viscosity coefficient narrows
the part of the phase space which is occupied by the mode of viscous
sound. So the conditions arise for the non-viscous sound propagation. At high wave numbers non-viscous sound decays moderately due
to Joule heat dissipation.

Introduction.

In strongly collisional plasma the Landau dumping does not to
be important.So ion sound wave may propagate in isothermal plasma in
spite of the equality of electron and ion temperatures.Sound propagation in dense media is governed by the hydrodynamics relations
which include the transport coefficients. The ion viscosity is the
most important. Calculations demonstrate the good comparison with
the experimental data [1]. In [1] the light scattering by fluctuations of isothermal argon plasma has been measured. The both maxima of
dynamic structure factor $S(q,\omega)$ have been registered which correspond to the sound mode.

However the argon plasma in [1] répresents the kind of
gaseous weakly non-ideal plasmas. Strongly coupled plasma is characterized by influence of strong interparticle correlations. The su-
perdense plasma reminds not gas but liquid. It is well known it may
be crystallized under the cooling (or further compression),e.g.
[2,3].At these circumstances one may indicate the change of the mechanism of viscous transfer of momentum.Every ion oscillates inside
of the Wigner-Seits cell,so the vacansion mechanism of the momentum

"

of the Wigner-Seits cell,so the vacansion mechanism of the momentum transfer takes the primary importance.As the consequence of it the viscosity coefficient increases sharply and the diapason of wave numbers of viscous sound waves becomes narrow.It results in the change of mechanism of propagation and decay of sound waves.The non-viscous sound propagation is accompanied by the absence of local thermodynamic equilibrium on the space scale of short wave lengths. For such waves the dominant reason of the decay becomes dissipation of the Joule heat due to eddy displacement currents.

PLASMA PARAMETERS.

Three coupling parametersare proper to the plasma considered: the ion-ion,the electron-ion, the electron-electron interaction parameters

$$\Gamma_{zz} = Z^2 e^2 / \bar{r} T, \quad \Gamma_{ze} = Z e^2 / \bar{r} T, \quad \Gamma_{ee} = Z^{1/3} e^2 / \bar{r} T \qquad (1)$$

Here Z is the ion charge, $\bar{r} = (4\pi N/3)^{-1/3}$ is the average inter-ion distance,N is the ion particle density. If Z is much in excess of unit, $Z \gg 1$, the following inequalities hold

$$\Gamma_{zz} \gg \Gamma_{ze} \gg \Gamma_{ee} \qquad (2)$$

So the plasma will be considered which contains multiply   charged ions. Inter-electron correlations are almost absent ($\Gamma \ll 1$), electron-ion correlations are moderate ($\Gamma_{ze} \stackrel{<}{\sim} 1$) and inter-ion correlations are very strong, $\Gamma_{ze} \gg 1$. If electron subsystem is degenerated the parameters Γ_{ze} and Γ_{ee} given by

$$\Gamma_{ze} = Z^{2/3} r_s / a_o , \quad \Gamma_{ee} = r_s / a_o \qquad (1')$$

Here $r_s = (4\pi N_e/3)^{-1/3}$ is the average inter-electron distance, N_e is the electron particle density. The hierarchy of inequalities (2) is retained.

The strongly coupled plasma with multiply charged ions may be generated by the action of intense power pulses upon the condensed matter.For example,due to the absorption of intense laser pulses and another swift heating processes, due to high-velocity impacts, in self-compressed electric discharges of powerful pulses of current

ION-SOUND IN NON-IDEAL GASEOUS PLASMAS

It is necessary for the subsequent reasoning to discuss briefly the
ion sound propagation in moderately non-ideal plasmas, where $\Gamma_{ze} < 1$.
The coefficient of ion sound dumping (in [exp(-yx)])is given by the
following expression

$$\gamma = \omega^2 (2v_s^3)^{-1} [4\eta/3\rho + (kr_d)^2 v_s^2/4\pi\sigma] \qquad (3)$$

Here $\omega = kv_s, \omega$, k ,v_s are the frequency,the wave number and velocity of sound, η is the viscosity coefficient, $\rho = MN$ is the density
of matter, M is the ion mass, σ is the plasma electric conductivity,
$r_d = (T/4\pi e^2 N_e)^{1/2}$ is the electron Debye length. The expression for γ
accounts for the viscous dissipation – the most important origin of
decay – and the Joule heat

$$\gamma = \gamma_V + \gamma_J$$

The last is of no importance ordinarily.

The field of electron temperature is smoothed due to   strong
electron heat conduction. The following inequality is not   fulfilled
for extremely small k only

$$\lambda\gamma_T \sim \lambda k\varkappa_e (v_s\rho)^{-1} \sim kr_d (M/m)^{1/2} (\Gamma_{ze}^{3/2}\Lambda)^{-1} \geq 1$$

Here $\lambda = 2\pi/k$ is the wave length of sound. γ_T is the contribution
of electron heat conduction to sound decay, M/m is the ion-electron
mass ratio, Λ is the Coulomb logarithm which one can find in the expression for the frequency of electron-ion collisions.

So sound for ions is adiabatic process. However for electrons
it is isothermal. Due to viscosity prevalence in the dumping factor
(3) the condition $\lambda\gamma_V$ indicates the maximal value of wave number

$$k_{max} \simeq \bar{r}^{-1} \Gamma_{zz}^2 \Lambda$$

With the density growth Γ_{zz} steams to the unit.The same relates to
the Λ due to the lowering of the role of far collisions. At this limit $k_{max} \simeq \bar{r}^{-1}$ So sound waves take now all the possible part of
the phase space [4].

Plasma becomes liquid-like if $\Gamma_{zz} \gg 1$. At $\Gamma_m \simeq 160$ plasma may crystallize. Viscosity coefficient for liquid state is written below as for ordinary liquid which is cooled significantly

$$\eta = \eta^*(Z^2e^2M/\bar{r})^{1/2}\bar{r}^{-2} \qquad (4)$$

Viscosity coefficient (4) has been calculated by molecular dynamic simulation [5]. It was concluded that the dimensionless viscosity $\eta \sim \Gamma_{zz}^{1/3}$ if $\Gamma_{zz} \gtrsim 40$. So at high Γ_{zz} η increases proportionally to $\Gamma_{zz}^{17/6}$. In [6] the generalized kinetic equation for the overcooled liquid $(\Gamma_{zz} > \Gamma_m)$ has been analyzed. The farther increase of η with the growth of Γ_{zz} has been demonstrated.

It is naturally to suggest that in the crystal domain viscosity coefficient must grow exponentially. Strong inter-ion correlation leads to cell-like structure of matter. Every ion oscillates inside of its cell. The frequency of oscillations equals to ionic plasma frequency $\omega_{pi} \sim (Z^2e^2/M\bar{r}^3)^{1/2}$. The amplitude is small in comparison with the cell radius. The displacement of ion into another cell becomes possible if a vacancy is present in the neighborhood. It requires long time interval to find out the vacancy

$$\tau \sim \omega_{pi}^{-1} \exp(W/T)$$

Certainly $W \gg T$ where W is the work which is necessary for vacancy creation.

Thus the viscous transfer of momentum becomes vacansion. Write down viscosity coefficient in the following way

$$\eta = (Z^2e^2M/\bar{r})^{1/2}\bar{r}^{-2} \exp(a\Gamma_{zz}/\Gamma_m) \qquad (5)$$

Here a is the constant of order of unit, accurate value of which is unknown now.

To elucidate the structure of exponent index we suppose that coupling becomes stronger as a result of cooling. As to density let it be constant, say $N = N_1$. For simplicity let also Z to be constant at this cooling. It means that thermal ionization does not proceeds and the cold ionization level (due to strong compression) is high enough.

At that case one may rewrite the coupling parameters which stand in exponent index

$$\Gamma_{zz} = Z^2e^2/\bar{r}_1T , \qquad \Gamma_m = Z^2e^2/r_1T_m$$

Here T_m is the melting temperature, $r_1 = (4\pi N_1/3)^{-1/3}$. Now viscosity coefficient takes the following form

$$\eta = (Z^2 e^2 M/\bar{r})^{1/2} \bar{r}^{-2} \exp(aT_m/T) \qquad (6)$$

The exponent index must contain W . Nobody knows its value. However it is proportional to the bond energy. In known liquids the bond energy is proportional to critical temperature. However (6) contains melting temperature. It do not seems to be strange. The empirical relation exists for many metals: $T_c/T_m \simeq 7$ [7]. Examples: Cs – 7,1, Na – 7,6, Fe – 5,5, W – 6,3, U – 8,9.

Eq. (5) gives viscosity coefficient in a qualitative way. It doesnot pretend any numerical accuracy.

ION SOUD DUMPING IN STRONGLY COUPLED PLASMA

In highly Z-plasma pressure of electrons determines to a considerable extent total pressure of plasma. Under condition $\Gamma_{ee} \ll 1$ electron subsystem represents nearly ideal gas. So sound velocity is given by

$$v_s \simeq [(dp/d\rho)_T]^{1/2} \simeq (ZT/M)^{1/2}$$

The value of Z we consider to be very large as it is necessary for the consistent theory. In real system inter-ion correlation influences velocity values even at high Z-numbers. At the given paper we shall not account for it. It does not change qualitative conclusions. The qualitative results only are presented now.

For determination of maximal value of wave number of viscous sound the condition $\lambda \gamma_v = 1$ may be used. In liquid states η $\Gamma_{zz}^{17/6}$. It leads to

$$k_{max} \sim \bar{r}^{-1} Z^{1/2} \Gamma_{zz}^{-5/6} \qquad (7)$$

Values of k_{max} decrease with the growth of coupling parameter Γ_{zz}. At the region of solid states maximal wave numbers decrease exponentially

$$k_{max} \sim \bar{r}^{-1} \exp(-a\Gamma_{zz}/\Gamma_m) \qquad (8)$$

The possibility of existence of viscous sound is limited by (7) and (8). The region of viscous sound is indicated approximately (region 1) at Fig. 1.

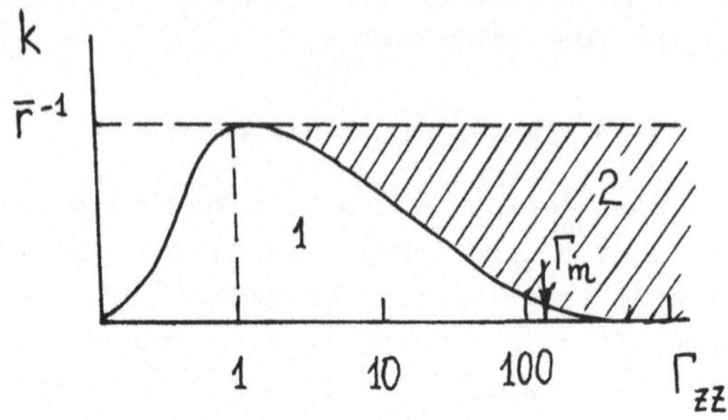

Fig.1. Diagram $k_{z\bar{z}}\ \Gamma$.

The propagation of viscous sound does not disturb local thermodynamic equilibrium. The condition $k = k_{max}$ is identical to condition $\omega\tau = 1$ which distinguish the domains of viscous and nonviscous sound. Viscous sound does not exists outside of region 1. Non-viscous sound may propagate if $\omega\tau > 1$ (region 2). The fulfillment of this inequality is necessary for propagation of zero-sound in the Fermi-liquid [8] and ultrasonic waves in cryogenic metals [9].

Under the unequally $\omega\tau > 1$ the ion subsystem dynamics cannot support local thermodynamic equilibrium. Propagation of sound waves results in the non-equilibrium of distribution function of particles over velocities. The function is elongated in direction of sound propagation. Electron-ion collisions tend to equilibrate the distribution. It leads to energy dissipation. The mechanism has to be considered on kinetic level [9]. This dissipation competes with the Joule one. For the decay coefficient we have

$$\gamma = \gamma_p + \gamma_J = \omega^2(2v_s^3)^{-1}[N_e T\tau_{ei}\rho^{-1} + (kr_d)^2 v_s^2/4\pi\sigma] \qquad (9)$$

Here τ_{ei} is the characteristic time for electron-ion interaction. Approximately $\tau_{ei} = \bar{r}/v_e$, v_e is the thermal electron velocity. For small wave numbers the ratio $\gamma_J/\gamma_p \sim (k\bar{r})^2\Gamma_{ee}^{-2}$ may be small, so the electron-ion collision dissipation prevails. For the maximal k the Joule dissipation dominates if $\Gamma_{ee} \ll 1$. The inequality $\lambda\gamma_J < 1$ indicates that the decay of non-viscous sound is weak.

In significant diapason of plasma parameters electrons are degenerated. One may write down for these conditions the similar relations and make the same conclusions.

Non-viscous sound deserves more accurate investigation. The same one may relate to viscosity of strongly coupled plasma.

REFERENCES

[1] Y.Q.Zhang, A.N.Mostovych, A.W.DeSilva, Phys. Rev. Lett. **62**, 1165 (1989)

[2] V.E.Fortov, I.T.Iakubov, Physics of Non-Ideal Plasma (Hemisphere, N.Y., 1989)

[3] Strongly Coupled Physics, S.Ichimaru, Ed. (Elsevier, Amsterdam, 1990)

[4] M.I.Berkovskii, A.A.Valuev, Yu.K.Kurilenkov, High Temperatures. 27, N 2 (1989)

[5] B.Bernu, P.Vieillefosse, Phys.Rev.A.**18**,2345 (1985)

[6] S.Ichimaru, H.Iyetomi, S.Tanaka, Phys.Rep.**149**,91 (1987)

[7] A.R.Ubbelohde, Melting and Crystal Structure (Clarendon, Oxford.1985)

[8] E.M.Lifshits, L.P.Pitaevskii, Statistical Physics, Part 2 (Nauka,Moscow,1978)

[9] A.B.Pippard, Phil.Mag.**46**,1104 (1955)

INTERPOLATION FORMULAE
FOR THE TRANSPORT COEFFICIENTS
IN DENSE PLASMAS

H. Reinholz, R. Redmer, G. Röpke,
University Rostock, Department of Physics,
Universitätsplatz 3, O-2500 Rostock 1,
Federal Republic of Germany

Abstract

The transport coefficients of dense plasmas are treated within linear response theory. Within a virial expansion, the respective coefficients can be related to many-particle effects such as dynamic screening or self-energy. Considering the explicit results for the transport coefficients in Born approximation, for the classical domain, and for the strong-coupling regime, interpolation formulae are derived that reproduce the data available for dense plasmas within the experimental errors.

1. Introduction

The transport properties of dense plasmas (electrical and thermal conductivity, thermoelectric power) are governed by the Coulomb interaction and are thus universal functions of the density of electrons n_e and of the temperature T. In the recent years, experimental efforts have been made to measure the electrical conductivity up to high values of n_e [1, 2, 3, 4, 5]. The interesting non-ideal behaviour of the electrical conductivity has been investigated by means of quantum-statistical approaches [6, 7, 8, 9, 10, 11]. Especially, different interpolation formulae have been constructed which are valid in a wide range of the temperature-density domain [8, 9, 12, 13].
In the case of a high-density, strongly coupled plasma, the concept of binary collisions between the plasma constituents cannot be applied, and we have to take into account many-particle effects such as degeneration and Pauli-blocking, dynamic screening and self-energy, structure factor and multiple scattering, the Debye-Onsager relaxation effect, bound-state formation and pressure ionization, etc. However, up to now there has been no rigorous quantum-statistical approach which accounts for all these many-particle effects mentioned above. We will treat the transport properties of dense plasmas within the framework of the linear response theory given here in the extended formulation of Zubarev [14], see also [10, 11].
The derivation of interpolation formulae for the transport coefficients implies the correct treatment of limiting cases. We will perform here a low-density expansion which reads for the electrical conductivity

$$\sigma(n,T) = \frac{(k_B T)^{3/2}(4\pi\epsilon_0)^2}{m_e^{1/2} e^2}\sigma^*(\Gamma,\Theta) = \left[A(T)\ln n + B(T) + C(T) n^{1/2}\ln n + \cdots\right]^{-1} . \quad (1)$$

The virial coefficient $A(T)$ is given by the Spitzer result [15] that can be obtained within a T-matrix approximation for both the electron-ion and electron-electron collision integral in the kinetic equation [16]. The evaluation of $B(T)$ requires the treatment of dynamic screening [17], and $C(T)$ is determined by self-energy effects and nonequilibrium two-particle correlations [11]. Introducing the dimensionless parameters [8]

$$\Gamma = \frac{e^2}{4\pi\epsilon_0 k_B T} \cdot (4\pi n/3)^{1/3} \quad , \qquad \Theta = \frac{2m_e k_B T}{\hbar^2} \cdot (3\pi^2 n)^{-2/3} \quad , \tag{2}$$

we distinguish the weakly nonideal, quasiclassical region $\Gamma < 1$, $\Gamma^2\Theta \gg 1$, the Born limit for weak nonideality $\Gamma < 1$, $\Gamma^2\Theta \ll 1$, and the strongly coupled case $\Gamma \geq 1$, where new effects as bound-state formation (for $\Theta \geq 1$) or structure factor (for $\Theta \leq 1$) may become of importance.

2. Linear response theory for the transport properties

A modified Liouville-von Neumann equation for the nonequilibrium statistical operator ρ can be derived employing a generalized Gibbs state for the "relevant" part of ρ. Within the frame of linear response theory, a generalized linear Boltzmann equation can be derived which allows for a direct comparison with respective expressions of standard kinetic theory. Applying a finite set of "relevant" operators P_m to characterize the nonequilibrium state, the conductivity is given by [10, 18],

$$\sigma = -\frac{e^2}{\Omega} \begin{vmatrix} 0 & N_m \\ N_n & D_{nm} \end{vmatrix} / |D_{nm}| \quad , \tag{3}$$

with the correlation functions $N_m = \frac{1}{m_e}(P_0, P_m)$, $D_{nm} = \langle \dot{P}_n(\eta); \dot{P}_m \rangle$, where

$$(A, B) = \int_0^\beta d\tau \, Tr\{\rho_0 A(-i\hbar\tau B\} \quad ,$$
$$\langle A(\eta); B \rangle = \int_{-\infty}^0 dt \, e^{\eta t}(A(t), B) \quad , \tag{4}$$
$$P_m = \sum_k \hbar k_z \left(\frac{\beta\hbar^2 k^2}{2m_e}\right)^m a_k^\dagger a_k \quad .$$

The time dependence of the operators and ρ_0 are determined by the system Hamiltonian. Eq. (3) is valid for arbitrary degrees of degeneration. Due to the relation between correlation functions in thermal equilibrium and thermodynamic Green functions which can be evaluated by means of efficient diagram techniques, a systematic treatment of many-particle effects in a strongly coupled plasma can be performed, see [11, 17].

For the nondegenerate case, the values $N_m = N\Gamma(m + \frac{5}{2})/\Gamma(\frac{5}{2})$ are immediately obtained ($N = n\Omega$, particle number). Furthermore, the correlation functions D_{nm} are given by a ladder sum of diagrams that is related to the T-matrix of the two-particle scattering process. Calculating these T-matrices for electron-ion and electron-electron scattering by means of the scattering phase shifts for a statically screened (Debye) potential, and determining the electrical conductivity (3) within at least a three-moment expansion, i.e. considering P_0, P_1, and P_2, the Spitzer result [15] is reproduced as the low-density asymptote for the electrical conductivity, see [16],

$$\sigma_{Sp} = 0.591 \frac{(4\pi\epsilon_0)^2 (k_B T)^{3/2}}{e^2 m_e^{1/2}} \left[\frac{1}{2} \ln \left(\frac{3}{2}\Gamma^{-3}\right) \right]^{-1} \quad . \tag{5}$$

For the degenerate domain, the usual Ziman formula [22] can be derived from eq. (3) already within a one-moment expansion, i.e. considering only P_0, see [19, 20, 21],

$$\sigma_{Zi}^{-1} = \frac{\Omega m_e^2 N}{12\pi^3\hbar^3 e^2 n_e^2} \int_0^\infty dq \, q^3 \, f\left(\frac{q}{2}\right) S(q) \left|\frac{V_{ei}(q)}{\varepsilon(q)}\right|^2 \quad , \tag{6}$$

where $f(q)$ denotes the Fermi distribution function, $S(q)$ is the ionic static structure factor, $V_{ei}(q)$ is the electron-ion pseudo-potential, and $\varepsilon(q)$ being the electronic static dielectric function, resp.

3. Interpolation formula for the electrical conductivity

The force-force correlation functions D_{nm} can be evaluated numerically for arbitrary degrees of degeneration and coupling strengths as well. However, in order to construct interpolation formulae for the transport coefficients, we consider for weak nonideality $\Gamma < 1$ the Born limit $\Gamma^2 \Theta \ll 1$ and the classical limit $\Gamma^2 \Theta \gg 1$, as well as the strongly coupled case $\Gamma \geq 1$.

A. Born limit $\Gamma^2 \Theta \ll 1$

The correlation functions $D_{nm} = D_{nm}^{ei} + D_{nm}^{ee}$ in Born approximation are given by Lenard-Balescu collision integrals which read, e.g., for D_{00}

$$D_{00}^{LB,cd} = 2\pi\beta\hbar \sum_{k,p,q} \int_{-\infty}^{\infty} d\omega \left| \frac{V_{cd}(q)}{\varepsilon_{RPA}(q,\omega)} \right|^2 \times \tag{7}$$

$$\times \delta(E_k - E_{k+q} - \omega)\, \delta(E_p - E_{p-q} + \omega)\, f(k)(1 - f(k+q))\, f(p)(1 - f(p-q)) \quad .$$

Inserting the RPA dielectric function for the nondegenerate case [24], the following low-density expansion can be obtained from the numerical solution of eq. (7) for $\Gamma \leq 1$, $\Theta \geq 1$, see [17],

$$D_{nm}^{LB,ei} = d\,(n+m)! \left[\tfrac{1}{2} \ln \tfrac{\Theta}{\Gamma} + \tfrac{1}{2} A(n+m) - 0.1895 \right] ,$$
$$D_{nm}^{LB,ee} = d\,\sqrt{2}\, b_{nm} \left[\tfrac{1}{2} \ln \tfrac{\Theta}{\Gamma} + \tfrac{1}{2} f_{nm} + 0.0810 \right] ,$$
$$A(m) = 1 + \tfrac{1}{2} + \tfrac{1}{3} + \cdots + \tfrac{1}{m} , \quad A(0) = 0 , \tag{8}$$
$$b_{0m} = b_{m0} = 0 , \quad b_{11} = 1 , \quad b_{12} = b_{21} = \tfrac{11}{2} , \quad b_{22} = \tfrac{157}{4} , \quad \cdots ,$$
$$f_{11} = 0 , \quad f_{12} = f_{21} = \tfrac{2}{11} , \quad f_{22} = \tfrac{48}{157} , \quad \cdots , \quad d = \tfrac{4}{3} \sqrt{2\pi}\, \frac{m_e^{1/2}}{(k_B T)^{3/2}} \left(\frac{e^2}{4\pi\epsilon_0} \right)^2 n N .$$

B. Classical limit $\Gamma^2 \Theta \gg 1$

For the classical domain, the correlation functions D_{nm}^{cd} were obtained from the solution of the Boltzmann collision integral that is given by transport cross sections Q_{cd}^T according to [16, 23]

$$D_{nm}^{B,ei} = \frac{8\hbar n^2 \Omega}{3\sqrt{\pi}\beta} \left(\frac{\beta\hbar^2}{2m_e} \right)^{n+m+5/2} \int_0^\infty dk\, k^{2n+2m+5}\, exp\left(-\frac{\beta\hbar^2 k^2}{2m_e} \right) Q_{ei}^T(k) ,$$
$$D_{nm}^{B,ee} = \frac{8\sqrt{2}\hbar n^2 \Omega}{3\sqrt{\pi}\beta} \left(\frac{\beta\hbar^2}{m_e} \right)^{7/2} \int_0^\infty dk\, k^7\, R_{nm}\, exp\left(-\frac{\beta\hbar^2 k^2}{m_e} \right) Q_{ee}^T(k) , \tag{9}$$
$$R_{0m}(x) = 0 , \quad R_{11}(x) = 1 , \quad R_{12}(x) = R_{21}(x) = \tfrac{7}{2} + x^2 , \quad R_{22}(x) = \tfrac{77}{4} + 7x^2 + x^4 .$$

The transport cross sections Q_{cd}^T were determined within a T-matrix calculation with respect to a statically screened (Debye) potential [16]. Again, we obtain the following low-density expansion for $\Gamma \leq 1$, $\Theta \geq 1$,

$$D_{nm}^{B,ei} = d\,(n+m)! \left[-\tfrac{1}{2} \ln(6\Gamma^3) + A(n+m) - 0.2680 - \frac{0.0905}{\Gamma^2 \Theta} - \frac{0.0196}{\Gamma^4 \Theta^2} \right] ,$$
$$D_{nm}^{B,ee} = d\,\sqrt{2}\, b_{nm} \left[-\tfrac{1}{2} \ln(6\Gamma^3) + f_{nm} + 0.2320 - \frac{0.3620}{\Gamma^2 \Theta} - \frac{0.2359}{\Gamma^4 \Theta^2} \right] . \tag{10}$$

Employing the convergent collision integral [25, 26] $D_{nm} = D_{nm}^B + D_{nm}^{LB} - D_{nm}^L$, where D_{nm}^L is the Landau collision integral, the following low-density expansion can be given for the electrical conductivity [17]

$$\frac{1}{\sigma} = \frac{d}{e^2 n N} \left(-0.2544 \ln(6\Gamma^3) + 0.5288 \right) . \tag{11}$$

Taking n in m^{-3}, T in K, and the electrical conductivity σ in $1/(\Omega m)$, the respective virial coefficients in eq. (1) read

$$A(T) = -32.8013\, T^{-3/2} , \quad B(T) = 98.4040\, T^{-3/2} \left(\ln T + 10.6177 \right) . \tag{12}$$

This formula fits the numerical results better than 1% for typical plasma conditions of $T = (15 - 30) \times 10^3\,K$, and $n = (1 - 10) \times 10^{17}\,cm^{-3}$.

C. Strong coupling effects $\Gamma \geq 1$

The correct Born approximation for strongly coupled, degenerate plasmas $\Gamma \geq 1$, $\Theta \leq 1$ was considered by Ichimaru and Tanaka [8] by taking into account the structure factor $S(q)$ and local field effects in the dielectric function $\varepsilon(q)$ for calculating the electrical conductivity. Lee and More [9] have applied the relaxation time approximation for the determination of the transport properties of strongly coupled plasmas in both electric and magnetic fields at arbitrary degrees of degeneration. Rinker [12] has derived interpolation formulae for the electrical conductivity from an improved Ziman formula that is valid in a wide region of the density-temperature plane for the elements of the periodic table.

Taking T in K and σ in $1/(\Omega m)$, we propose here the following fit formula for the electrical conductivity, see also [13, 23],

$$\sigma_{FIT} = a_0 \cdot T^{3/2} \left(1 + \tfrac{b_1}{\Theta^{3/2}}\right) \left[\ln(1 + A + B) - C - \tfrac{b_2}{b_2 + \Gamma\Theta}\right]^{-1}, \tag{13}$$

$$A = \Gamma^{-3}\,\frac{1 - a_4/\Gamma^2\Theta}{1 - a_2/\Gamma^2\Theta - a_3/\Gamma^4\Theta^2}\,\left[a_1 + c_1 \ln(c_2\Gamma^{3/2} + 1)\right]^2$$

$$B = \frac{b_2(1 + c_3\Theta^{4/5})}{\Gamma\Theta(1 + c_3\Theta)}\,,\quad C = \frac{c_4}{\ln(1 + \Gamma^{-3}) + c_5\Gamma^2\Theta}\,.$$

The parameters a_i are fixed by the low-density limit, and the values for the b_i are due to the strong degenerate limit. Only the parameters c_i were fitted to the numerical data for the correlation functions. The whole set of parameters reads $a_0 = 0.03064$, $a_1 = 1.1590$, $a_2 = 3.136$, $a_3 = 3.0369$, $a_4 = 3.6590$, $b_1 = 0.3831$, $b_2 = 3.274$, $c_1 = 1.5$, $c_2 = 6.2$, $c_3 = 0.3$, $c_4 = 0.6$, $c_5 = 0.1$.

Comparison between eq. (13) and experimental values for the conductivity can be performed with data for strongly coupled plasmas such as given in [1]. The results are shown in Table I. Taking into account an experimental error of about 30%, theory and experiment are not in contradiction. However, the experimental values are systematically smaller for the lower-density plasmas with $\Gamma \sim 0.2 - 0.7$. For the higher-density plasmas, they are higher than the theoretical ones. For comparison, we present also the results of Ichimaru and Tanaka [8] in Table I which include, especially, the effects of structure factor. They found also an overall agreement with the experimental values of [1] within the error bars of about 30%.

The recent experimental result of Shepard et al. [5] for a partially degenerate, strongly coupled plasma (polyurethane) is shown in addition. The strong deviation of the theoretical values is due to the neglection of bound-states (atoms, molecules, molecular ions). The influence of bound-states on the transport properties of dense plasmas can be investigated systematically within the given linear response approach [10, 23]. A better agreement with the experimental value of Shepard et al. [5] is obtained within the approaches of Rinker [12] and Djurić et al. [27] which are based on the Ziman formula and employ usual methods of the theory of liquid metals.

In Table I, we compare also between eq. (13) and experimental values for the conductivity of non-degenerate ($\Theta \gg 1$) and weakly nonideal ($\Gamma \ll 1$) plasmas [28, 29]. The theoretical results are systematically higher than the experimental ones and the deviations (50-100%) observed are due to the neglection of partial ionization.

The present approach can be immediately generalized by including further effects as bound-state formation [10, 23], the Debye-Onsager relaxation effect [11], or structure factor [19], which affect the thermoelectric properties in different regions of the density-temperature domain strongly. Furthermore, interpolation formulae were developed also for the other transport coefficients such as the thermoelectric power or the thermal conductivity, see [20, 23].

Table I Electrical conductivity of strongly coupled plasmas.
a: [1], b: [28], c: [5], d: [29], σ_{RRR}: present approach, σ_{IT}: [8].

	T (10³ K)	n (10²⁵ m⁻³)	Γ	θ	σ_{exp}	σ_{RRR} 10²/(Ωm)	σ_{IT}
Ar (a)	22.2	2.8	0.368	56.9	190	226	200
	20.3	5.5	0.505	33.2	155	232	203
	19.3	8.1	0.604	24.4	170	241	209
	19.0	14.0	0.736	16.7	255	272	234
	17.8	17.0	0.838	13.7	245	274	232
Xe (a)	30.1	25.0	0.564	17.9	450	457	442
	27.5	59.0	0.822	9.24	680	544	506
	27.0	79.0	0.922	7.47	740	599	546
	26.1	140	1.150	4.93	690	769	657
	25.1	160	1.260	4.34	780	832	660
	24.6	200	1.380	3.66	1040	957	728
	22.7	200	1.500	3.38	930	967	694
Ne (a)	19.8	1.1	0.303	94.6	130	173	148
	19.6	1.9	0.367	65.0	165	187	160
Air (a)	11.0	0.13	0.267	218	60	67	53
Ar (b)	16.4	0.059	0.128	551	83	93	78
		0.101	0.165	385	79	102	83
		0.131	0.180	324	76	106	86
		0.154	0.190	291	64	108	88
Xe (b)	12.4	0.062	0.185	403	46.4	70	55
		0.112	0.226	272	43.8	75	59
		0.125	0.234	252	41.1	76	60
	12.6	0.072	0.192	371	48.0	73	57
		0.126	0.230	255	46.3	78	62
		0.140	0.239	238	43.5	79	65.5
Polyurethan(c)	≈116	≈6000	0.6... 1.8	1.2... 1.9	166... 1333	5021	9270
H(d)	15.4	0.10	0.175	364	62.5	95	77
	18.7	0.15	0.165	337	91.3	125	103
	21.5	0.25	0.170	276	114.6	156	131

References

[1] Yu. V. Ivanov, V. B. Mintsev, V. E. Fortov, and A. N. Dremin, Zh. Eksp. Teor. Fiz. **71**, 216 (1976) [Sov. Phys.-JETP **44**, 112 (1976)].

[2] V. K. Gryaznov, I. L. Iosilevskii, Yu. G. Krasnikov, N. N. Kuznetsova, V. I. Kucherenko, G. B. Lappo, B. N. Lomakin, G. A. Pavlov, E. E. Son, and V. E. Fortov, *Thermophysical Properties of the Working Medium of Gas-Phase Nuclear Reactors* (Atomizdat, Moscow, 1980).

[3] K. Günther and R. Radtke, *Electrical Properties of Nonideal Plasmas* (Birkhäuser, Basel, 1984).

[4] W. Ebeling, V. E. Fortov, Yu. L. Klimontovich, N. P. Kovalenko, W. D. Kraeft, Yu. P. Krasny, D. Kremp, P. P. Kulik, V. A. Riaby, G. Röpke, E. K. Rozanov, and M. Schlanges, *Transport Properties of Dense Plasmas* (Birkhäuser, Basel, 1984).

[5] R. L. Shepard, D. R. Kania, and L. A. Jones, Phys. Rev. Lett. **61**, 1278 (1988).

[6] H. Minoo, C. Deutsch, and J. P. Hansen, Phys. Rev. **A 14**, 840 (1976).

[7] D. B. Boercker, Phys. Rev. **A 23**, 1969 (1981);
D. B. Boercker, F. J. Rogers, and H. E. DeWitt, Phys. Rev. **A 25**, 1623 (1982);
F. Perrot and M. W. C. Dharma-wardana, Phys. Rev. **A 36**, 238 (1987).

[8] S. Ichimaru and S. Tanaka, Phys. Rev. **A 32**, 1790 (1985);
S. Ichimaru, H. Iyetomi, and S. Tanaka, Phys. Rep. **149**, 91 (1987).

[9] Y. T. Lee and R. M. More, Phys. Fluids **27**, 1273 (1984).

[10] G. Röpke and F. E. Höhne, phys. stat. sol. (b) **107**, 603 (1981);
F. E. Höhne, R. Redmer, G. Röpke, and H. Wegener, Physica **128 A**, 643 (1984);
V. Christoph and G. Röpke, phys. stat. sol. (b) **131**, 11 (1985).

[11] G. Röpke, Phys. Rev. **A 38**, 3001 (1988).

[12] G. A. Rinker, Phys. Rev. **B 31**, 4207 (1985); 4220 (1985); Phys. Rev. **A 37**, 1284 (1988).

[13] G. Röpke and R. Redmer, Phys. Rev. **A 39**, 907 (1989).

[14] D. N. Zubarev, *Nonequilibrium Statistical Thermodynamics* (Plenum, New York, 1974).

[15] L. Spitzer and R. Härm, Phys. Rev. **89**, 977 (1953).

[16] F. Sigeneger, S. Arndt, R. Redmer, M. Luft, D. Tamme, W. D. Kraeft, G. Röpke, and T. Meyer, Physica **152 A**, 365 (1988);
S. Arndt, F. Sigeneger, F. Bialas, W. D. Kraeft, M. Luft, T. Meyer, R. Redmer, G. Röpke, and M. Schlanges, Contr. Plasma Phys. **30**, 273 (1990).

[17] R. Redmer, G. Röpke, F. Morales, and K. Kilimann, Phys. Fluids B: Plasma Phys. **2**, 390 (1990).

[18] G. Röpke, Physica **121 A**, 92 (1983).

[19] C.-V. Meister and G. Röpke, Ann. Phys. (Leipzig) **39**, 133 (1982).

[20] H. Reinholz, PhD-Thesis (Rostock, 1989, unpublished).

[21] R. Redmer, H. Reinholz, G. Röpke, R. Winter, F. Noll, and F. Hensel, J. Phys.: Condensed Matter **4**, 1659 (1992).

[22] J. M. Ziman, Phil. Mag. **6**, 1013 (1961).

[23] H. Reinholz, R. Redmer, and D. Tamme, Contr. Plasma Phys. **29**, 395 (1989).

[24] Yu. L. Klimontovich and W. D. Kraeft, Tepl. Vys. Temp. **12**, 239 (1974) [Sov. Phys.-High Temp. **12**, 212 (1974)].

[25] H. A. Gould and H. E. DeWitt, Phys. Rev. **155**, 68 (1967).

[26] R. H. Williams and H. E. DeWitt, Phys. Fluids **12**, 2326 (1969).

[27] Z. Djurić, A. A. Mihajlov, V. A. Nastasyuk, M. Popović, and I. M. Tkachenko, Phys. Lett. **A 155**, 415 (1991).

[28] M. M. Popović, Y. Vitel, and A. A. Mihajlov, in: Strongly Coupled Plasmas, S. Ichimaru, Ed. (Elsevier, Yamada, 1990), p. 561.

[29] R. Radtke and K. Günther, J. Phys. D: Applied Phys. **9**, 1131 (1976).

STRONG ELECTRIC FIELDS AND NONLINEAR CURRENT-VOLTAGE
CHARACTERISTIC NON-IDEAL PLASMAS IN LIQUID
SEMICONDUCTORS AT HIGH TEMPERATURES AND HIGH PRESSURES

A. Andreev
Joffe Institute
26 Polytekhnicheskaya, 194021 St. Peterburg (Leningrad), Russia

V. Alekseev
Branch of Atomic Energy Institute
142092, Moscow region, Troitsk, Russia

Abstract

This paper shows the results of observation of switching effect
in liquid semiconductors at high pressures.

The study of electric field influence on the semiconductor melts
conductivity provides new possibilities for investigations of
conductivity mechanisms at the transition from liquid semiconductor
state to dense non-ideal plasma state at high pressure. Such pressure
permits to support the necessary density of melts at high
temperatures. This high pressure technique up to 550 atm makes it
possible to get stable and reliable results. It allows us to observe
the stable relaxation oscillations in liquid Se with frequency
~10Mhz[1].The first observations of switching effect under electric
field influence have been related to glass [2-4] and liquid [5-6]
semiconductors. This effect reveals itself in abrupt conductivity
change of some orders at the supplement of strong (~10^5v/cm)
fields.The investigations of the Volt-Ampere characteristics (VAC)
was conducted on pulses within the limits of 250 nsec to 2 μsec for a
gap with melt, the distance between the graphite electrodes being
80-100μm.The typical VAC for liquid Se (410°C) and Sb_2S_3 (570°C) are
given in photos (Fig.1,2).Two positive branches are seen:one is
corresponding to the high-ohmic state and another - to the low-ohmic
state. The order of differential resistance for Se'is 10^6.The third
branch connecting the ends of curves for the high-ohmic and low-ohmic
states rests unseen on the oscillogramme a cause of high speed of the
switching process. A generalized picture of the switching process is
given at Fig.3.
The Fig.1,2 show the high-ohmic branch nonlinearity. The dependence

σ(E) is plotted in Fig.4 in the Poole-Frenkel coordinate. It is not difficult to sea a good σ(E) alignment and therefore the thermo-electron ionization mechanism is observed in the prebreakdown region. It is an important result since it permits to suppose an electronic (nonthermal nature) switching mechanism in liquid Se, at least at temperatures up to 500-600°C. This result is well demonstrated by comparison of BAC at different temperature. The temperature dependence of the switching voltage field was investigated by two methods a pulse one a sinusoidal one, at sound frequency sinusoidal regime an agreement between the temperature dependence $E_{*(t)}$ and that predicted by thermal theory - i.e. E_* is equal to half of the activation energy (regard Fig.5). The temperature dependence of such kind isn't revealed in pulse regime. However at melt overheating $E_{*(t)}$ begins to decrease as temperature increases and the electronical mechanism of the switching is been replaced by the thermal one. The ab-branch in-clination is determined by the

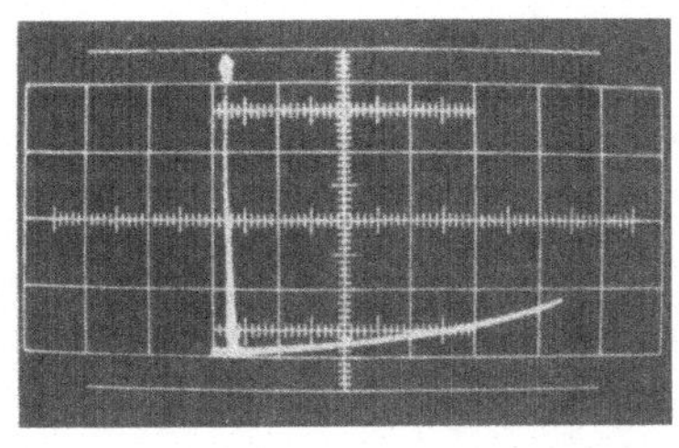

Fig.1.Voltampere characteristic (VAC) of Se at 410°C. The switching field is $3*10^4$v/cm The current of the pulse is 3 A.

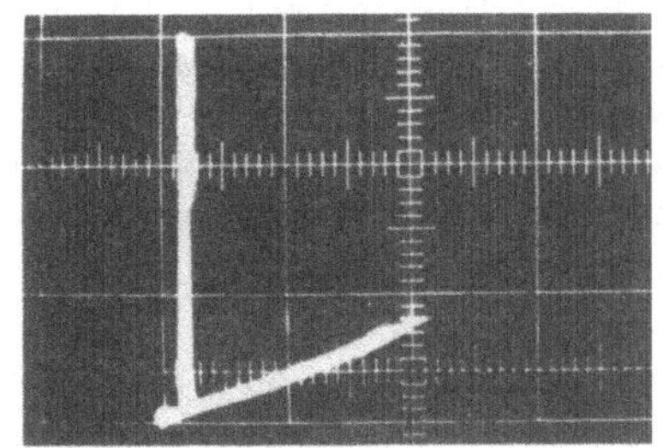

Fig.2. VAC of Sb_2S_3 at 460°C.

resistance of the energy source. A high-conducting state "bc" is characterized by non-ohmic current dependence on voltage in sense that BAC doesn't passe through zero, the voltage approaching to zero value. Some "voltage V_g of holding" the high-conductivity state is observed. The current carrying nature in the high-conductivity state is of a special interest. The dynamical resistance of this current $\Delta V/\Delta I \to 0$. The current density reachs 1÷3 thousands ampere per 1 cm^2. It is evedent that it is a plasma cord of high density. The "bc" branch nonlinearity of BAC indicates injection character of the current. The plasma volume resistance doesn't limit the injection.

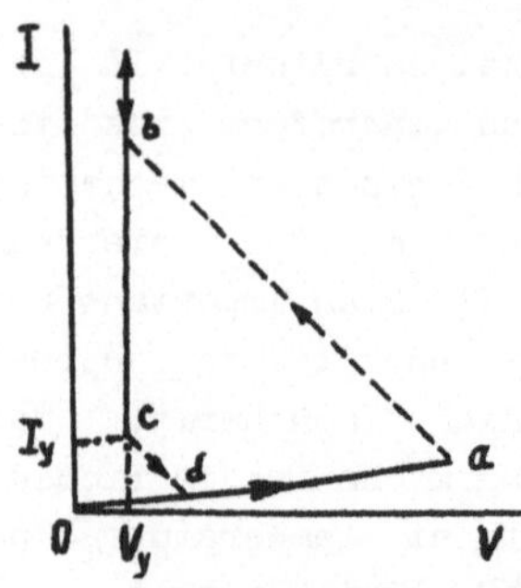
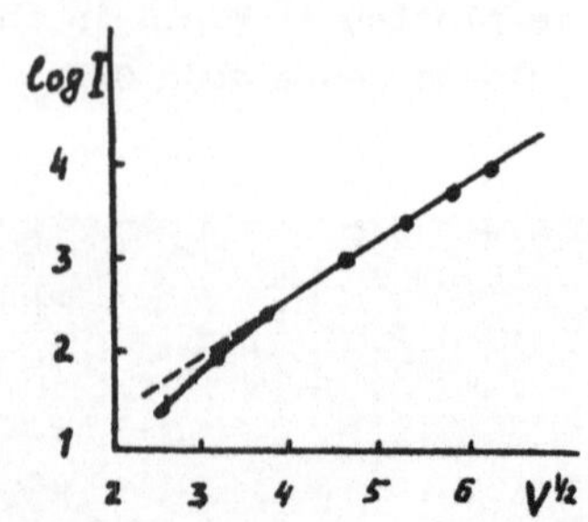

Fig.3. General VAC switching effect
    oa – the high-ohmic branch;
    ab – the breakdown in the low-
        ohmic state;
    bc – the low-ohmic branch;
    cd – the relaxation to the
        initial conductivity value.

Fig.4. High-field dependence of
    the current of the voltage.
    The current and are given
    in relative units

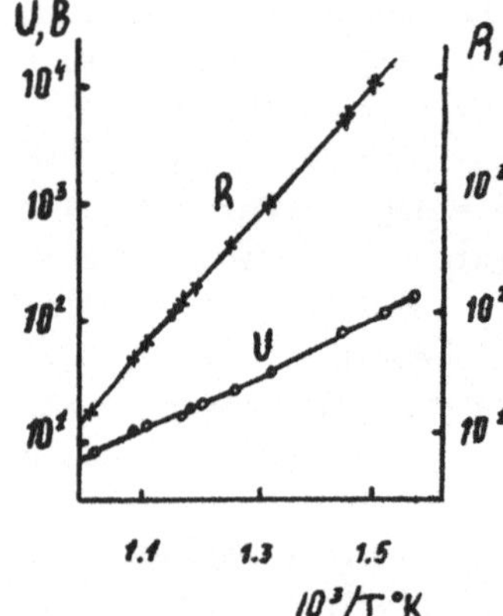

Fig.5. Temperature de-
pendence of resestivity
(R) and Voltage swit-
ching (U) in $Se_{0.91}Te_{0.09}$

The plasma is close to nonideal. At current decrease in conductive state below some definite value I_y of "the current of holding" the switching to an high-ohmic state occurs along the load – line "cd".

If the external chain parameters provide the current decrease below the current of holding it is possible to observe relaxation oscillations.Fig.6 shows the typical picture of such generation.The oscillation amplitude consist of kilovolts, the current – about hundreds milliampere.

The frequence is determined by parameters of RC-chain. Upper frequence edge is limited by switching front speed. The equipment with increase time permission about $0.5*10^{-9}$s being used ,there was no possibility to find the switching front speed. However it should be noticed that this remark is related to liquid Se at temperature up to ~500°C. At higher themperatures the switching characteristic changes as specific conduction grows.

Therefore:

1.The speed of switching from the highohmic to the lowohmic

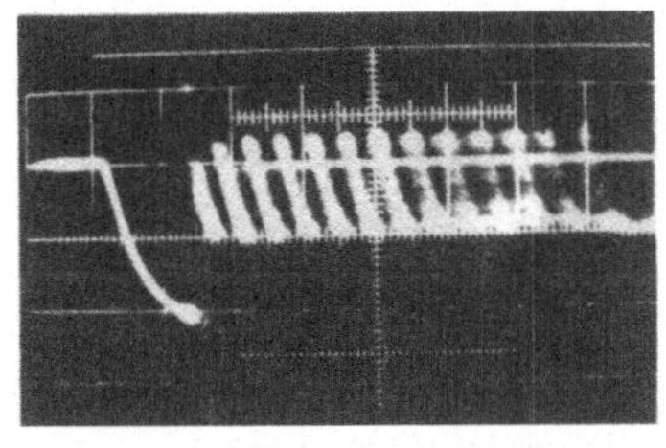

Fig.6.Relaxation oscillation in liquid Se with the frequency 10 MHz at 230°C

conduction decreases with the increase of temperature.

2.The critical field of switching decreases with the increase of temperature.

3.The order of conductivity changes at the switching decreases.

4.VAC become more "loop-like".

In the first approximation the critical temperatures (at which the switching vanishes) are governed by the criterion of the the thermal mechanism $E_g \gtreqless 4$ kT, where E_g is the activation energy, the absolute conductivity magnitude approaching to $10^{-1}-1.0$ $Ohm^{-1}cm^{-1}$. Attention should be paid to the fact that this conductivity value is just adjacent to the region of the liquid semiconductor-metal transition at analogoes pressures. The Se data in the measurements [7-9] may be used as an example. Probably these conductivity values are threshold ones for the Joule's effect self-heating.

References

[1].A.A.Andreev, V.A.Alekseev, E.A.Lebedev,   M.Mamadaliev,B.T.Meleh, A.R.Regel, Yu.F.Ryzhkov, Fiz.Tekhn.Poluprovodnikov   6,661 (1972).

[2].B.T.Kolomiets & E.A.Lebedev, Radiotekhn. Electron. 8, 2097 (1963).

[3].S.R.Ovshinsky, Phys.Rev. 21, 1450 (1968).

[4].H.Fritzohe and S.R.Ovshinsky J. Non-Crystalline Solids 2, 393 (1970).

[5].S.Busch, H.J.Guntherodt,H.U.Kunzi and A.Schweiger, Phys.Letters 33A, 64 (1970).

[6].A.R.Regel,   A.A.Andreev   and   M.Mamadaliev,   Fiz.   Tekhn. Poluprovodnikov 5, 2187 (1971); J.Non-Crystalline Solids 8-10, 455 (1972).

[7].V.A.Alekseev,V.G.Ovcharenko, Yu.F.Ryzhkov and A.A.Andreev, V-th Int. Confer. on Amorphous and Liquid Semiconductors (Nauka, Leningrad, 1975), p.395.

[8].V.A.Alekseev, Yu.F.Ryzhkov, V.G.Ovcharenko and M.V.Sadovsky, Sov.Phys. JETP Lett. 24, 214 (1976); Phys.Lett. 65A, 173 (1978).

[9].H.Hoshino, R.W.Schmutzler and F.Hensel, Ber. Bunsenges. Rhysic Chem 80,27 (1976).

MEASUREMENT OF THE ELECTRICAL CONDUCTIVITY
OF A STRONGLY COUPLED METAL PLASMA

A. W. DeSilva[†] and H.-J. Kunze
Ruhr-Universität   Bochum,
4630 Bochum, Germany

Abstract

Strongly coupled plasmas at near solid density have been created by vaporizing and heating thin metal wires in glass capillaries by means of a short pulse of current. The plasma is confined until the induced pressure pulse in the capillary wall propagates to the outer wall and reflects back to the inner wall. The conductivity of the plasma may be deduced from the observed resistance, and the temperature is measured spectroscopically. Plasmas having densities in the range 1-2 g/cm^3, and temperature in the range 6-10 x 10^3 K have been produced, and the measured conductivities compare well with predictions of theory.

Introduction

We are investigating a method for creating strongly coupled metal plasmas, at very high density, that allows independent and accurate measurements of the ion density, electron temperature, and electrical conductivity, over a wide range of densities and temperatures. The plasma is created by evaporating a metal wire inside a glass capillary. The glass envelope confines the metal plasma for a short time, within which the measurements are completed. The technique of vaporizing a metal wire inside a glass capillary tube by a pulsed electric current is not new, but does not appear to have been applied for this purpose.

Exploding wire plasmas have been widely studied, and strongly coupled plasmas are produced in these discharges. However, in the case of unconfined wire explosions, the density of the resulting plasma is ill-defined, making precise measurement of the properties of the dense plasma difficult. In addition, instabilities tend to produce axially nonuniform plasmas, making interpretation of conductivity measurements uncertain[1].

Wire explosions confined by glass capillaries [2], water [3,4], and plastics [5] have been reported, but these works generally focused on the later phases of the discharge, after considerable expansion of the initial current channel had occurred.

Strongly coupled plasmas created by high current discharges in alkali metals and noble gases in capillaries have been studied [6,7,8] for plasma densities in the range of 10^{20} cm^{-3}, and a measurement of the electrical conductivity of the plasma resulting from evaporating material from the inner wall of a capillary in polyethelene has been reported [9].

Copper resistivity at elevated temperature has been measured by exploding copper wires encased in plastic [10], and by a transient technique involving the measurement of the rate of diffusion of magnetic fields into copper in the presence of a very rapidly rising field [11].

† Permanent address: Laboratory for Plasma Research, University of Maryland, College Park, MD 20742, USA.

Experiment

In the technique reported here, a metal wire is threaded through a glass capillary, with the wire diameter selected to produce the desired final ion density after the vapor fills the tube bore. When a rapidly rising electrical current is driven into the wire, it melts and vaporizes in a very short time (a few hundred nanoseconds). This timescale is faster than the observed growth time of instabilities in unconfined wire explosions [12]. The wire vapor fills the capillary, and remains confined until the capillary itself disintegrates.

The rapidly rising pressure of the vapor will create a compression wave in the glass, that propagates radially out at roughly sound speed. Previous studies of strong compressions in glass [13] and quartz [14] have indicated that for pressures up to about 100 kBar, the compressibility of glass is nearly constant and equal to 1.75×10^{-3} $kBar^{-1}$, and that the shock speed is nearly constant at about 5.4×10^3 m/sec., up to about 70 kBar[13]. When the shock wave reaches the outer wall of the glass, it reflects as a tension wave, and one then expects that the outer wall will spall off [15]. The metal plasma, however, cannot know of the finite size of the tube until the reflected wave has returned to the inner wall, at which time confining effects of the wall are lost. The total transit time out and back for the 6mm (outside) diameter capillaries we have typically used is about 1.2 μs.

Pressures generated in the capillary by the heated plasma are of the order of 1-10 GPa, or 0.01-0.1 MBar, and rise in a time short compared to the transit time for sound across the capillary wall. Under influence of the plasma pressure, the inner diameter of the capillary will grow in time, due to the finite compressibility of the glass. This affects the calculation of both the plasma conductivity, and the density. In future work, the pressure and inner wall diameter will be directly measured, but this has not been done in the preliminary work reported here

A schematic diagram of the apparatus is shown in Fig 1. The glass capillary is typically 6 mm in diameter, has a bore of 0.45 mm, and is 22 mm in length. The capillary tube with wire inside is placed between brass electrodes, and the ends lightly clamped to contact the ends of the wire. A capacitor bank of 3.6 μF charged to 10-15 kV is discharged into the electrodes. The rate of rise of current is limited by the circuit inductance (150 nH) to about 5×10^{10} A/s. Full vaporization of the wire occurs in about 200 nsec, during the linearly rising phase of the current, at which time the rapidly rising resistance of the wire causes the current to decrease.

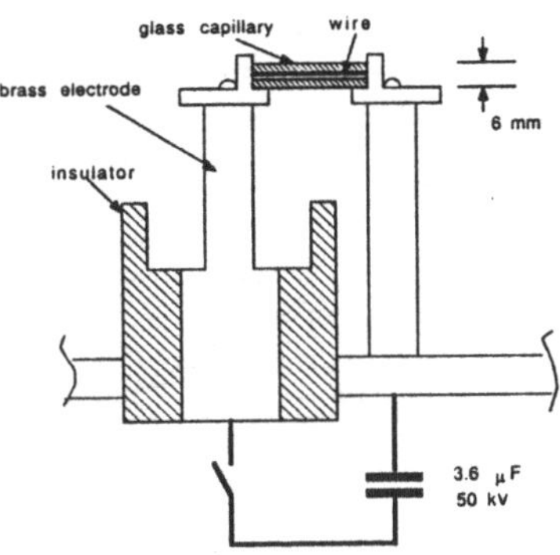

Fig. 1. Schematic

Voltage dividers (Fig. 2) are placed to measure the potential at each end of the wire, and are carefully positioned to avoid undue flux linkages. The signals from the two dividers are subtracted, and the resulting voltage recorded. This voltage may be written

$$V = IR + d/dt\,(LI) \tag{1}$$

where I is the current, L the inductance of the current channel, and R the resistance of the channel. L changes slightly during the rise of the current, but the term arising from IdL/dt is negligible. We record both the current and its derivative, and solve the above equation to find R.

The temperature of the plasma is determined from measurement of the absolute intensity of the continuum radiation in a band of 30 nm width centered on 500 nm, under the assumption that the radiation is blackbody. That the spectrum approximates that of a blackbody has been checked by measurements at several wavelengths in the visible.

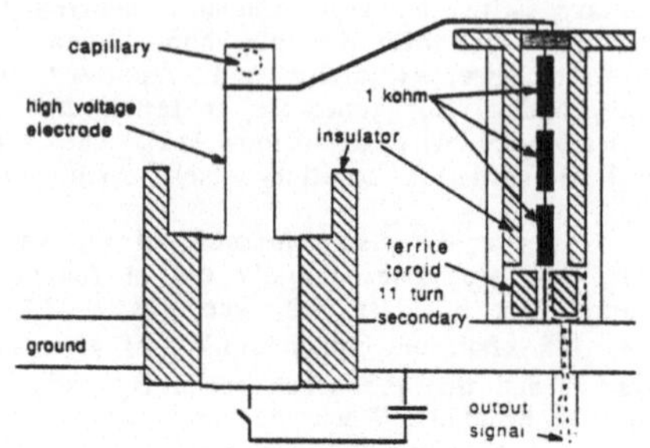

Fig. 2. Voltage divider

Results

Figure 3 displays the time record of a typical discharge, made using a copper wire 0.24 mm in diameter, in a capillary 0.45 mm diameter, with the capacitor charged to 16 kV. The third trace shows an attempt to determine the time of vaporization of the

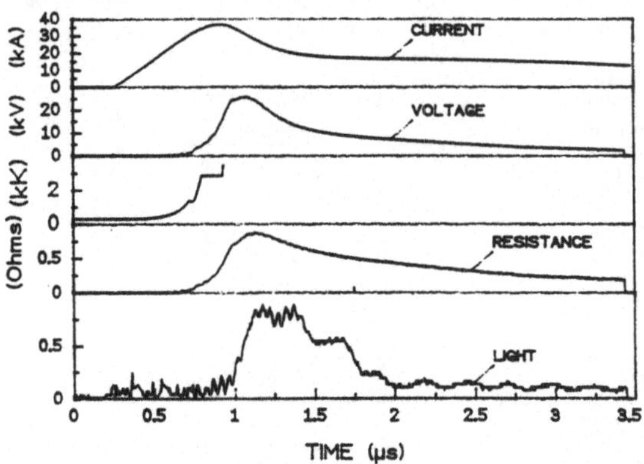

Fig 3. Current, voltage, calculated wire temperature, resistance of the plasma, and light intensity at 500 nm, as function of time, for copper wire. The inductive contribution Ldi/dt has been subtracted from the voltage trace.

wire. The wire temperature is calculated at five nanosecond intervals, from initiation of the discharge until the melting point, using handbook values of the temperature dependence of electrical resistivity and specific heat, and the measured current to determine energy input. As we have not used highly pure copper wires, impurities cause the resistivity to be 17% above the handbook value, and this constant factor is used in the calculations up to the melting point. In this regime, use of the measured potential in the calculation would require subtracting from the measured voltage a term representing Ldi/dt, and the small difference between these two quantities is subject to large error. After the melting point is reached, the measured voltage has risen to a high enough value that the power input is most accurately taken to be the product of measured current and voltage. The calculation carries the temperature rise through until vaporization is complete, using published values for the latent heats of fusion and vaporization, at atmospheric pressure, which leads to an overestimate of the time to vaporization.

The lower trace of Fig. 3 shows the light signal at 500 nm. The oscillations evident in the trace are believed to be of instrumental origin.

To obtain a value for the plasma conductivity, it is necessary to know the cross-sectional area of the plasma column. As discussed above, the inner diameter of the capillary grows due to the influence of the plasma pressure, but a direct measurement of the diameter has not yet been made. If we estimate the plasma pressure, the capillary diameter may be deduced from the known compressibility of glass[13]. In Fig 4 we display the conductivity, density, and temperature for the case of Fig 3, as a function of time, deduced under the assumption that the plasma pressure is equal to nkT. The temperature T is determined from the radiated power, and the density n from the known mass of the wire, also taking into account the growth of the capillary diameter. The trace begins at the time of full vaporization, and displays the 600 ns interval which is the estimated transit time for the shock to reach the outer wall. After that time, scattering of the emitted radiation by the disintegrating outer wall makes interpretation of temperature from the radiation difficult.

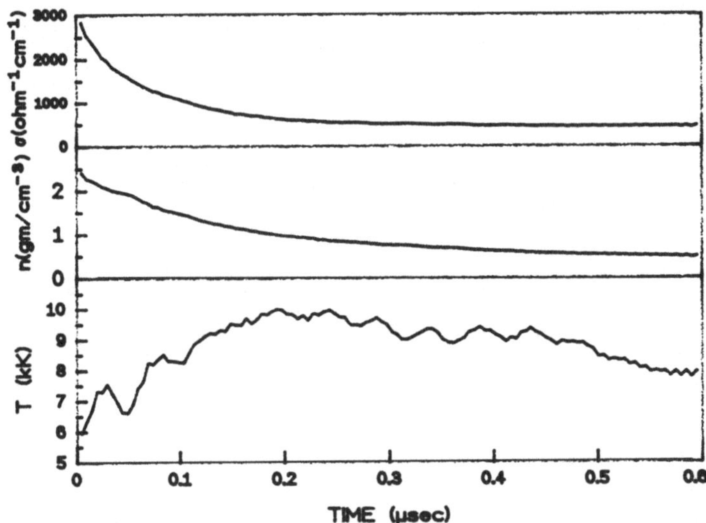

Fig. 4. Conductivity, density and temperature for the case shown in Fig 3.

The time at which the shock reaches the outer wall may be roughly identified by noting the time of obscuration of a He-Ne laser beam adjusted to pass at grazing separation from the outer tube wall. Such a measurement has been made, with the expected result that obscuration begins at about 600 ns.

Discussion

In Figure 5, the conductivity for the case shown in Fig. 4 is compared with the theoretical model for the conductivity of copper plasmas of Ebeling et al. [16].   Our measurements falls in the

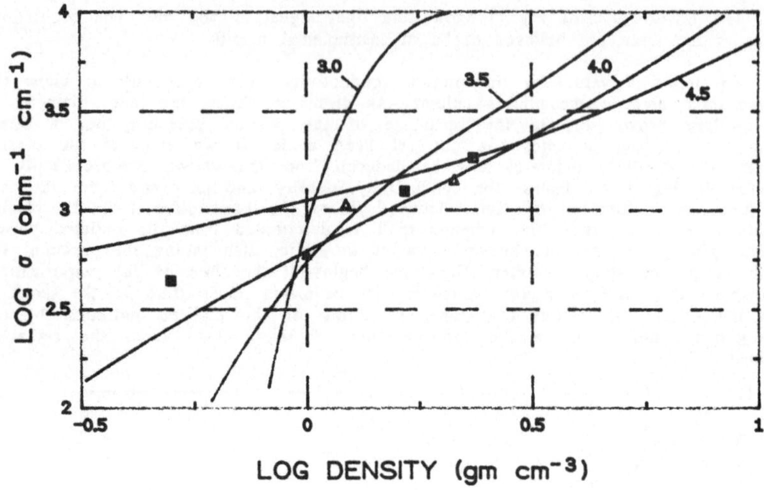

Fig 5. Conductivity as a function of density. Solid curves are theory (see text) for four temperatures, with parameter log T, and filled symbols represent data from Figure 4, for two temperatures ; ■:8000K; Δ:10,000K. Open symbols represent the same data, evaluated under the assumption P=0.1 nkT.

region where the theory curves cross, and the conductivity is therefore expected to be only a weak function of temperature. Experimental points are displayed for two measured temperatures. In order to show the sensitivity of this measurement to the pressure model assumed, we show also the result of assuming that the pressure is only one-tenth nkT. This has the effect of increasing both the deduced conductivity and density, so the measurements still fall close to the theory curves. The temperature measurement is also affected slightly by this change of assumed pressure.

The major source of uncertainty in this work is the lack of a good measurement of the capillary inner diameter. The diameter may be observed directly by means of streak photography, and will be measured as the next step in this study. This technique should then be capable of allowing measurement of conductivity over a wide range of densities and temperatures.

Acknowledgement

We are indebted to R. Durand and E. Schmieder for help with the data analysis. This work was supported by the Deutsche Forschungsgemeinschaft.

204

References

[1] K. S. Fansler and D. D. Shear, in _Exploding Wires_, edited by W. G. Chace and H. K. Moore, V.4, p185.

[2] C. P. Nash, R. P. DeSieno and C. W. Olsen, in _Exploding Wires_, edited by W. G. Chace and H. K. Moore, V.3, p231.

[3] R. R. Buntzen, in _Exploding Wires_, edited by W. G. Chace and H. K. Moore, V.2, p 195.

[4] J. A. Kersavage, in _Exploding Wires_, edited by W. G. Chace and H. K. Moore, V.2, p225.

[5] I. M. Fyfe and R. R. Ensminger, in _Exploding Wires_, edited by W. G. Chace and H. K. Moore, V.3, p257. See also F. D. Bennett, Phys. Fluids 5, 102 (1961).

[6] S. G. Barol'skii, N. V. Ermokhin, P. P. Kulik, and V. A. Ryabyi, High Temperature 14, 626 (1976).

[7] P. P. Kulik, E. K. Rozanov and V. A. Ryabyi, High Temperature 15, 349 (1977).

[8] V. M. Adamyan, G. A. Gulyi, N. L. Pushek, P. D. Starchik, I. M. Tkachenko, and I. S. Shvets, High Temperature 18, 186 (1980).

[9] R. Shepherd, D. Kania, L. A. Jones, M. Maestas, D. Nothwing, and K. Stetler, Bull. Am. Phys. Soc. 32, 1818 (1987).

[10] W. G. Chace, M. A. Levine, and C. V. Fish, in _Exploding Wires_, edited by W. G. Chace and H. K. Moore, V.4, p51.

[11] H. Knoepfel and R. Luppi, in _Exploding Wires_, edited by W. G. Chace and H. K. Moore, V.4, p233.

[12] W. Muller, in _Exploding Wires_, edited by W. G. Chace and H. K. Moore, V.1, p186.

[13] P. J. A. Fuller and J. H. Price, Brit. J. Appl. Phys. 15, 751 (1964).

[14] J. Wackerle, J. Appl. Phys. 33, 922 (1962).

[15] Ya. B. Zel'dovich and Yu. P. Raizer, _Physics of Shock Waves and High-Temperature Hydrodynamic Phenomena_, Vol II, (Academic Press, New York, 1967), p762.

[16] W. Ebeling, A. Förster, V. E. Fortov, V. K. Gryaznov and A. Ya. Polishchuk, _Thermophysical Properties of Hot Dense Plasmas_, Band 25, (B. G. Teubner Verlagsgesellschaft, Stuttgart, 1991), p274.

RECENT DEVELOPMENT OF EXPERIMENTS ON NON-IDEAL PLASMAS WITH THE KIEL BALLISTIC COMPRESSOR

H.J. Kusch
Institute for Experimental Physics
University of Kiel
Olshausenstr. 40, W-2300 Kiel, Germany

Abstract

Recent results of research work on high pressure plasmas obtained by the Kiel ballistic compressor are reported: 1. determination of electrical conductivity, 2. environmental effects on Hall sensor measurements for the piston approach, 3. homogeneity checks of the compressed plasma near the wall, 4. determination of the equation of state from sound velocity measurements, and 5. line broadening studies in highly compressed rare gas and hydrogen-atom plasmas.

Introduction: The Experimental Device

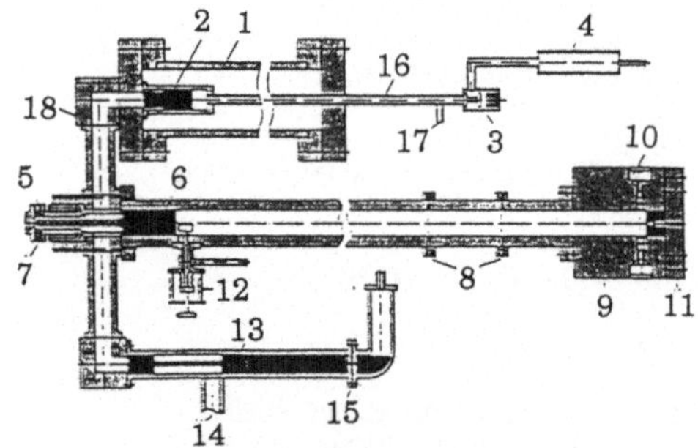

Highly compressed plasmas are produced in a ballistic compressor; the technical data are:
Driver gas: 68,484 cm^3, 200 bar max
Tube: diameter: 8,55 cm
       length: 1,150 cm
Piston: 8,157 g
Maximum plasma conditions:
10,000 bar, 15,000 K

Fig. 1 The Kiel Ballistic Compressor

(cross-sectional view).

For details see [1].

The driver gas in 1 is released by a stepped piston device 2 operated by an electromagnetic valve 3 and a vacuum tank 4. The gas accelerates the piston 6 towards the test gas section 9, 10, 11. The piston motion is reduced to one stroke only by an oil-hydraulic gas exhaust device 13,14,15. For further details see [1].

Determination of the electric conductivity of highly compressed plasmas

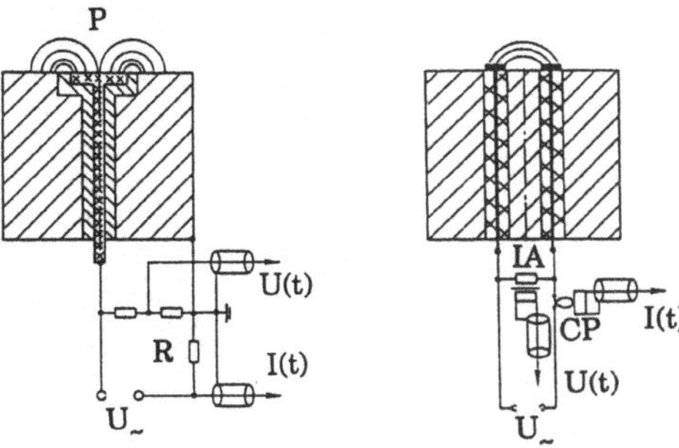

Fig. 2 Electrical conductance of high pressure plasmas
obtained by ballistic gas compression in electrode
configurations. Left: Central electrode towards
grounded endplug with electric circuitry (grounded).
Right: two electrodes (insulated to ground) with
electric circuitry (ungrounded). IA: isolation amplifier
CP: current probe.

a) Electrical conductivities from electrode configurations - either grounded or ungrounded-. Plasma conductance from simultaneous measurements of I (t) and voltage drop U (t), DC and AC $\leq$ 100 kc/s, Fig. 2. Field geometry factors from model measurements with electrolytes, mercury, or - with less accuracy - mixtures of quartz powder with graphite or copper powder. For current I between electrodes 1,2 is:

$$I = \int_F \vec{j} \, d\vec{f} = \sigma \int_F \vec{E} d\vec{f} \quad (1)$$

$$= \sigma U \int_F \nabla\varphi(1) \, d\vec{f}, \quad (2)$$

$$\Psi = \int_F \nabla\varphi(1) \, d\vec{f} \quad (3)$$

is a geometry factor (for unit voltage drop). In modelling with σ_0 is:

$$\Psi = \frac{1}{\sigma_o} (I/U)_o \quad (4)$$

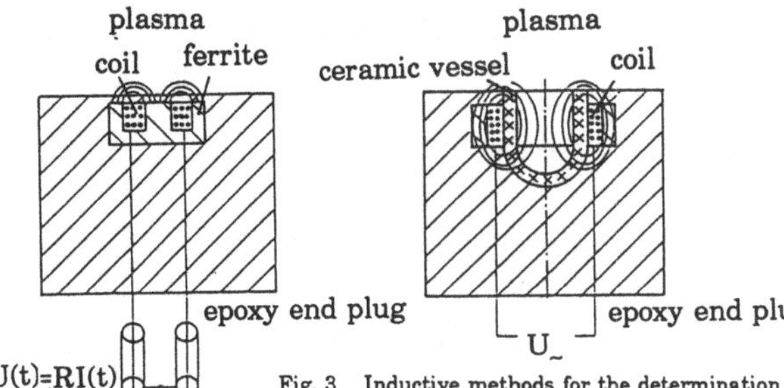

Fig. 3 Inductive methods for the determination of
the electrical conductivity of high pressure
plasmas. Left: ferrite pot core in epoxy endplug with electrical HF-circuitry. Right:
Cylindric coil configuration with plasma
inside causing dissipation of HF-field energy.

and for the plasma follows:

$$\sigma_{pl} = \frac{1}{\psi} (I/U)_{pl} \, . \quad (5)$$

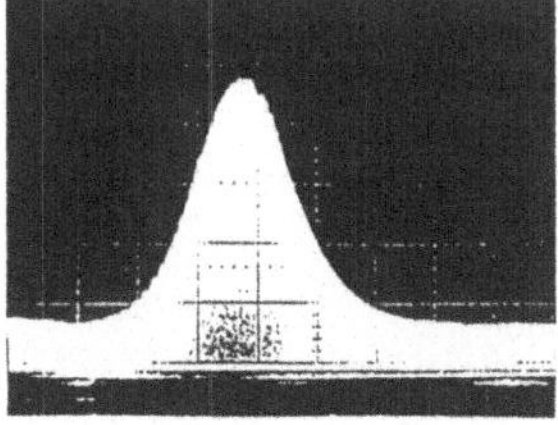

Fig. 4 Current I(t) during
compression phase due
to HF-energy dissipation.
Frequency: 2,5 Mc/s,
500 μsec/unit.

b) Inductive methods:
1. pot core in epoxy end plug with $v < v_{res}$ increase of coil current due to neighboring conductance. Calibration by salt solutions, mercury, powder mixtures. Experimental result: conductivity of the plasma too high. With test gas helium and simultaneous electrode configuration measurements

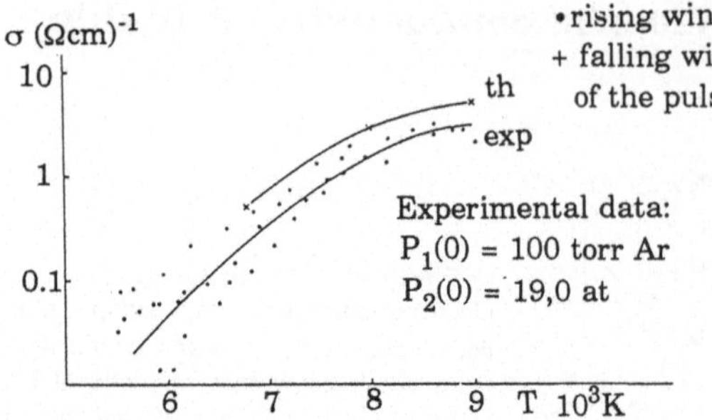

Fig. 5 Experimental values for the electrical conductivity of high
pressure argon plasmas vs. T compared with theoretical values
based on the Frost/Phelps cross-section Q_{eo} (T)

$\rightarrow$ piezomagnetic effect χ (p).σ_{pl}
= signal Ar (p) - signal He (p)
or correction by statical hydraulic
values.
Improvement: nickel ferrite
1) $\rightarrow$ powder iron: χ (p) $\approx$ 0
small μ $\rightarrow$ small σ-signal
2) $\rightarrow$ nickel free ferrite: χ (p) $\approx$
0 large μ $\rightarrow$ large σ-signal
3) $\rightarrow$ vitreous ferrite: χ (p) $\approx$ 0
large μ $\rightarrow$ large σ-signal
4) $\rightarrow$ ferrite-less coil.

Piston motion determination in ballistic gas compression

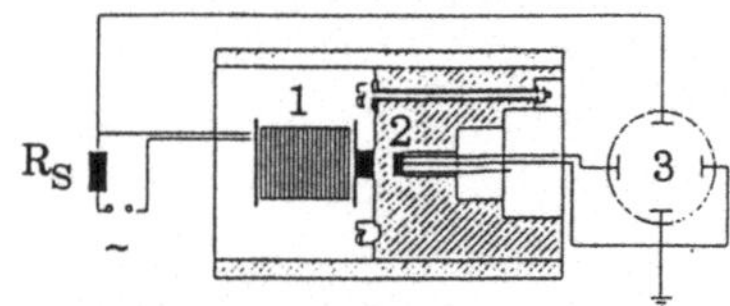

Fig. 6 Monofrequency determination of Hall voltage
yielding damping and phase shift.
1. coil powered from sine-wave generator via
low inductance shunt R_s; 2: Hall sensor.

From the piston motion x (t) thermodynamic
parameters of the test gas plasma can be
obtained by the equation of motion.
Maximal compression x_{min}:
1.   telescope sensor   [2]
2.   copper rod deformation   [3]
3.   voltage maximum of Hall sensors [4,5]

Discrepancy: statical calibration $\leftrightarrow$ dynamic Hall
voltage

Piston position x (t):
1.   eddy current sensor   [6]
2.   light gap methods   [3]
3.   Hall sensoric methods   [4,5]

a) Monofrequent determination of Hall voltage
Fourier components, Fig. 6.

b) Piston approach vs. time from copper pin
measurements, Fig. 7.

Dynamic calibration:

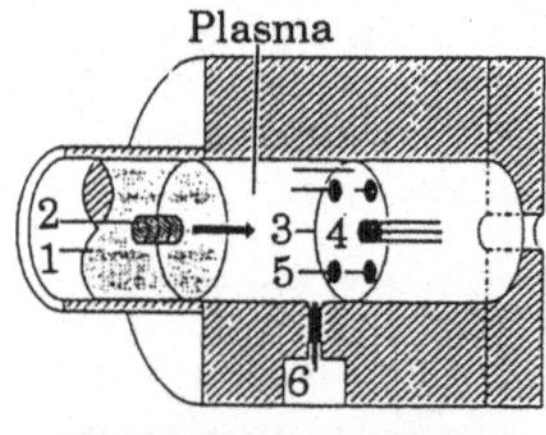

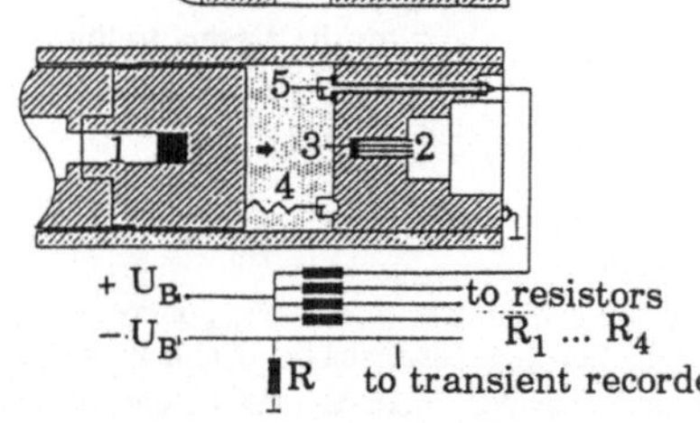

Fig. 7 Determination of the piston approach
vs. time using the contact between
piston front and copper pins and of
the Hall voltage vs. time.

$$H(\omega) = A(\omega) \, e^{j\varphi(\omega)} \rightarrow B(j\omega) = \int_{-\infty}^{-\infty} B(t) \, e^{j\omega t} dt \qquad (6)$$

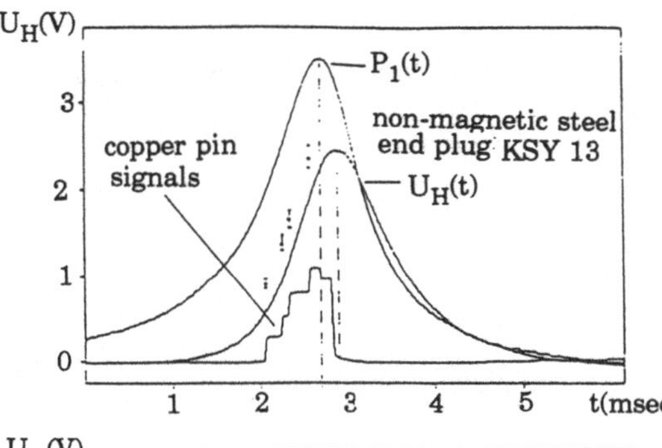

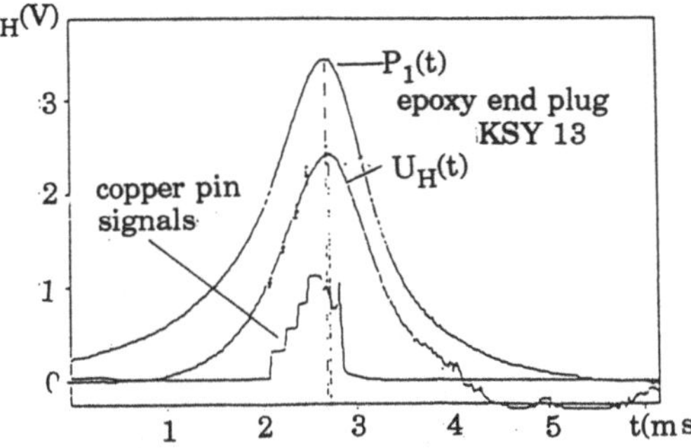

Fig. 8 The pressure pulse $P_1(t)$, the Hall voltage $U_H(t)$ and the electrical signals due to contact between copper pins and piston front. Upper picture: Hall sensor surrounded by non-magnetic steel. Lower picture: Hall sensor surrounded by epoxy-quartz. The Hall values ꞌ are taken from static calibration.

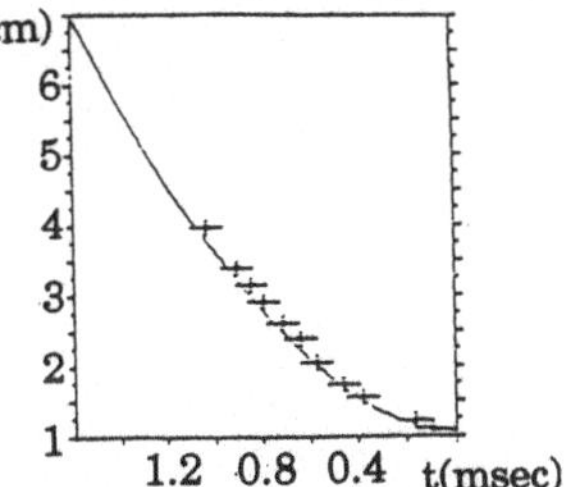

Dynamic calibration --- exp. values

Fig. 9 The piston approach x(t) vs. t. Solid line: Fourier transform results from Hall voltage $U_H(t)$. : experimental values by copper pin contact.

Change due to conductive environment, thickness d:

$$A(j\omega) = e^{\alpha d}e^{-j\beta} \qquad (7)$$

$$\rightarrow B_d(j\omega) = B(j\omega)\,A(j\omega) \qquad (8)$$

$$\rightarrow B_d(t) = \frac{1}{2\pi}\int_{-\infty}^{+\infty} B_d(j\omega)e^{j\omega t}dt \qquad (9)$$

α damping, β phase shift from measurements Fig.6.

Fig. 9 shows agreement between Fourier transform results and dynamic values.

c) Calibration with rectangular magnetic pulses.

$$u_H(t) = s(t) \text{ response } S(t) \qquad (10)$$

$$S(t) = \frac{ds}{dt} \rightarrow H(\omega) = \int_{-\infty}^{+\infty} S(t)e^{j\omega t}dt \qquad (11)$$

1. Modulation of the magnetic field using a rotating sector, Fig. 10. Results Fig. 11a, 11b.

2. Rectangular magnetic pulses discharging a LC-condensor bank through a coil. High frequency results obtained by this method are shown in Fig. 13a, 13b.

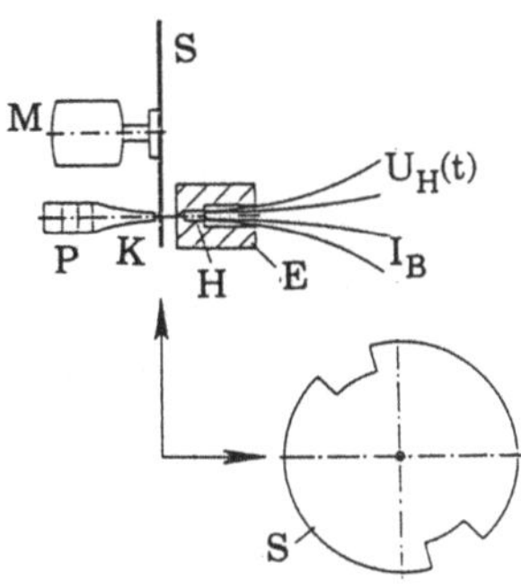

Fig. 10 Production of rectangular magnetic pulses by a rotating sector S (iron). M: motor 25/50 c/s, P: CoSm-permanent magnets, K: soft iron cone, H: Hall sensor in endplug E.

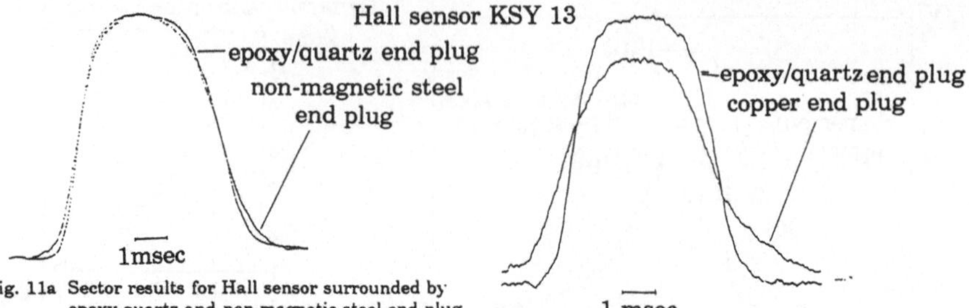

Fig. 11a Sector results for Hall sensor surrounded by epoxy-quartz and non-magnetic steel end plug.

Fig. 11b Sector results for Hall sensor surrounded by epoxy-quartz and copper end plug.

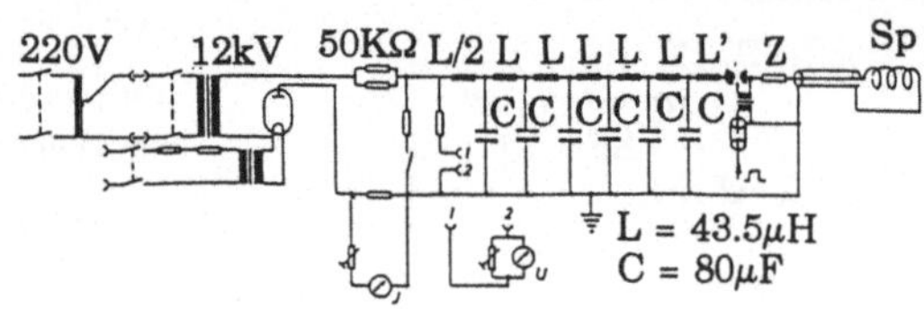

Fig. 12 LC-capacitor bank producing rectangular pulses of the magnetic field strength if discharged through coil Sp. Z is the characteristic impedance (carbon rods).

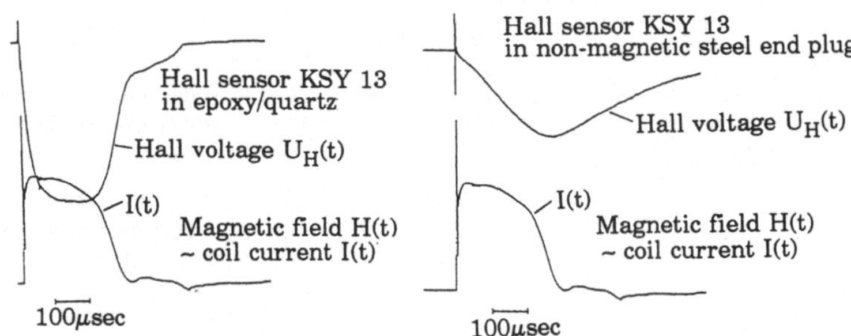

Fig. 13 a, b. Hall sensor excited by rectangular magnetic field. a. (left): lower trace - magnetic field ~ coil current. upper trace - epoxy-quartz end plug b. (right): lower trace - magnetic field ~ coil current. Upper trace - non-magnetic steel end plug.

Boundary layer measurements of the compressed plasmas

In the theoretical treatment of adiabatic gas compression dissipative processes, as gas friction and gas losses, are taken into account, and with rising compression values heat losses due to conduction, convection and radiation cannot be neglected. These processes give rise to temperature gradients towards the wall, which influence the evaluation of plasma parameters or atomic data, where mostly the plasma column is considered as an homogeneous layer. Boundary layers are studied by optical - spectroscopic methods.

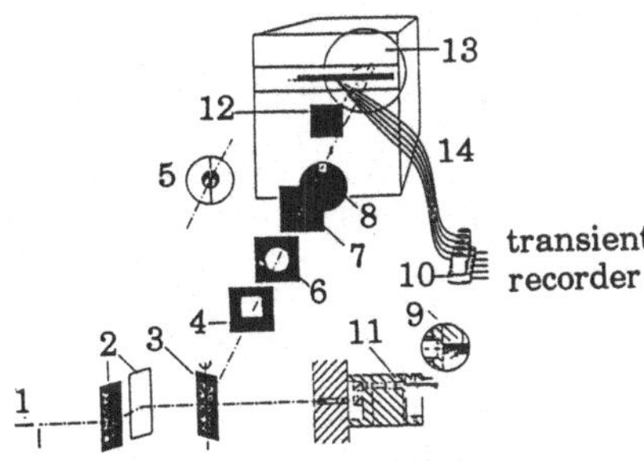

Fig. 14 Spectroscopic device for plasma emission parallel to the wall. 1: calibration carbon arc, 2: deflecting mirrors, 3: turnable mirror, 4: filter, 5: entrance slit, 6: focussing lens, 7: pin hole aperture, 8: rotating sector for calibration light pulses, 9: cross-section of the end plug near the wall, 10: photo-multipliers, 11: end plug, 12: plane grating, 13: focal plane of the spectrometer, 14: 11-channel light fiber system.

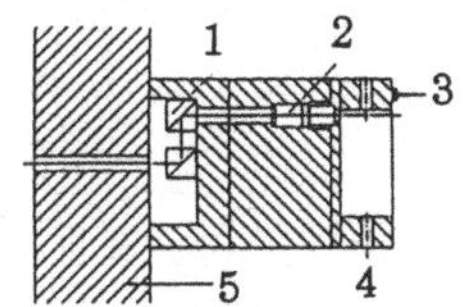

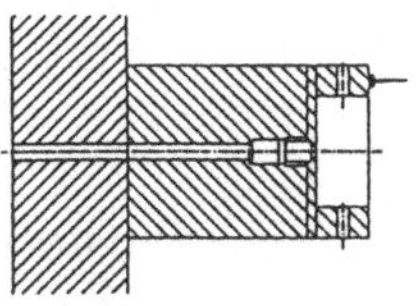

Fig. 15 End plugs for plasma emission parallel to the wall (upper picture) and parallel to the axis.
1: deflecting mirrors,
2: sapphire-quartz-windows,
3: copper pin for X_{min},
4: radial bores for pressure and side-on radiation measurements.

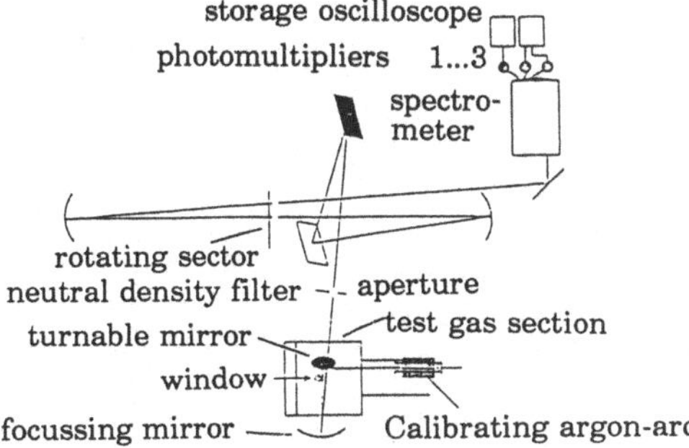

Fig. 16 Side-on spectroscopy of high pressure plasmas. U: turnable mirror for light flux from the plasma and the calibration light source (argon cascade arc), respectively.

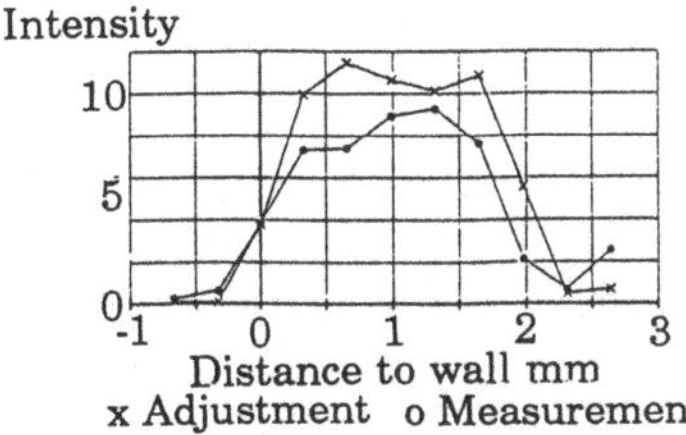

Fig. 17 Intensity distribution vs. distance to the wall. Adjustment: light source inside the compressor tube. Measurement: light emission during compressor operation. Note: no change of the outer light fiber channels due to perturbations by shocks.

The plasma emission parallel to a cylindrical wall was recorded using a 11-channel light pipe device (Fig. 14). The parallel light beams from the end plugs (Fig. 15) are imaged to a pin hole aperture 7 by a lens 6; image ratios can be varied by different focal lengths. Absolute intensities at 5400 Å are determined using the anode of a carbon arc 1 as a standard, accounding for absorption losses of the mirrors in the end plug by the additional mirrors 2. The PM-signals were stored on a 12-channel transient recorder. Side-on spectra (Fig. 16) for the determination of plasma temperatures were recorded on three channels of a multichannel-spectrometer. The total equipment was mounted on a vibration-isolated desk to secure the optical adjustment before the shot (laser beam from the rear side of the compressor or lamp near to the end plug), see Fig. 17. The optical layer was determined by a Hall sensor monitored by the copper rod (3 in Fig. 15). From the continuous

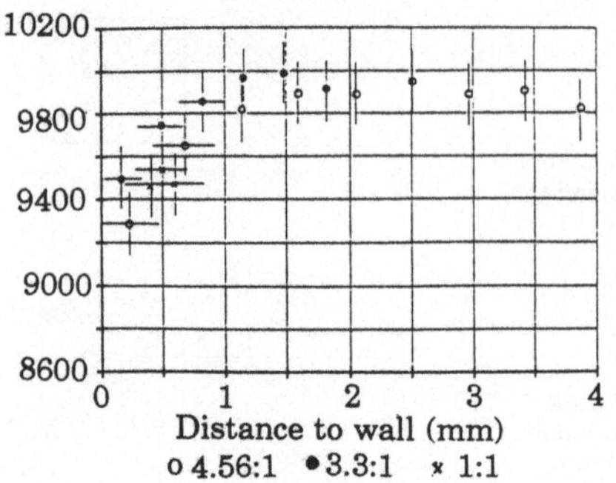

Fig. 18 Temperature distribution of highly compressed argon plasma (maximum compression) vs. distance from the wall, obtained from shots with different ratios of imaging.

emissison of the plasma calibrated by an argon-arc and pressure values from a KISTLER transducer temperature profiles near the wall were obtained, see Fig. 18. The influence of Fresnel diffraction of the convergent light beam to the focus pin hole on the resolution was studied by Kichhoff's integral, and of deviations from straight line light propagation due to refractive index gradients by Fermat's principle and an approximative solution of the Euler-Lagrange equation; both influences are less than one light fiber channel. Energy losses from the compressed gas (either argon or helium with 5% argon) were studied on the basis of: 1) thermal conduction, 2) convective heat losses, 3) radiative losses, estimating boundary layer values.

Equation of state obtained from the velocity of sound in highly compressed plasmas

Discontinuities on the rising wing of pressure pulses during ballistic gas compression caused by sound wave propagation between piston front and end plug can be used to determine sound velocity values S at different plasma parameters; thus the equation of state can be easily obtained. From general theory follows:

$$S^2 = (K_s nm)^{-1} \tag{12}$$

$$\text{with} \quad K_s^{-1} = -V\left(\frac{\partial P}{\partial V}\right)_T \tag{13}$$

inverse compressibility
In the Debye approximation is:

$$K_s^{-1} = kT\left(n - \frac{\kappa_D^3}{16\,\pi}\right) \tag{14}$$

where κ_D is the reciprocal Debye length.

Finally results:

$$p = -\int S^2\, nm\, \frac{dV}{V} \tag{15}$$

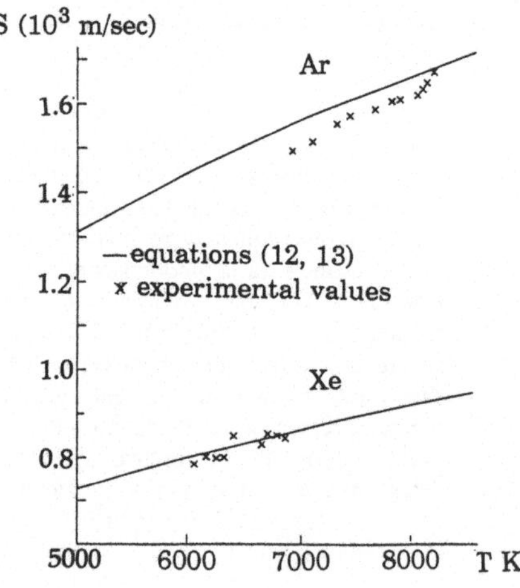

Fig. 19 The velocity of sound for argon and xenon plasmas vs. T(K): comparison between experimental values and theoretical estimations.

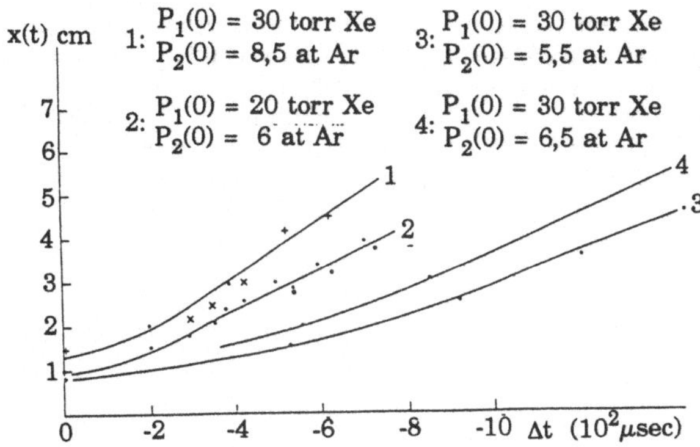

Fig. 20 The piston approach x(t) for different compressor operation parameters, obtained from copper pin- and Hall-sensor-measurements.

For argon and xenon as test gas $S(t) = S(p,T)$ is determined from the steps on the rising pressure pulse wing, the piston approach x(t) (Fig. 20) from Hall sensoric or galvano-contact data (copper pin - piston front), and the number density n (p,T) from the plasma composition. Plasma temperatures were obtained from continuous emission measurements at 3 UV-wave lengths and pressure values from piezoelectric transducer signals.

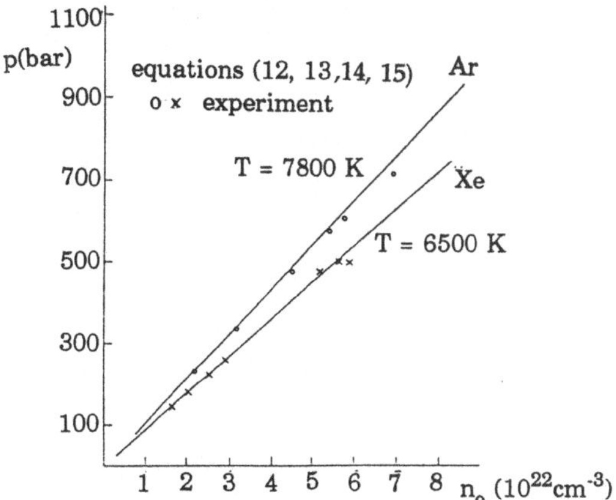

Fig. 21 The equation of state for argon and xenon high pressure plasmas. Solid line: theoretical calculations, o, x: experimental values from sound speed determination.

Fig. 21 shows the equation of state $p(n_0, T)$, which in the regime of small ionization degrees and medium pressures, i.e. small Debye and neutral interaction correction, is the straight line of ideal gases. Measurements at enhanced plasma conditions approaching non-ideal plasma states are in progress.

Ballistic compressors operated at moderate conditions have been often used to determine line broadening paramters, mostly line shifts caused by foreign atom impacts [7].

Spectroscopic studies with the Kiel Ballistic Compressor

1. Preliminary time-resolved spectra of argon and xenon as test gas were obtained by a drum-camera equipped spectrograph. The rare gas lines are strongly broadened and shifted in the early phases of compression by resonance and van der Waals interaction, and with increasing electron density the broadening and shift by charged particles is the dominant effect. Plasma temperatures were obtained by side-on spectroscopy (see Fig. 16) and the perturber number densities calculated from the plasma composition. Improvement shall be made using a vidicon-multichannel device for spectra recording and interferometrie techniques for the number densities determination.

2. Impurity metal lines (either added as a compund to the test gas or caused by evaporation of the walls) are broadened and shifted by the test gas atoms (argon, helium, xenon). In case of using hydrogen as test gas, van der Waal interaction with hydrogen atoms can be studied.Preliminary time-resolved spectra were exposed; photoelectric multichannel recording is in preparation too, as well as the interferometric procedure for atom density determination. Plasma temperatures were obtained from x(t) and p(t) assuming adiabatic compression; from the entropy conservation:

$$S_{H_2}[p_1(0),\ T_1(0)] = S_H[p_1(t),\ T_1(t)] + S_{H_2}[p_1(t),\ T_1(t)] \tag{16}$$

follows $T_1(t)$ and furthermore n_H and n_{H_2}

References:

[1] H.J. Kusch, Journ.Phys. E, 18,654 (1985)

[2] G.L. Hammond, G.T. Lalos, NOLTR 71-228 (White Oak, Silver Springs,Md,USA)

[3] H. Kasimir, H.J. Kusch, Journ.Phys. D, 19, 513 (1986)

[4] H.J. Kusch, 5th Internat. Workshop on Nonideal Plasmas, Wustrow (1988)

[5] M. Wirsig, H. Schneidenbach, Contrib. Plasma Phys. 29, 545 (1989)

[6] M.J. Lewis, B.P. Roman, G. Rouel, B.E. Richards
     Karman Inst. Fluid Dynamics, preprint 71-7 (1971) Rhode-St.Genese(Belgium)

[7] Shang Yi Chen, P.K. Henry, Journ. Quant. Spectr. Rad. Transfer, 13, 385 (1973)

ELECTRIC FIELD DYNAMICS AT A CHARGED POINT
STRONG COUPLING LIMIT

JAMES W. DUFTY[1]

Department of Physics, University of Florida

Gainesville, FL 32611, USA

Abstract

The dynamics of electric fields at a charge Q_o imbedded in a one component plasma of charges Q is considered for conditions of large Q_o/Q and large plasma parameter Γ. The complete response of the charge Q_o to the surrounding plasma is formulated in terms of path integrals over the probability density functional for the electric field histories. Two applications of this formulation illustrating problems of both transport and plasma spectroscopy are noted. A set of Langevin equations for the position and velocity of the charge Q_o, and the total electric field at the charge are studied in the large charge, strong coupling limit. The analysis suggests a Gaussian, non-Markovian process for these variables. The resulting joint distribution of fields at two times and the generating functional for fields at many times are given. Finally, it is proposed that the set of observables can be extended slightly to yield an equivalent Gaussian-Markov process.

INTRODUCTION

An ion of charge Q_o and mass m_o is placed in a one component plasma (OCP) of ions with charge and mass, Q and m. The entire system is taken to be charge neutral (uniform background) and at equilibrium. The objective here is to develop a complete description of the effects of the OCP on the imbedded charge. This objective encompasses static and dynamic effects, linear and nonlinear response, radiative and transport properties. Clearly such a program is too ambitious for progress beyond the formal level in general. However, it is noted that significant simplifications occur for conditions of strong coupling and large Q_o/Q. These simplifications lead to a tractable stochastic model that meets the stated objective.

Most theoretical studies of tagged particle motion in a many body system are based on kinetic theory (e.g., time correlation function/Green's function methods). While such methods are highly developed for linear response and one- or two-particle properties, they are not easily implemented for nonlinear coupling effects that involve many-particle or many-time properties. For such problems a quite different point of view is considered[1,2] that focuses directly on the dynamics of a few

[1]Research supported by National Science Foundation grant PHY-8822581

relevant many-particle properties, rather than a detailed confrontation of the complete underlying microscopic N+1 particle dynamics.   In the present case that property is the total electric field of the OCP at the ion, which represents the entire coupling of the ion to its environment.   The formulation described below is exact, but its value lies primarily in suggesting approximations appropriate for complex problems.

The center of mass degrees of freedom obey Newton's equations,

$$\frac{\partial \vec{r}(t)}{\partial t} = \vec{v}(t) \quad , \quad \frac{\partial \vec{v}(t)}{\partial t} = (Q_o/m_o)\vec{E}(t) \tag{1.1}$$

where $\vec{r}(t)$ and $\vec{v}(t)$ are the position and velocity of the ion, and the microscopic electric field is given by,

$$\vec{E} = \sum_{i=1}^{N} \vec{e}(\vec{q}_i - \vec{r}) + \vec{E}_b \tag{1.2}$$

Here, $\vec{e}(\vec{q}_i - \vec{r})$ is the Coulomb electric field at the ion due to the ith particle, and $\vec{E}_b$ is the field of the uniform neutralizing background charge.   If there is internal electronic structure to this ion, the main coupling of the internal coordiantes to the OCP is through a dipole interaction, $\vec{E}(t) \cdot \vec{d}$.   The electric field is therefore the only relevant plasma variable determining the effects of the plasma on the ion for both its radiative and transport properties.   To solve Eqs. (1.1) it is necessary to know the field at all times over some time interval of interest, $0 \leq t \leq T$. Every allowed function, $\vec{\varepsilon}(t)$, specified over this interval will be referred to as a field "history".   The approach taken here is first to select a possible field history for given inital conditions, calculate the property of interest, and assign a probability for the chosen history.   A final summation over all such choices provides the average value for the property considered.   In this approach, the central unknown is the probability density functional for the electric field histories.   A formal definition is obtained in terms of the corresponding joint probability density for the field to have specified values at M successive times,

$$P_M[\varepsilon;T] \equiv < \prod_{p=0}^{M} \delta(\vec{\varepsilon}(t_p) - \vec{E}(t_p)) > \quad , \quad t_p = pT/M \tag{1.3}$$

$$P[\varepsilon;T] \equiv \lim_{M \to \infty} P_M[\varepsilon;T] \tag{1.4}$$

The brackets in (1.3) denote an equilibrium ensemble average, and in the following we choose $t_{i+1} > t_i = 0$.   The average of a functional of the field history, $F[\varepsilon;T]$, is then given by,

$$<F> = \lim_{M \to \infty} <F_M> \equiv \int \mathcal{D}[\varepsilon] \, P[\varepsilon;T] \, F[\varepsilon;T] \tag{1.5}$$

where $\mathcal{D}[\varepsilon]$ denotes functional integration. This equation defines the average as a functional "path" integration over all histories, weighted by the probability density P. Virtually all properties of interest regarding the ion can be cast in this form: the microscopic property is calculated for a given field, followed by a weighted path integral over all possible choices.

Two Examples

Two examples are given to show the utility of such a formulation.[3] The first is the generating functional for time correlation functions of the ion's displacement,

$$F[\lambda; T] \equiv \left\langle \exp\left(i \int_o^T dt \; \vec{\lambda}(t) \cdot [\vec{r}(t) - \vec{r}(0)] \right) \right\rangle \tag{2.1}$$

The correlation functions are obtained from $F[\lambda; T]$ by functional differentiation,

$$\left\langle \prod_{p=1}^{M} [\vec{r}(t_p) - \vec{r}(0)] \right\rangle = \prod_{p=1}^{M} \frac{\delta}{\delta i \lambda(t_p)} F[\lambda; T] \Bigg|_{\lambda=0} \tag{2.2}$$

or, alternatively, the associated cumulants are given by,

$$\left\langle \prod_{p=1}^{M} [\vec{r}(t_p) - \vec{r}(0)] \right\rangle_c = \prod_{p=1}^{M} \frac{\delta}{\delta i \lambda(t_p)} \ell n \left(F[\lambda; T] \right) \Bigg|_{\lambda=0} \tag{2.3}$$

In addition, the correlation functions involving the ion velocity can be obtained from suitable time derivatives of (2.2). The representation of (2.1) in terms of electric field histories follows directly from integration of Eqs. (1.1),

$$F[\lambda; T] = \int d\vec{v} \; \phi(v) \int \mathcal{D}[\varepsilon] \; W[\varepsilon; v, T] \exp\left(i \int_o^T dt \vec{\lambda}(t) \cdot \left[\vec{v} t \right. \right.$$

$$\left. \left. + (Q_o/m_o) \int_o^t d\tau \; (t-\tau)\vec{\varepsilon}(\tau) \right] \right) \tag{2.4}$$

Here, $\phi(v)$ is the usual Maxwell-Boltzmann distribution and $W[\varepsilon; v, T]$ is the conditional probability for electric field histories, given the initial velocity $\vec{v}$, (analogous to (1.3)),

$$W[\varepsilon; v, T] \equiv \left\langle \prod_{p=0}^{M} \delta(\vec{\varepsilon}(t_p) - \vec{E}(t_p)) \; \delta(\vec{v} - \vec{v}(0) \right\rangle / \phi(v) \tag{2.5}$$

As expected, the probability density for electric field histories completely determines all aspects of the center of mass motion for the ion.

As a second example, consider an ion whose internal degrees of freedom are characterized by a dipole moment operator, $\vec{d}$. Emission and absorption spectra are broadened both by the ion's center of mass motion (Doppler broadening) and the coupling of the dipole moment to the plasma electric field (Stark broadening). These line shapes are determined from the Fourier transform of the dipole auto correlation function,

$$C(t) = < \vec{D}(t) \cdot \vec{d}(0) \ e^{i\vec{k} \cdot [\vec{r}(t) - \vec{r}(0)]} > \qquad (2.6)$$

$$-i\hbar\frac{\partial}{\partial t}\vec{D}(t) = \left[H_a, \vec{D}(t)\right] + \left[\vec{E}(t) \cdot \vec{d}, \vec{D}(t)\right] \quad , \quad \vec{D}(0) = \vec{d} \qquad (2.7)$$

Clearly, for given $\vec{E}(t)$ all effects of the plasma on the line shape can be calculated explicitly. Introducing the conditional probability functional the dipole autocorrelation function can be written,

$$C(t) = \int d\vec{v}\phi(v) \int \mathcal{D}[\varepsilon] \ W[\varepsilon; v, T] \ C[\varepsilon; v, t] \ S[\varepsilon; t] \qquad (2.8)$$

The functions $C[\varepsilon; t]$ and $S[\varepsilon; t]$ represent the Doppler and Stark broadening effects, respectively, for an atom in a specified field, $\vec{\varepsilon}(t)$,

$$C[\varepsilon; t] \equiv \exp\left(i\vec{k} \cdot \left[\vec{v}t + (Q_o/m_o) \int_o^t d\tau \ (t-\tau)\vec{\varepsilon}(\tau)\right]\right) \qquad (2.9)$$

$$S[\varepsilon; t] \equiv \mathrm{Tr}_{int} \ \rho_{int} \ \vec{D}[\varepsilon; t] \cdot \vec{d}(0) \qquad (2.10)$$

The trace in (2.10) extends over the internal degrees of freedom for the atom, with density operator ρ_{int}. The dipole operator $\vec{D}[\varepsilon; t]$ obeys Eq. (2.7) with $\vec{E}(t)$ replaced by the specified field $\vec{\varepsilon}(t)$, so that the analysis is reduced to an atomic physics problem for each field history. Both Doppler and Stark broadening effects are determined by the same field, so the representation (2.6) enforces explicitly the self-consistency in their calculation.

The calculation of line shapes in this representation then is carried out as follows. An initial velocity and field history are chosen; the Doppler and Stark broadening effects are calculated for these fixed conditions (an atomic physics problem); the process is repeated with different choices of $\vec{v}$ and $\vec{\varepsilon}(\tau)$ selected from the distributions, $\phi(v)$ and $W[\varepsilon; v, t]$. This formulation of the problem is essentially the same as that used in current methods for molecular dynamics simulation of spectral line shapes. In these methods, the field histories are generated by computer simulations of the OCP and used as input to the atomic physics problem (2.7). The sampling of field histories from the molecular dynamics data is effectively equivalent to reconstruction of the probability density functional.

218

The explicit calculation of the probability functional $P[\varepsilon;t]$ or the corresponding conditional functional $W[\varepsilon;v,t]$ is clearly a very difficult problem. To describe the limiting case for which simplifications are expected, a formal Langevin description is first introduced. These are exact equations for the position, velocity, and electric field that can be obtained by the projection operator methods of Mori and Zwanzig.[4] The details of their derivation are not important for the considerations here, and only the results are quoted. Also, in the following we use units such that the time is measured in units ω_p^{-1} and distance in units of r_o, where $\omega_p^2 = 4\pi n Q^2/m$ and $4\pi n r_o^3/3 = 1$. The resulting dimensionless Langevin equations are,

$$\frac{\partial}{\partial t}\vec{r}(t) = \vec{v}(t) \quad , \quad \frac{\partial}{\partial t}\vec{v}(t) = (Q_o m/3Qm_o)\,\vec{E}(t) \tag{3.1}$$

$$\frac{\partial}{\partial t}\vec{E}(t) + (Q_o/Q)\,\vec{v}(t) + \int_o^t d\tau\,v(t-\tau)\,\vec{E}(\tau) = \vec{f}(t) \tag{3.2}$$

Equations (3.1) simply restate the kinematic relation of $\vec{r}$ and $\vec{v}$, and Newton's second law. The many-body difficulties are contained in (3.2) in the non-local relaxation kernel $v(t)$ and the source term $\vec{f}(t)$. These two quantities are constrained by the exact conditions,

$$\langle\, f_i(t)E_j(0)\,\rangle = 0 \quad , \quad \langle\, f_i(t)f_j(0)\,\rangle = \delta_{ij}\,v(t)\,(Q/Q_o\Gamma) \tag{3.3}$$

$$v(0) = 4(Q_o m/Qm_o)\int_o^\infty dr\,r^{-4}g(r) \tag{3.4}$$

The Langevin equations are a contraction of the dynamics of a many-body system to that for a few degrees of freedom, $\vec{r}$, $\vec{v}$, $\vec{E}$. Of course, it is not possible to have a closed description in terms of this contracted set so the effects of the remaining degrees of freedom are contained in the source term, $\vec{f}(t)$. If this term were absent, there would be a unique "deterministic" solution, $\vec{E}_d(t)$, for given initial conditions,

$$\vec{E}_d(t) = \alpha(t)\,\vec{E}(0) \tag{3.5}$$

$$\left(\frac{\partial^2}{\partial t^2} + \sigma^2\right)\alpha(t) + \int_o^t d\tau\,v(t-\tau)\alpha(\tau) = 0 \quad , \quad \sigma^2 \equiv (Q_o/Q)^2(m/3m_o) \tag{3.6}$$

More generally, $\vec{f}(t)$ generates a statistical distribution of solutions for given initial conditions (i.e., each $\vec{f}$ determines a different solution). The amplitude of the correlation function in (3.3) is a measure of the width of this distribution. If this distribution is not too broad, most solutions are close to $\vec{E}_d(t)$ and $\vec{f}(t)$

represents a residual source of "noise" in the system. Our first observation is that the amplitude for the noise decreases with decreasing $(Q/Q_o\Gamma)$. A simple estimate gives $\nu(0) \sim (m/m_o)(Q/Q_o)^2\Gamma^{-3}$ and consequently $\langle f^2 \rangle \sim (m/m_o)(Q/Q_o)^3\Gamma^{-4}$ (the dependence on Γ is somewhat weaker for $Q = Q_o$). This decrease in the noise amplitude is illustrated in Figure 1 for $Q = Q_o$. The conclusion is that the formal Langevin description becomes more relevant for strong coupling and/or large charge ratio. Even for weak noise, the distribution of field histories about $\vec{E}_d(t)$ depends in detail on the statistical properties of the noise, $\vec{f}(t)$.

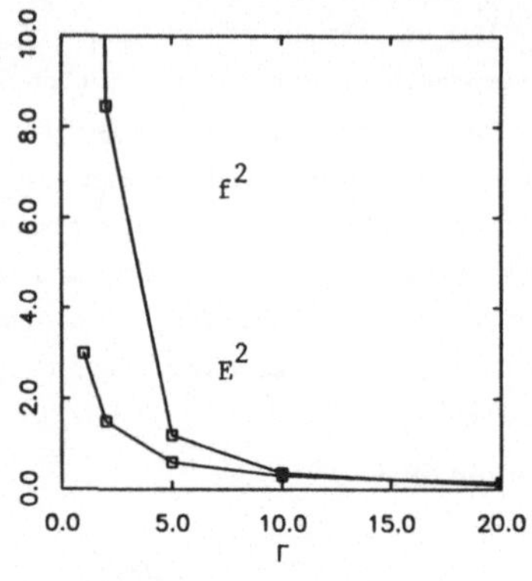

Figure 1: $\langle f^2 \rangle$ and $\langle E^2 \rangle$

Generally, this entails specification of all higher order fluctuations in $\vec{f}(t)$. An exception is the case of noise that is Gaussian distributed. In this case, the statistical features of $\vec{f}(t)$ are entirely determined by the co-variance, $\langle \vec{f}_i(t)\vec{f}_j \rangle$, which is known once $\nu(t)$ is specified (see Eq. (3.3)). The linearity of the Langevin equations then implies that the variables $\vec{r}(t)-r(0)$, $\vec{v}(t)$, and $\vec{E}(t)$ also have Gaussian distributions in this case. Finally, the equilibrium distributions of $\vec{v}$ and $\vec{E}$ must also be Gaussian.

The argument for Gaussian statistics at strong coupling and large Q_o/Q is qualitatively as follows. The large charge ratio creates an excluded region around the ion due to Coulomb repulsion. The amplitude for the field fluctuations is given exactly by $\langle E^2 \rangle = 3 \ (Q/Q_o\Gamma)$. For strong coupling and large charge ratio, these fluctuations are small. Consequently, the electric field is due to increasingly distant charges which can lead to only but small incremental changes in the field. These are the typical characteristics of a Gaussian process for which transitions occur only between nearby "states". This is in contrast to fields at a neutral point, where nearby perturbers can induce large changes in the field during a short encounter. These latter events are better described by a Kubo-Anderson or Kangaroo process, and are excluded here by the Coulomb repulsion. The linearity of the Langevin equations implies that Gaussian noise implies that the variables $\vec{r}$, $\vec{v}$, and $\vec{\varepsilon}$ have Gaussian distributions (for given initial values), and that the stationary distributions must be Gaussian. It is known from computer simulation that the joint probability density for $\vec{r}(t)-\vec{r}(0)$ is well approximated by a Gaussian at strong coupling.[5] The equilibrium distribution $\phi(v)$ is always Gaussian (classical mechanics) but in general the field distribution $Q(\varepsilon) \equiv \langle \delta(\vec{\varepsilon}-\vec{E}) \rangle$ is not, <u>except</u> for conditions of large $Q/Q_o\Gamma$. Figure 2 shows $P(\varepsilon) \propto \varepsilon^2 Q(\varepsilon)$ in the Gaussian limit and simulation results (---) for the marginal conditions of $Q = Q_o$ and $\Gamma = 10$.

It remains to determine the response function $\nu(t)$. This can be done by relating it to the electric field correlation functions[3],

$$< \vec{E}(t)\cdot\vec{E} > = \alpha(t) <E^2> \qquad (3.7)$$

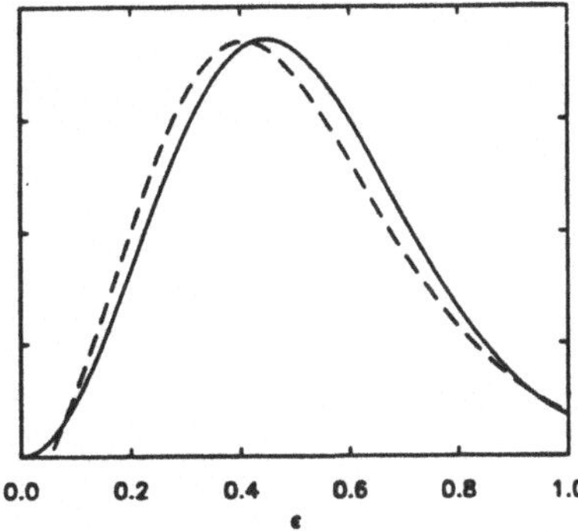

Figure 2: E field distribution

with $\alpha(t)$ given by (3.6). An approximate model for $\nu(t)$ can be obtained by requiring that it yield the first two time derivatives of the electric field autocorrelation function (equivalently, the first four time derivatives of the velocity autocorrelation function), and that it yield the correct self-diffusion coefficient (time integral of the velocity autocorrelation function). These requirements are met by the simple approximation[2],

$$\nu(t) = \nu(0) \; e^{-t/\tau} \quad , \quad \tau = D \; \Gamma/\nu(0) \qquad (3.8)$$

where $\nu(0)$ is given by Eq. (3.4). The self-diffusion coefficient D is assumed known from either theory or simulation. Figure 3 shows a comparison of the results from this model with computer simulation for the case of $Q = Q_o$ at $\Gamma = 1$, 3, and 5. Clearly, the agreement is quite good.

In summary, the discussion of this section suggests that at strong coupling and large charge ratio, the variables $\vec{r}$, $\vec{v}$, and $\vec{\varepsilon}$ can be described by the Langevin equations (3.1) - (3.4) with $\nu(t)$ given by (3.8) and a Gaussian distribution for $\vec{f}(t)$.

Gaussian non-Markovian process

The simplifications of the last section allow explicit calculation of the probability functional for electric field histories. Consider first the solution to the Langevin equations, which are written in matrix form as,

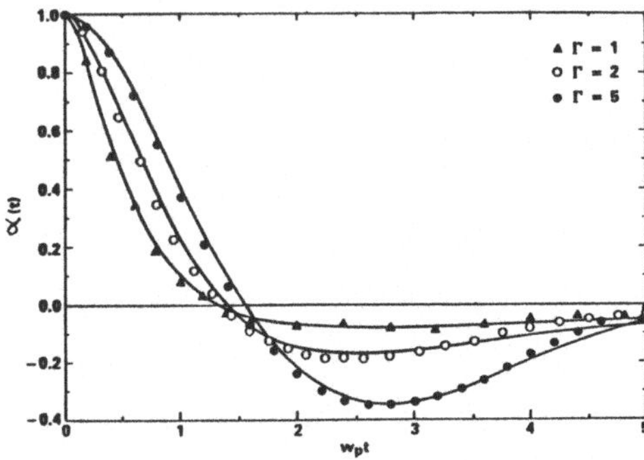

Figure 3: Field autocorrelation function

$$y(t) \leftrightarrow \left\{ \vec{r}(t), \vec{v}(t), \vec{E}(t) \right\} \quad , \quad \frac{\partial}{\partial t} y(t) + L y(t) = f(t) \tag{4.1}$$

$$y(t) = G(t) y(0) + \int_0^t d\tau \, G(t-\tau) \, f(\tau) \tag{4.2}$$

The simplest application of these results is to the calculation of the joint probability density for y at t and y_0 at 0,

$$P(y,t;y_0,0) \equiv \; < \delta(y-y(t))\delta(y_0-y(0)) \; >$$

$$= (2\pi)^{-9} \int d\lambda \; < \delta(y_0-y(0)) \exp\left[i\lambda \int_0^t d\tau \, G(t-\tau) \, f(\tau) \right] > \exp\left[i\lambda[y - G(t)y_0] \right]$$

Since the noise $f(\tau)$ is Gaussian and independent of y_0 the average is easily performed to give,

$$P(y,t;y_0,0) = P(y_0)(2\pi)^{-9} \int d\lambda \; \exp\left[i\lambda[y - G(t)y_0] - H(t)\lambda^2 \right] \tag{4.3}$$

where $P(y_0) = \; < \delta(y_0-y(0)) \; >$ is the distribution of initial values, and $H(t)$ is,

$$H(t) = \frac{1}{2} \int_0^t d\tau \int_0^t d\tau' \; G(t-\tau) \, <f(\tau)f(\tau')> \, G^T(t-\tau') \tag{4.4}$$

The autocorrelation function for $f(\tau)$ is given by (3.3) so that $H(t)$ can be calculated explicitly, and the Gaussian integral in (4.3) is easily performed. For example, the joint distribution for electric fields alone is found to be,

$$Q(\vec{\varepsilon},t;\vec{\varepsilon}_0,0) = Q(\varepsilon_0)\left[1-\alpha^2(t)\right]^{-3/2} Q\left(\frac{\vec{\varepsilon}-\alpha(t)\vec{\varepsilon}_0}{[1-\alpha^2(t)]^{1/2}}\right) \tag{4.5}$$

$$Q(\varepsilon) \equiv \left[3/2\pi<E^2>\right]^{3/2} \exp\left(- \frac{3\varepsilon^2}{2<E^2>}\right) \tag{4.6}$$

and $\alpha(t) \equiv <\vec{E}(t)\cdot\vec{E}>/ <E^2>$ is computed from equations (3.5) and (3.6).

The probability functional $P[y;T]$ is calculated in an analogous way,

$$P[y;T] \propto P(y_0)\int D[\lambda] \; \exp\left[i \int_0^t d\tau \, \lambda(\tau)[y(\tau) - G(\tau)y_0] \right.$$

$$\left. - \int_0^t d\tau \int_0^t d\tau' \; \lambda(\tau)H(\tau,\tau')\lambda(\tau') \right] \tag{4.7}$$

$$H(t,t') = \frac{1}{2} \int_o^t d\tau \int_o^{t'} d\tau' \; G(t-\tau) \; \langle f(\tau)f(\tau')\rangle \; G^T(t'-\tau') \tag{4.8}$$

Performing the functional integration over $\lambda(t)$ yields a Gaussian functional for the distribution of histories $y(t)$ about the most probable value $G(t)y_o$.

GAUSIAN MARKOVIAN PROCESS

The above Gaussian process is that it is not Markovian since the electric field correlation function of Figure 3 does not have an exponential decay. Instead, the model chosen here yields a superposition of three exponentials,

$$\langle \vec{E}(t)\cdot\vec{E}\rangle = \langle E^2\rangle \sum_{\alpha=1}^{3} A_\alpha \exp(-\lambda_\alpha t) \tag{5.1}$$

Furthermore, this correlation function is not positive definite. As a consequence, the approach of $Q(\vec{\varepsilon},t;\vec{\varepsilon}_o,0)$ to its asymptotic equilibrium value is non-monotonic. This is illustrated in Figure 4 for the case of $\vec{\varepsilon}$ parallel to $\vec{\varepsilon}_o$ at $\Gamma = 10$ and $Q = Q_o$.

The desirable feature of a Markov process is that the prediction of field histories over any chosen interval Δt is independent of the previous history. Such a Markovian description can be introduced for the problems considered here by expanding the set of variables. The form (5.1) suggests that three new variables ξ_α representing the OCP degrees of freedom should be considered such that the electric field has the representation,

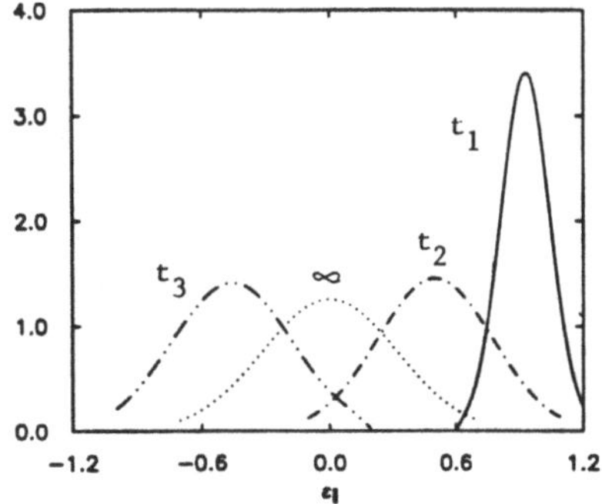

Figure 4: Joint field distribution

$$\vec{E}(t) = \sum_{\alpha=1}^{3} \vec{B}_\alpha \, \xi_\alpha(t) \tag{5.2}$$

The associated Langevin equations are of the form,

$$\frac{\partial}{\partial t}\vec{r}(t) = \vec{v}(t) \quad , \quad \frac{\partial}{\partial t}\vec{v}(t) = (Q_o m/3Qm_o) \sum_{\alpha=1}^{3} \vec{B}_\alpha \, \xi_\alpha(t) \tag{5.3}$$

$$\frac{\partial}{\partial t}\xi_\alpha(t) + \mathcal{L}_{\alpha\beta}\xi_\beta(t) = \phi_\alpha(t) \tag{5.4}$$

$$\langle \phi_\alpha(t)\phi_\beta(0)\rangle = \left[2\mathcal{L}_{\alpha\sigma}\langle\xi_\sigma\xi_\beta\rangle\right]^H \delta(t) \tag{5.5}$$

where the superscript H in (5.5) denotes the Hermitian part of the matrix. In contrast to the corresponding equations of section 3, these Langevin equations are local in time and have white (δ-correlated) noise amplitudes. The matrix $\mathcal{L}_{\alpha\beta}$ and the coefficients B_α must be selected to reproduce (5.1). In this way, the relevant collective property of the OCP is no longer the electric field, but three variables which determine the field. A more detailed analysis of these variables and their interpretation will be given elsewhere.

SUMMARY

The problem of calculating center of mass an internal state dynamics of an ion in a plasma has been formulated in terms of the electric field at the ion. Such dynamics is completely determined if the electric field history is known. The probability density functional for the distribution of all possible histories plays a central role in this formulation. Conditions of strong coupling and large charge ratio Q_o/Q appear to admit simplifications that make calculation of this and related functionals practical. The statistical features are then determined from a Gaussian process whose covariance is described by an accurate model for the electric field autocorrelation function. Further simplification to a Gaussian-Markov process can be affected by the formal introduction of three variables representing the OCP. Future studies of this approach will include applications to spectral line broadening and an investigation of the underlying assumption by comparison with electric field histories generated by computer simulation.

REFERENCES

[1] A. Brissaud and U. Frisch, JQRST $\underline{11}$, 1767 (1971); J. Dufty, in *Spectral Line Shapes*, B. Wende, ed. (W. de Gruyter, NY, 1981)

[2] D. Boercker, C. Iglesias, and J. Dufty, Phys. Rev. A$\underline{36}$, 2254 (1987)

[3] J. Dufty in *Spectral Line Shapes 6*, Fromhold and Keto, eds. (AIP Conf. Proc. 216, NY, 1990)

[4] See for example, B. Berne in *Statistical Mechanics, Part B*, B. Berne ed. (Plenum, NY, 1977)

[5] B. Bernu, J. Stat. Phys. $\underline{21}$, 447 (1979)

ABOUT THE INFLUENCE OF DENSE PLASMAS ON SPECTRAL LINE SHAPES OF DIFFERENT ATOMIC RADIATORS — THEORY

L. Hitzschke
Zentralinstitut für Elektronenphysik,
Hausvogteiplatz 5—7, O—1086 Berlin, Germany

Abstract

The influence of dense plasmas on profiles of isolated spectral lines has been investigated on the basis of a recently developed consequent many-particle approach. As the most important result of this approach the electron-atom interaction became dynamically screened. Calculations have been performed for the Xe I 467.1 nm and Cs I 621.3 nm lines which react quite sensitive on this effect. A comparison with corresponding experiments reveals that only by taking into account the dynamic character of screening for both lines the experimental behaviour can be explained.

Introduction

It has been known for a long time that with increasing electron densities in plasmas many-particle effects are revealing. This concerns not only thermodynamic and transport properties but also optical phenomena [1,2].
A lot of theoretical and experimental investigations has been done to study the influence of collective effects on spectral line shapes [3]. Concerning the theory, in principle a quantum statistical many-particle problem has to be considered. In fact, with increasing electron density the well-known binary collision approximation (BCA) fails to work and collective degrees of freedom of the plasma as, e.g. plasma oscillations, become important.
Recently a many-particle approach to the theory of spectral lines has been developed using the Green's function concept [1,2,4-7]. As a starting point the relationship between optical properties and the dielectric function has been chosen. A diagram technique has been used to select the relevant approximations within a perturbative expansion. All the effort towards the inclusion of many-particle effects has been

taken, of course, to investigate dense plasma systems. The aim of the present contribution is devoted to a detailed study of those systems in order to answer the question about the appearance of collective degrees of freedom. Therefore, after providing the necessary theory, results of calculations for the Xe I 467.1 nm and Cs I 621.3 nm lines are discussed and compared with corresponding experiments. The two lines have been chosen due to their quite sensitive reaction on the effects under consideration.

Theory

For the small plasma parameters considered here, it is possible to decouple the ion and electron subsystems by treating the ion-electron correlation within a microfield distribution where the ion field is shielded by electrons in the Debye-Hueckel form [8].
The objective of this paper is to investigate the electron contributions to the shift and broadening of spectral lines in dense plasmas. The latter are obtained from the imaginary part of the corresponding polarization function [1]. However, for a comparison with experiment the full line shape function $P(\lambda)$ has to be calculated. It reads [3]

$$P(\lambda) = \frac{1}{\pi} \int_0^\infty d\beta \; W(\beta) \; \frac{w^{(e)}}{(\Delta\lambda - d^{(e)} - C\beta^2)^2 + (w^{(e)})^2} \cdot \tag{1}$$

Here $\Delta\lambda$ is the difference between the actual and the unperturbed wavelength of the transition

$$\Delta\lambda = \lambda - \lambda_{if}. \tag{2}$$

The static Stark coefficient C is given by

$$C = C_i - C_f \tag{3}$$

and $W(\beta)$ stands for the microfield distribution function, where $\beta = E/E_0$ denotes the normalized and E_0 the Holtsmark (normal) field strength. In this paper the microfield distribution according to Hooper [9] has been chosen. Further, we will not treat the case of hydrogen here as that has been done elsewhere [7]. Therefore we neglect the problem of overlapping lines and expect the ions to produce quadratic Stark effects.
The electron shift $d^{(e)}$ and broadening $w^{(e)}$ of the line consist of

contributions which depend on the initial and final states of the transition

$$d^{(e)} = d_i^{SE} - d_f^{SE}, \tag{4}$$

$$w^{(e)} = w_i^{SE} + w_f^{SE} + w_{if}^{V}. \tag{5}$$

The electron shift and broadening result from self-energy (SE) and vertex contributions (V). They emerge from the interaction between the radiating atom and the plasma environment which can be treated using a perturbative expansion. For this a Green's function approach in connection with a diagram technique offers the opportunity of a systematic and complete consideration of all contributing terms. As relevant parameters the perturber density and the atom-perturber interaction appear. Due to the fact that our main concern is directed to weakly non-ideal plasmas, it is reasonable to restrict the theory to modifications in the line shape which are linear in the density of the perturbing electrons. Therefore, solving under these assumptions the corresponding Bethe-Salpeter equation , corrections to the isolated two particle energy appear owing to [1]: (a) phase space occupation due to statistical correlations, (b) exchange and (c) dynamic self-energy and (d) a dynamically screened effective potential. Further it has been shown [5] that in general contributions from (a) and (b) are negligibly and the corrections of type (c) and (d) are compensating each other up to a definite measure. So, the (SE)-contributions result in an expression which can be explained in the way that the initial electron-atom interaction has been replaced by a dynamically screened one. Finally, the same approximation has to be applied with respect to the vertex contributions. The final results read [1,5]

$$d_n^{SE} + i w_n^{SE} = -\sum_\alpha \int \frac{d\vec{k}}{(2\pi)^3} V(k) \int_{-\infty}^{\infty} \frac{d\omega}{\pi} \, \text{Im} \, \varepsilon^{-1} (\vec{k}, \omega + i\,0^+) \tag{6}$$

$$\cdot \left[1 + n_B(\omega)\right] \frac{|M_{n\alpha}(\vec{k})|^2}{E_n - E_\alpha - (\omega + i\,0^+)},$$

$$w_{if}^{V} = -2 \int \frac{d\vec{k}}{(2\pi)^3} M_{ii}(\vec{k}) \, M_{ff}(-\vec{k}) \, V(k) \cdot \int_{-\infty}^{\infty} d\omega \, \text{Im} \, \varepsilon^{-1}(\vec{k}, \omega + i\,0^+) \left[1 + n_B(\omega)\right] \delta(\omega) \tag{7}$$

where in principle in (6) α is running over the whole two-particle spectrum. (We have put $\hbar=1$.) Further, within the frame of our approximations the dielectric function of the electron gas $\varepsilon(\vec{k},z)$ being the only ingredient within (6) and (7) still carrying many-particle informations has been taken in random-phase-approximation (RPA)

$$\varepsilon(\vec{k}, \omega + i\, 0^+) = 1 - 2 \int \frac{d\,\vec{p}}{(2\pi)^3}\ V(k)\ \frac{f_e(E_{\vec{p}}) - f_e(E_{\vec{p}-\vec{k}})}{E_{\vec{p}} - E_{\vec{p}-\vec{k}} - (\omega + i\, 0^+)} \cdot \tag{8}$$

The expression

$$M_{n\alpha}(\vec{k}) = i \int \frac{d\,\vec{p}}{(2\pi)^3}\ \psi_n^*(\vec{p}) \left[\psi_\alpha(\vec{p}) - \psi_\alpha(\vec{p} + \vec{k}) \right] \tag{9}$$

denotes the isolated vertex function leading for small $\vec{k}$ to the dipole matrix element what can be proved easily. Further, ψ_n is to be identified with the wavefunction while E_n means the corresponding energy for the isolated atom. At last, we have introduced $V(k)=4\pi e^2/k^2$, $E_{\vec{p}}=\vec{p}^2/2m_e$ and

$$f_e(E_{\vec{p}}) = \left\{ \exp\left[\beta\, (E_{\vec{p}} - \mu_e) \right] + 1 \right\}^{-1}, \tag{10}$$

$$n_B(\omega) = \left[\exp(\beta\, \omega) - 1 \right]^{-1} \tag{11}$$

with μ_e standing for the chemical potential of the electrons, $\beta=1/k_BT$ and k_B being the Boltzmann constant.

The above approximation is restricted to weak electron-atom collisions. The consideration of strong collisions has to avoid a perturbative expansion with respect to the interaction. However, different from the case of weak collisions, collective effects as screening are without influence on them. Unfortunately, from theoretical point of view the general situation in literatur concerning tractable treatments of strong collisions is more or less unsatisfying although practicable improvements seem to be possible after recent investigations [10]. However, following the aim of the present paper which is to end up with a comparison with experiment, a model is needed to include strong collision interactions. That has been the reason for us to go back to a quite simple cutoff procedure. (For more details, see of [3].) In the face of this elementary approach the impression could come up as if all the effort to describe the week collision processes is going to become questionable . Fortunately however, these doubts are groundless [11].

Results

After a detailed analysis of the above equations the following conclusions can be drawn [11]:

1. Plasma screening for the shift and broadening contributions from level α to level n takes place if the energy distance between the two levels becomes comparable with the plasma frequency.
2. Within the screening process both the real and imaginary part of the dielectric function are of equivalent importance.
3. The influence of the plasma screening on level shift is much more pronounced than that on broadening.

Therefore, before plasma screening can be observed several conditions must be fulfilled. For the Xe I 467.1 nm line corresponding in Paschen notation to the $3p_8-1s_5$ transition all of them are realized. In fact, at first the line can be considered as an isolated one, the Stark effect of which is determined essentially by the upper level ($3p_8$). The shift and broadening of this level is dominated by contributions from the most neighboring levels with dipole allowed transitions to it ((4d), see Fig. 1). Within these again, in the BCA the $4d_4'-3p_8$ contribution with $\omega_{\alpha n}=0.004$ already amounts up to 80%. On the other hand, for reaching $\omega_{pl}=0.004$ the electron density must be about $2*10^{18}\,\mathrm{cm}^{-3}$. Therefore, beginning with $n_e-10^{18}\,\mathrm{cm}^{-3}$ the electron contribution especially to the shift should be influenced by dynamic plasma effects.

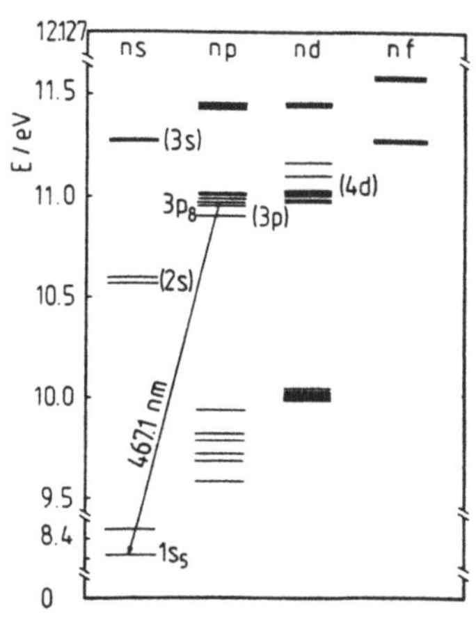

Fig. 1. Energy level diagram for Xe I

In Fig. 2 a comparison of the calculated line shifts with experimental results is given. The quite good correspondence of both over the whole range of electron densities indicates the applicability of the theoretical model. In detail, the agreement for small densities, the region of dominant binary collision contributions ($n_e<2*10^{17}\,\mathrm{cm}^{-3}$), illustrates the practicability of the cutoff approach for strong collisions. With increasing densities the plasma influence is rising. However, due to the compensation process caused by the growing ion contributions which keeps the line shift proportioanal to n_e the latter can only be seen only in the experiment at densities higher

10^{18}cm^{-3}. Finally, for $n_e>2*10^{18}$cm^{-3} the ion contributions to the line shift dominate. Unfortunately, for this densities only a few experiments are available. Although the experimental and theoretical results agree quite well some more measurements would be desirable. Nevertheless, it is obvious that both the Debye cutoff as well as the full

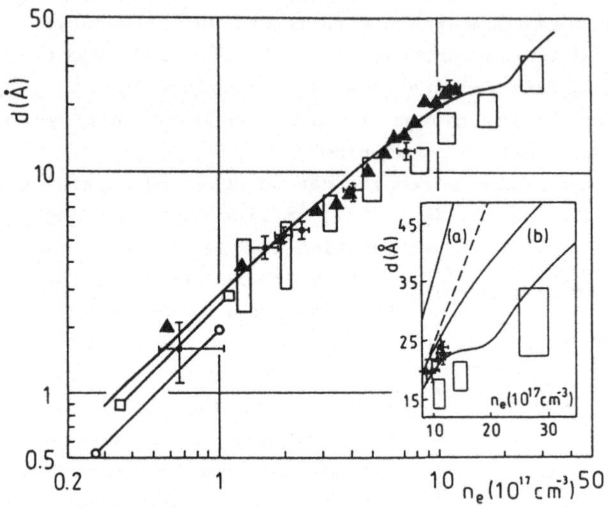

Fig. 2. Shift of the Xe I 467.1 nm line. $\square$, $\boxplus$, $\boxplus$, $\diamondsuit$—$\diamondsuit$, $\bigcirc$—$\bigcirc$: [12-16], respectively; ——— this paper, RPA for the electron contribution; ——— (a) this paper, BCA; ——— (b) this paper, cutoff approach; ----- linear extrapolation of [15].

Debye approximation are not able to describe the plateau region of the shift at about $n_e=1.5*10^{18}$cm^{-3} because it can only arise if the electron contribution is decreasing within it. (Compare Figs. 3 and 4. Note that the Debye approximation results after replacing $|\varepsilon(\vec{k},\omega)|^{-2}$ by $k^2/(k^2+k_D^2)$ with k_D being the inverse Debye radius while the Debye cutoff [3] follows after introducing a cutoff for the $\vec{k}$-integrals in (6) and (7) at $k_{min}=k_D/1.1$ [11].)
The above results confirm the conclusions concerning the dynamic character of the plasma influence especially on spectral line shifts. This influence is characterised by an effective screening of contributions having an energy distance to the perturbed level of about ω_{pl}. For a further study of this effect, the Cs I 621.3 nm line (transition $8d_{5/2}-6p_{3/2}$) seems to be very suitable. In fact, for the upper level $8d_{5/2}$ we now have a pair of main contributions resulting from the $9p_{3/2}$ and 6f which dominate the electron contribution to shift and

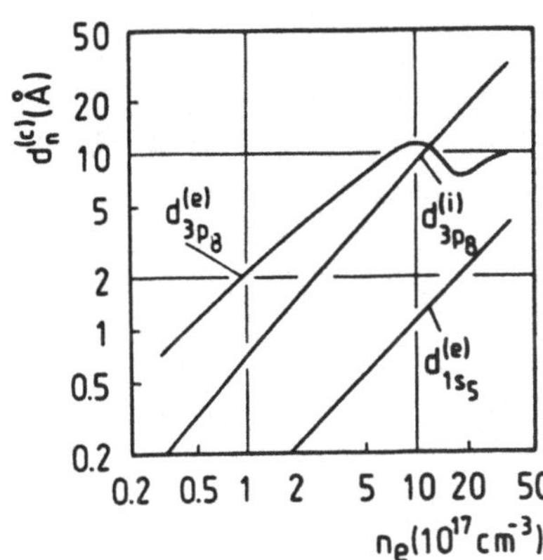

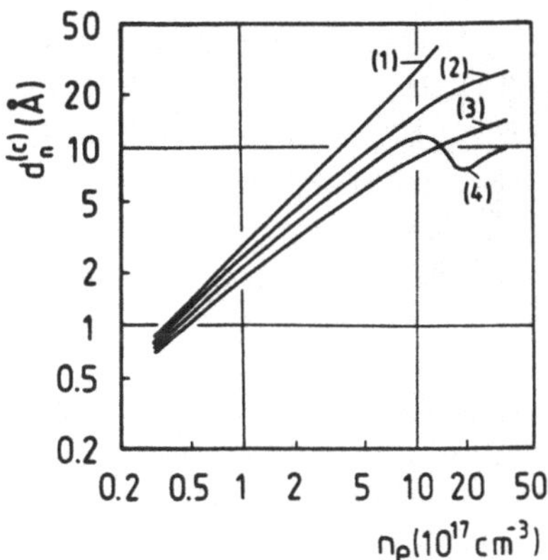

Fig. 3. Contributions to the Xe I 467.1 nm line shift. (Note that the ion contribution $d_{1s_5}^{(i)}$ is negligible

Fig. 4. Different approximations for the electron contribution to the shift of the upper level ($3p_8$). (1) BCA; (2) cutoff at k_{min}; (3) Debye approximation; (4) RPA

broadening within the BCA. But different to the broadening, for the shift these terms are competitive leading to blue and red level shift contributions, respectively (see Fig. 5). Within the BCA the blue shifting one dominates with 25% of itself. However, at electron densities of about 10^{17}cm^{-3} $\omega_{9p_{3/2}, 8d_{5/2}}$ becomes comparable with ω_{pl}. Therefore, at these densities we expect the blue shift contribution to be screened out. As the result, the blue electron shift for small densities should turn over into a red shift. According to the given theory this electron shift reversal should be a very sensitive proof for the different theoretical approaches. The results of the calculations are included in Fig. 5 of the subsequent contribution. It is to be seen there that in the ex-

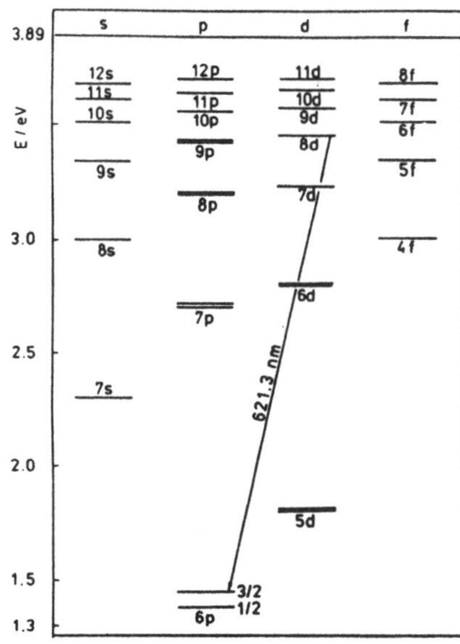

Fig. 5. Energy level diagram
for Cs I

periment the expected shift reversal really takes place. But among the
theoretical models only the RPA one is able to describe it. In fact,
all the other do not even deliver the shift reversal from the blue
into red. Further, within the experimental and theoretical uncertainties the RPA results agree quite well with the experiment over the
whole range of densities. (For more details, see [11].)

References

[1] W.D. Kraeft, D. Kremp, W. Ebeling, G. Röpke, *Quantum Statistics
     of Charged Particle Systems* (Akademie-Verlag, Berlin, 1986)

[2] G. Röpke and L. Hitzschke, in: *Spectral Line Shapes V*, ed. by
     J. Szudy (Ossolineum Publishing House, Wroclaw, 1989), p. 49

[3] H.R. Griem, *Plasma Spectroscopy* (Mc Graw-Hill, New York,
     1964); *Spectral Line Broadening by Plasmas* (Academic,
     New York, 1974)

[4] G. Röpke, T. Seifert, and K. Kilimann, Ann. Phys. (Leipzig)
     38, 381 (1981)

[5] L. Hitzschke, G. Röpke, T. Seifert, and R. Zimmermann,
     J. Phys. B: At. Mol. Phys. **19**, 2443 (1986)

[6] L. Hitzschke and G. Röpke, Phys. Rev. A **37**, 4991 (1988)

[7] S. Günter, L. Hitzschke, and G. Röpke, Phys. Rev. A **44**,
     10 (1991)

[8] C.A. Iglesias and J.W. Dufty, in: *Spectral Line Shapes II*,
     ed. by K. Burnett (de Gruyter, Berlin, 1983), p. 55

[9] C.F. Hooper, Phys. Rev. **165**, 215 (1968)

[10] S. Günter (unpublished)

[11] L. Hitzschke, to be submitted to Phys. Rev. A

[12] H. Heß, L. Hitzschke, E. Metzke, R. Niepraschk, M. Wirsig,
      in: *Contributed Papers of the XIIth Symposium on Physics of
      Ionized Gases*, ed. by M.M. Popovic (Insitute of Physics
      Belgrade, Belgrade, 1984), p. 453; H. Heß, Contr. Plasma
      Phys. **26**, 209 (1986)

[13] M. Kettlitz, R. Radtke, R. Spanke, and L. Hitzschke,
      J. Quant. Spectrosc. Radiat. Transfer **34**, 275 (1985)

[14] M. Kettlitz and R. Radtke, within this volume

[15] P. Klein, and D. Meiners, J. Quant. Spectrosc. Radiat.
      Transfer **17**, 197 (1977)

[16] Truong-Bach, J. Richou, A. Lesage, and M.H. Miller,
      Phys. Rev. A **24**, 2550 (1981)

ABOUT THE INFLUENCE OF DENSE PLASMAS ON SPECTRAL LINE SHAPES OF DIFFERENT ATOMIC RADIATORS — EXPERIMENT

M. Kettlitz, R. Radtke
Zentralinstitut für Elektronenphysik
Hausvogteiplatz 5-7, O-1086 Berlin,
Germany

Abstract

Electric pulse discharges have been used to generate Xe plasmas with temperatures of the order of 10^4 K and electron densities between 10^{17} and 10^{18} cm^{-3} as well as Cs plasmas with temperatures of about 4000 K and electron densities between 10^{16} and 10^{17} cm^{-3}. Spectroscopic measurements have been performed in the visible range to show the influence of the particle interaction in weakly nonideal plasmas on the line shift behaviour. The shifts of the investigated Xe I 467.1 nm and Cs I 621.3 nm lines were found to react sensitively to an increase of the electron density.

Introduction

In 1985 we published shift data of the Xe-line at 467.1 nm [1] within an electron density range between 10^{17} and 10^{18} cm^{-3}. There have been stated differences to other experiments performed using an adiabatic compressor [2]. In the latter experiment deviations have been found from the linear behaviour of the line shift as the electron density was increased to values up to $3*10^{18}$ cm^{-3}. On the other hand, in our Cs experiments we observed also an unexpected behaviour at high particle densities [3]. While the line at 635.5 nm was shifted with increasing electron density the position of the line at 621.3 nm seemed to be almost unchanged. This behaviour was possible to be explained by a dynamic screening of the atom-electron interaction [4]. However, there was still the necessity to throw more light on this subject. That's why new measurements under slightly changed conditions have been performed to extend the electron density range to higher values and to obtain some more reliable values.

Experimental set-up

For the generation of the plasmas we used pulsed wall-stabilised electric arcs which are fairly well understood [5] and allow precise measurements over a wide range of conditions. The xenon plasmas have been contained within an L-shaped quartz tube with a length of 10 cm (fig. 1). The inner diameter was reduced to 3 mm in difference to our earlier experiments where it had been 6 mm. At one end of the tube a quartz rod was mounted for a plasma end-on observation. The optical quality of the lamp was excellent and allowed spectroscopic measurements both end-on along the lamp axis and side-on normal to the lamp. As has been demonstrated in [5], the radial profiles of pulsed Xe-arcs are so flat that no corrections for boundary layers were necessary if the measurements were carried out

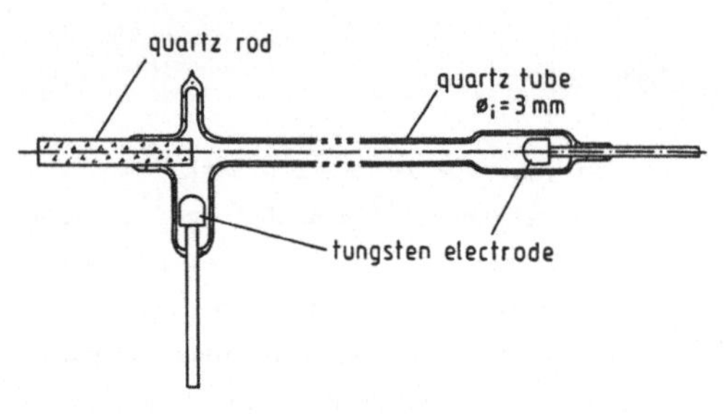

Fig. 1. Xe discharge lamp

side-on. The initial pressure in the lamp was about 400 Torr.

For Cs we have chosen a different experimental set-up. The extreme reactivity of the alkaline vapours obliged the employment of discharge tubes consisting of polycrystalline aluminia (fig. 2). This material is highly resistant to these vapours. However, it has the disadvantage of being only translucently not transparently. That is the reason why a sapphire rod was sealed into an end-cap for observing the undisturbed radiation of the arc - without any scattering effects. Cold boundary layers have almost been eliminated

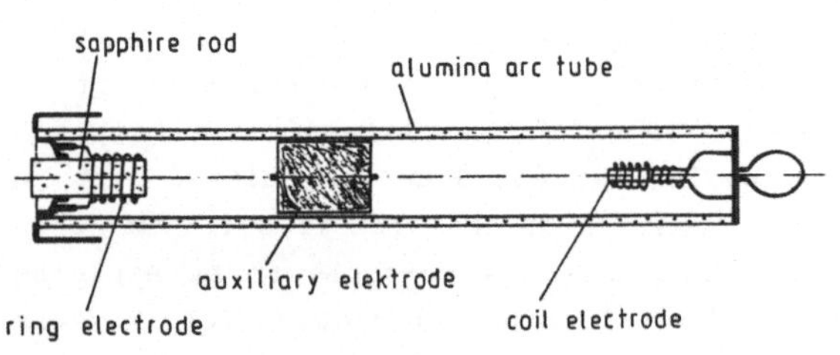

Fig. 2. Cs discharge lamp

in mounting the tungsten electrode around that window. So, the whole
tube reached a length of 11 cm and an inner diameter of 7 mm. Inside
the tube an auxilary electrode with small tungsten ends has been
placed. For allowing spectroscopic end-on observations at variable
optical path length without changing the plasma parameters it has
been made magnetically movable. The cesium discharge tube has been
operated within a vacuum chamber to protect the seal and the end caps
from reactions with oxygen at higher temperatures.

The lamps were part of a discharge circuit with a pulse forming net-
work (pfn) as the energy storage system. The impedance of the pfn is
matched to that of the discharge lamp. In that way current pulses
with a quasi-stationary period of about 1 ms have been produced. The
ignition of the discharges differed depending on the investigated
objects. For Xe we used the following method. The high voltage peak
of an ignited spark gap is coupled inductively into the discharge
circuit causing the breakdown of the discharge distance between the
electrodes of the lamp. In that way a conducting channel is formed
resulting in a high current discharge of the pfn.

The Cs lamp has been operated in the so-called simmermode which keeps
the discharge tube in the conducting state by means of a dc discharge. Therefore, we did not need a spark gap for the ignition. On
the other hand, the dc discharge has been used for delivering the
desired initial particle number densities. In that mode the stored
energy can easily be fed into the lamp after opening an electronic
switch. The applied current pulses had a height of up to 400 A. All
our measurements have been performed in the quasistationary period of
the pulses.

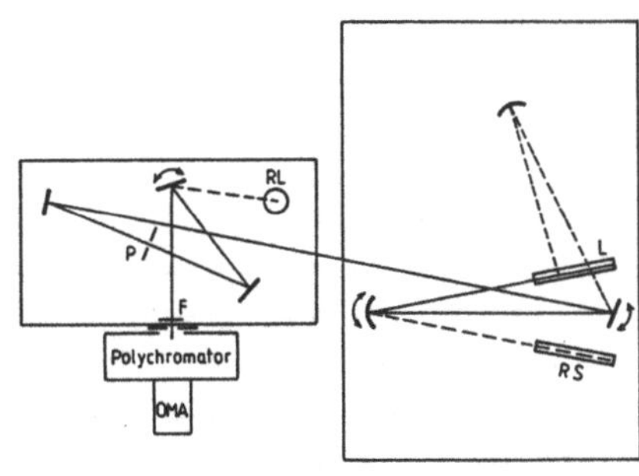

Fig. 3. Optical set-up

Figure 3 shows the optical arrangement for the measurements of the radiation intensity. The arc was imaged with a focussing mirror onto the entrance slit of a polychromator set-up with a 1300-lines/mm grating. In the focal plane of the polychromator the detector of a pulsed optical multichannel analyzer (OMA) was mounted which enabled us to obtain spectra in the visible range with a spectral resolution of 0.04 nm. The recorded spectra have been digitised and afterwords analysed by customary codes.

To provide an absolute intensity calibration of the optical detection system a pulsed Xe - radiation standard (RS) was used producing a blackbody intensity at 12000 K for wavelength larger than 300 nm [6]. After each experiment an independet wavelength calibration has been performed with different low pressure lamps (RL).

Plasma diagnostics

For both elements we have measured the radiation intensity

$$I_\lambda(T) = B_\lambda(T) \left[1 - \exp\left(-\varkappa_\lambda l\right)\right] \tag{1}$$

of several arcs differing in the power input. From the spectral distribution of the radiation emitted from two different optical lengths $l_{1,2}$ - in the case of Xe in the direction of the axis and normal to it; in the case of Cs at two different positions of the auxiliary electrode - the absorption coefficient $\varkappa$ has been determined from the ratio

$$\mu(\varkappa) = \frac{I_{\lambda,1}(T)}{I_{\lambda,2}(T)} = \frac{1-\exp\left(-\varkappa_\lambda l_1\right)}{1-\exp\left(-\varkappa_\lambda l_2\right)} \tag{2}$$

The temperature we obtained from the blackbody function

$$B_\lambda(T) = I_{\lambda,2}(T)/1 - \exp\left(-\varkappa_\lambda l_2\right) \tag{3}$$

where the subscript 2 refers to the measurement at the large optical path length.

The particle number densities have been received in different ways for both facilities. For Xe the electron densities were obtained via a Mach-Zehnder interferometer by recording the temporal change of the refractive index from the stationary state maintained during the last part of the pulse to the cold gas. The measurements have been carried out at two wavelength (λ = 633 and 1150 nm). This method leads to uncertainties in the electron density of 5 - 14 % corresponding to the variation of conditions in our experiment.

For Cs we recorded spectra of the continuous radiation in the wavelength region around 420 nm. The electron densities are calculated on
the basis of the measured absorption coefficient and temperature using the expression for the recombination continuum and the ξ-function
as given by Hofsaess [7] :

$$\varkappa^{b-f} = n_a \cdot \varkappa_{at} \cdot \xi(\lambda,T) \cdot \left[1-\exp\left(-C_2/\lambda T\right)\right] \tag{4}$$

where $\varkappa_{at}$ denotes the absorption coefficient per atom :

$$\varkappa_{at} = C \cdot T \cdot \lambda^3 \cdot \exp\left(-E_i/kT\right) \cdot \left[\exp\left(C_2/\lambda T\right)-1\right] \cdot \gamma/Z_0 \tag{5}$$

Here n_a is the neutral particle density, Z_0 the partition function of
neutrals, E_i the ionization potential, γ the statistical weight of
the ion ground state and $C = 6.62*10^{-6}$ $(m^{-1}K^{-1})$.

Results and discussion

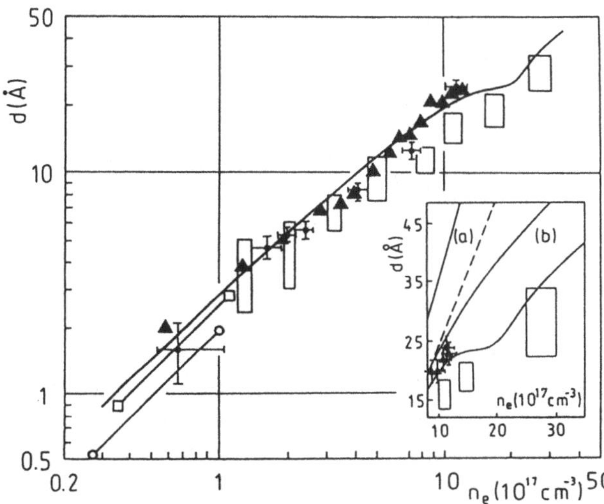

Fig. 4. Shift of Xe I 467.1 nm

▲ this paper, ┼ [1], ☐ [2], ─── [8],

◇─◇ [11], ○─○ [12]

The experimental results of our recent measurements in xenon have already been shown in fig. 2
of the previous contribution of L. Hitzschke in
this conference volume [8]; they are additionally
given in fig. 4. The first aim to fill the gaps
between the older measured values has been reached.
Secondly, the shift data at higher electron

densities could be confirmed. Unfortunately, we are not able to extend our measurements to electron densities above $1.2*10^{18}$ cm^{-3}. At larger values the line was not any longer discernible from the continuous ground. With this set-up we could not reach the interesting electron density region of about $2*10^{18}$ cm^{-3} where the theory predicts a plateau. Although the experimental results agree quite good with the theory there is still the need for some more measurements.

Figure 5 shows as an example the spectral intensity distribution for different cesium arcs of 0.5 cm path length around 621 nm. The growing power input leads to a growing intensity connected with an increasing electron density. Within this spectral region we are faced with an unpleasant fact. The comparatively high opacity of this line leads to reabsorption in the line centre causing problems in the direct determination of the line position. For this reason shift values were derived from an iterative procedure of fitting the line wings of the experimental line shape to a synthetic spectrum. It is seen that the in this way determined line position remains almost unchanged in comparison with the predictions for the uncorrelated electron gas - the bars mark the expected line shift for that case [9].

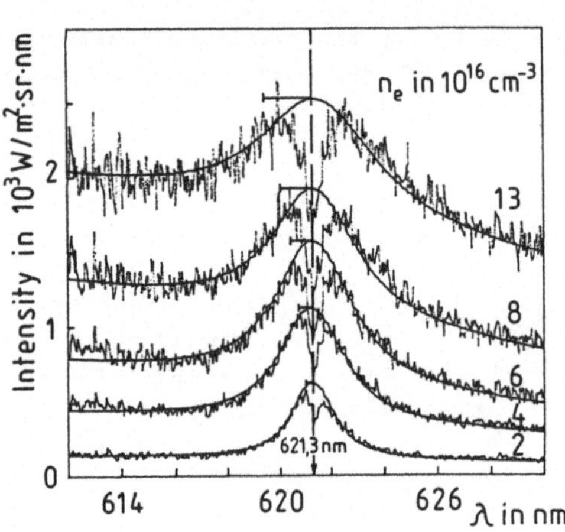

Fig. 5. Intensity distribution in Cs
around 621 nm

More precisely (fig. 6), with increasing density there is a shift reversal from the blue to the red of the Cs I 621.3 nm line (transition $8d_{5/2} - 6p_{3/2}$). The largest blue shift of 0.1 nm was obtained at electron densities of about $4*10^{16}$ cm^{-3}. After that, a shift reversal takes place leading to a red shift of this line of 0.2 nm at an electron density of $1.4*10^{17}$ cm^{-3}. This behaviour can be explained by the dynamic screening of the electron contribution to the line shift

- e.g. in our case : the blue shifting contribution is screened out with increasing n_e. Such a dynamical screening is taken into consideration within the random phase approximation (RPA) for the dielectric function. Within this approach the detailed surroundings of the levels for the regarded transition are considered in detail. This seems to be of importance if the energy distance between the perturbing and the perturbed levels compares to the plasma frequency. For more detailes see [10]. In fig. 6 a comparison of our experimental data with results of different theoretical approaches for the electron contribution to the shift are given. So, $d^{(g,RPA)}$ denotes the RPA, $d^{(g,Debye)}$ the Debye approximation, $d^{(g,cut-off)}$ the cut-off approach and $d^{(g,uc)}$ the uncorrelated case. Additionally, the corresponding ion shift contribution has been calculated [8]. Admittedly, the measured values try to avoid to meet the theoretical curves but fig. 6 suggests that only the RPA can be used in order to explain the shift reversal of this line.

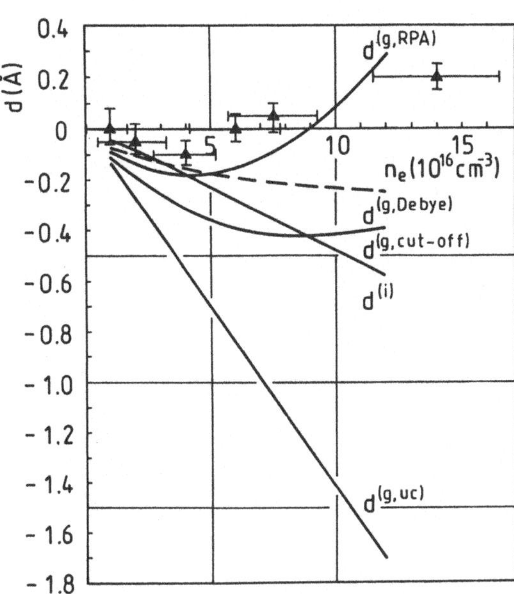

Fig. 6. Shift of Cs I 621.3 nm versus n_e
$d^{(g,...)}$ - full shift
$d^{(i)}$ - ion contribution

Moreover, after an analysis of the theoretical and experimental uncertainties the RPA agrees quite good with the experiment over the whole range of densities.

From the experimental point of view the relatively large error bars are a bit unsatisfactory. They originate from the chosen method of determining the electron density from the continuous absorption coefficient and therefore the main error is related to the uncertainty of

the measured optical lengths and the intensity ratios. For avoiding
these uncertainties in the electron density determination it would be
necessary to use more precise methods - that means for example laser
interferometry.

References

[1]   M. Kettlitz, R. Radtke, R. Spanke, L. Hitzschke, J.Q.S.R.T. **34**,
      275 (1985)

[2]   H. Heβ, L. Hitzschke, E. Metzke, R. Niepraschk, M. Wirsig,
      Contr. Pap. XII. Symp. on Phys. of Ionized Gases, Belgrade, 453
      (1984)

[3]   M. Kettlitz, R. Radtke, Contr. Pap. ICPIG, Belgrade, 350 (1989)

[4]   L. Hitzschke, Contr. Pap. ICPIG, Belgrade, 352 (1989)

[5]   K. Günther, R. Radtke, Electric Properties of Weakly Nonideal
      Plasmas (Birkhäuser, Basel, 1984)

[6]   K. Günther, S. Lang, R. Radtke, J. Phys. D **16**, 1235 (1983)

[7]   D. Hofsaess, Z. Physik A **281**, 1 (1977)

[8]   L. Hitzschke, in this conference volume

[9]   H.R. Griem, Spectral Line Broadening by Plasmas (Academic,
      N.Y., 1974)

[10] L. Hitzschke, to be submitted to Phys. Rev. A

[11] P. Klein, D. Meiners, J.Q.S.R.T. **17**, 197 (1977)

[12] Truong-Bach, J. Richou, A. Lesage, M.H. Miller, Phys. Rev. A **24**,
      2550 (1981)

IR Spectra of Dense Xenon and Argon Plasmas

V.E.Fortov
Intense Action Scientific Center - IVTAN, Moscow;
M.U.Kulish, V.B.Mintsev
Institute of Chemical Physics, Chernogolovka, Moscow region;
J.Ortner
Zentrum fur Wissenschaften Geratebau, Berlin;
I.M.Tkachenko
Odessa University, Odessa

Abstract.

The results are presented of experimental studies of IR radiation spectra of dense Xe and Ar shock compressed plasmas. The radiation cutoff is found at the wavelength $\lambda \simeq 1 \mu m$. The experimental data are interpreted within the elaborated model of generation of noble gas excimer molecules before the shock wave front.

Introduction.

The investigation of emissivity of strongly coupled plasmas is of essential physical importance, in particular, because it allows one to trace the influence of strong interparticle interactions on the processes in dense disordered media. Strict theoretical approaches developed only for dilute plasmas where the elementary processes are easily separated, and the influence of plasma surroundings is reduced to the broadening of spectral lines and the shift of the photo recombination limit. The increase of the plasma density leads to the decrease of population of upper energy levels, to the disappearance of excited states and deformation of the plasma electronic spectrum as a whole, thus making the application of traditional methods of description of such media impossible.

In this situation the experiment and model approximation acquire decisive importance.

The studies of emissivity of shock waves in noble gases carried out earlier provided information on the temperature, absorption and electronic concentration of dense plasmas [1]. These experiments also allowed one to elucidate the reflection properties [2] and the influence of strong Coulomb interactions on the broadening and shift of lines [3] in strongly coupled plasmas. The concentration of free carriers reaches the value of $n_e \geqslant 10^{21} sm^{-3}$, and the plasma frequency is as high as $w_p \geqslant 1.8 \cdot 10^{15} sec^{-1}$, so that the critical wavelength is in the near-infrared region, $\lambda_{cr} \leqslant 1 \mu m$. The radiation spectrum of shock-compressed plasmas proved to be similar to the black body one in the visible range.

In this circumstances the appearance is expectable of well known in the radiolocation and solid state physics peculiarities of emission and absorption properties of strongly coupled plasmas due to the sharp decrease of the plasma emissivity at wavelengths exceeding the critical one. The special investigations aimed at discovering these peculiarities were carried out.

The results are presented here of experimental studies of IR radiation spectra ($\lambda \approx 800 \div 1100$ nm) of xenon and argon dense plasmas at pressures p=(0.1÷10) GPa produced beyond the front of strong shock waves generated by explosion devices. One failed to describe the observed radiation "cutoff" at $\lambda \approx 1020$ nm in xenon and at $\lambda \approx 997$ nm in argon on the basis of plasma theories.

The model is presented of kinetic processes occurring while the strong radiating shock wave passes across the unperturbed compressed gas under investigation.

The discovered absorption features are shown to be explained by the formation in the resting gas under the action of the plasma strong UV radiation of excimer molecules (and other excited and ionized states) screening the radiant flux of the shock front in the IR range.

The generation of shock-compressed plasmas

The emissivity of dense plasmas of argon and xenon were measured with the explosive linear type generators. As it was shown in ref.[1], if the shock wave velocity $D \geqslant 6$ mm/μsec and the initial pressure $P_o \leqslant 1$ MPa one reaches the plasma temperature $T \approx 3eV$ and the electronic concentration $n_e \approx 10^{21} cm^{-3}$. The construction of experimental setup are given in Figs.1 a),b). The experimental assembly presented in Fig.1 b) is designed for the initial pressure below 10 MPa, and is structurally aligned with the explosive generators used to measure the radiant fluxes of plasmas produced of gases under normal initial conditions are given in Fig.1 a). For the mass velocity of 5.4 km/sec the design temperature of argon beyond the shock wave is 18600 K; the measured value of temperature in this case coincided with the design one within the experimental accuracy.

The hemispherical surface of the assembly was covered with the layer of the magnesium oxide to provide for the isotropic scattering of the plasma layer radiation within the volume occupied by the unperturbed

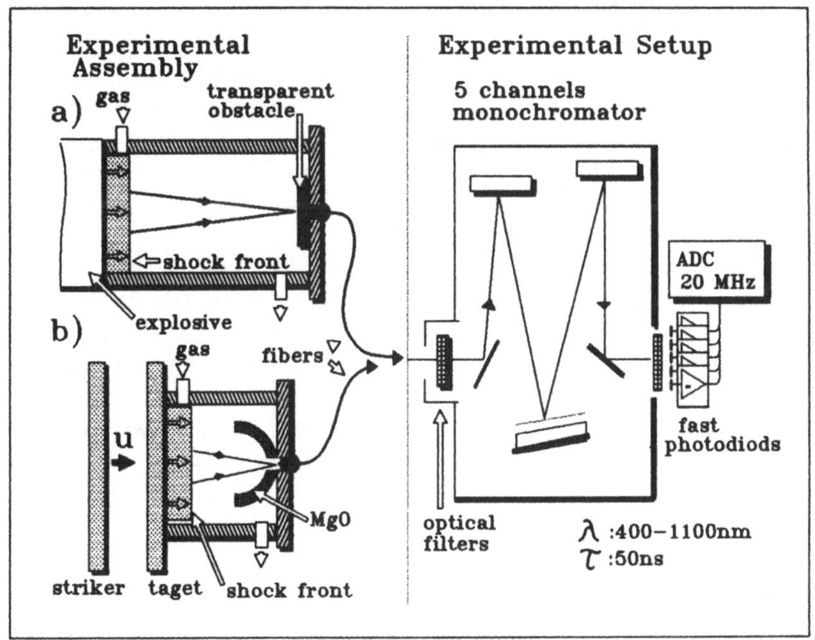

FIGURE 1.

gas. The coefficient of diffuse reflection from the surface covered by MgO is near unity (0.98-0.99). If there is no absorption by the cold gas, the measurements of the radiant flux in the assemblies with reflecting and blackened surfaces of the hemispheres allows one to determine the emissivity of the plasma layer surface directly [4]. For the experiments with xenon the striker and the target of the explosive generator were made of 1.0 mm thick steel. For the exploration of the argon plasma the target was composed of the 1.0 mm wide layer of steel and the 1.2 mm wide layer of aluminum. The ablation of the compound target in argon was generating the shock wave of a decaying profile and with the initial speed of 7.4 km/sec. The plasma radiation was tapped from the assembly through the diaphragm located in the hemisphere vertex along the ≈30m long quartz light pipe.

The methods of diagnostics

The experimental setup is shown in Fig.1. The plasma radiation along the quartz light pipe was advanced to the grating spectrograph. The exit slit of the spectrograph monochromator was replaced by the system

of slits aligned with the photoreceiver. The dimensions of the exit
slits matched the size of the receiving elements of the photoreceiver
1x1 mm. For three replaceable monochromator grids (1200, 600, 300
1/mm) and the focal distance of the camera lens the spectral width of
each measuring channel in 2.4 and 8 nm, respectively. The grid 600
1/mm for which the covered in one measurement spectral interval is of
$\simeq$120 nm with 5 channels was used in the majority of experiments. Fast
p-i-n photodiodes with the spectral sensitivity in the range of
400$\div$1100 nm were used as the photoreceivers. The signal from the
photodiodes was advanced through the amplifier with the dynamic range
of $\simeq$10^4 to the analog-digital converter with the clock frequency of up
to 20 MHz. The sensitivity of the spectral channels was determined in
advance of each experiment using the tungsten ribbon lamp with
$T_L \simeq$2800K. The precision of the calibration was increased by the
program-digital filtration of the noise of the lamp periodical signal.
A special series of experiments was carried out with the standard
explosive argon illuminant with T $\simeq$18600K.

The experimental results

The emission properties of plasmas were measured in the wavelength
range of 800$\div$1100 nm. The initial pressure of gas was chosen within
the limits of P_O=(0.1$\div$5.0) MPa, the pressure behind the shock wave
front reached the values of P=(1$\div$10) GPa at T$\simeq$(2.5$\div$3.2)$\cdot$10^{4}K. The
composition of the plasma was computed within the Debye approximation
in the grand canonical ensemble [1]. The characteristic experimental
data for the radiation intensity vs. time are given in Fig.2 at
P_O=0.1MPa the stationary value of intensity was reached in $\simeq$ 250 ns,

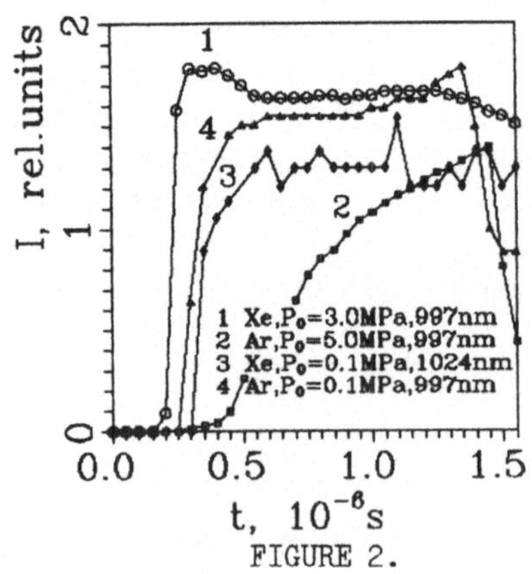

FIGURE 2.

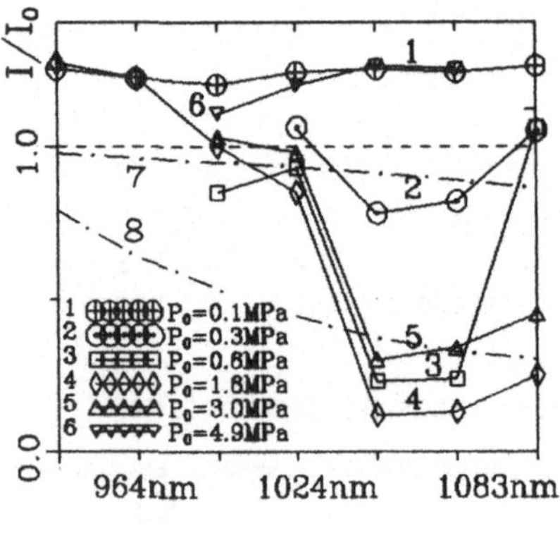

FIGURE 3.

and this value was equal to the intensity of the black body radiation
with the temperature equal to the temperature of the plasma with the
given shock wave speed (curves 3.4 in Fig.2). The growth of the
initial pressure causes the deviation of the plasma radiation from the
equilibrium one for the wavelength ≃1000 nm. The emission of the plane
stationary shock wave in argon was increasing in time as the wave was
approaching the transparent obstacle (see curve 2 in Fig.2). In
experiments with xenon the maximal intensity was observed during ≃100
ns after the exit of shock wave into the gas (curve 1 in Fig.2).
The results of measurements normalized to the design value of the
black body radiation intensity are presented in Figs.3,4. In Fig.3 for
xenon the appearance of a "gap" in the range of 1000-1100 nm is
observed. The comparative experiments using both reflective and
blackened hemispherical reflectors before the shock wave front did not
show the possible influence of reflection from the plasma layer
surface. The maximal absorption in xenon was registered at the initial
pressure of 1.6 MPa, when the radiation intensity made up the tenth
part of the black body radiation. In argon the absorption was observed
within a narrow band around $\lambda \approx 997$ nm. In this case the shock wave had
a decaying profile and far from the center of the absorption band the
decrease of the radiation intensity was observed along with its growth
in the center.
It is significant that the position of the "gap" is independent of the
electronic concentration. Indeed, if it were connected with the
radiation "cutoff" for wavelengths exceeding the critical one, then it
had to move to smaller wavelengths as the plasma concentration was
growing. The dot-and-dash lines in Fig.3 show the "expected" for

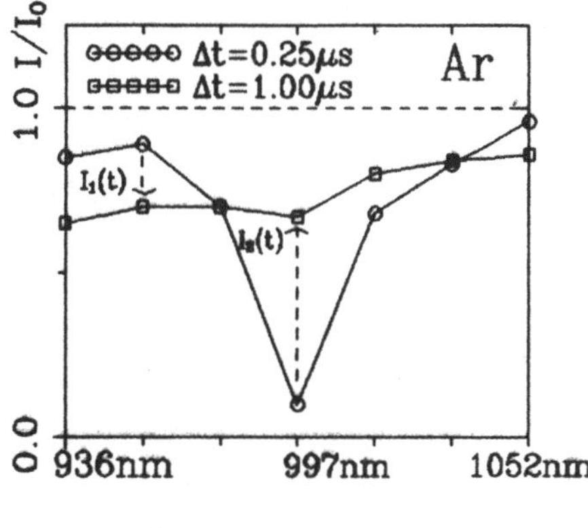

FIGURE 4.

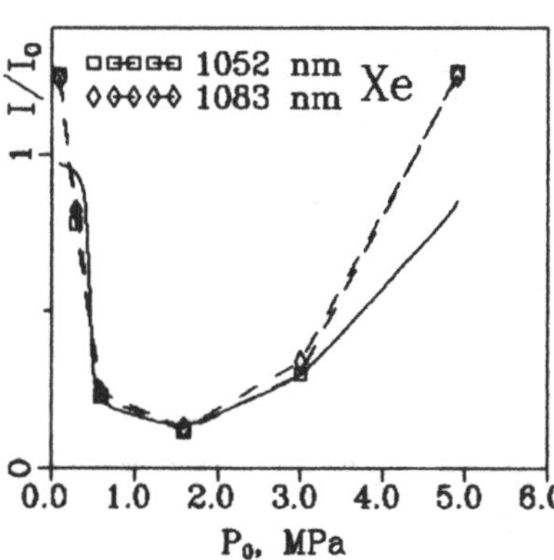

FIGURE 5.

$n_e=0.56\cdot10^{21}cm^{-3}$ (curve 7) and $n_e=1.2\cdot10^{21}$ cm^{-3} (curve 8) wavelength dependence of the normalized (to the black body radiation) plasma radiation. These curves were constructed on the basis of the measured in [2] values of the xenon plasma reflectivity in the case when $\nu/\omega=0.1$. A strong dependence is seen of the position of the absorption edge on the concentration free carriers.

Analogous to the actually observed dependencies in the IR range were observed in ref.[5] in noble gases exposed to the electronic beam.

The explanation of the observed "gap" and its initial pressure dependence is given below within the kinetic model based on the assumption that the essentially nonequilibrium slightly nonideal buffer plasma consisting of noble gas atoms (N=Xe, Ar), excited atoms N^*, excimers N_2^* and ions $N^+\cdot N_2^+$ is generated in the resisting gas by the shock compressed UV radiant flux.

The kinetic model

There is experimental evidence [6] that at elevated values of pressure the most effective of all kinetic processes [1,8] of excitation and ionization occurring in the buffer layer of unperturbed gas under the influence of the UV flux are:

$$\gamma_1 + 2N \Rightarrow N_2^*, \tag{1}$$

$$\gamma_2 + 2N \Rightarrow N_2^+ + e^- \tag{2}$$

Here and below γ_i stand for the UV photons of certain characteristic energy. Since the resonance radiation of excited atoms and molecules is trapped [1], one should account for the quenching of metastable states in collisions:

$$N^* + N \Rightarrow 2N^+ + \gamma_3, \tag{3}$$

$$N^* + e^- \Rightarrow N + e^- + \gamma_4, \tag{4}$$

$$N_2^* \Rightarrow 2N + \gamma_1', \tag{5}$$

and notice that the lowest states of excimers $^{1,3}\Sigma_u^+$ are not crossed by the repulsion curve [1]. In Eqs.(2), (4) e^- obviously stands for electrons.

If the characteristic time of the process Eq.(1) $\tau_0=(k_0n^2)^{-1}$ is small compared to those of radiational decay of excited atoms τ_1 and excimers τ_2, the net number density of ions (n^+, n_2^+) and excited states (n^*, n_2^*) in the buffer gas $\nu(t)=n^+(t) + n_2^+(t) + n_2^*(t)$ can be evaluated as

$$\nu(t) \approx \frac{c}{D} n_\gamma^{(0)} \left[\exp\left[-\frac{x-Dt}{D\tau_0} \right] - \exp\left[-\frac{x}{D\tau_0} \right] \right], \tag{6}$$

where n is the number density of ground state atoms calculated taking into account the second virial coefficient; c and D are the light and shock wave velocities, $n_\gamma^{(0)}$ is the concentration of γ_1 photons, x is the coordinate across the shock wave front, x>0 within the buffer gas. The estimates for the pumping parameter k_0 are given in Table 1.

Table 1

gas	state	K_0 $(10^{-34}$ cm^6/sec)
Xe	$^1\Sigma_n^+$	4.0
	$^3\Sigma_n^+$	0.4
Ar	$^1\Sigma_n^+$	2.0
	$^3\Sigma_n^+$	$6 \cdot 10^{-3}$

Then one can approximately resolve the nonlinear system of kinetic equations corresponding to the above processes to show that if the temperature of electrons $T_e \leqslant 3000K$,

$$n_2^*(t) = \frac{k\,n^2}{kn^2 + \Gamma'}\,\nu(t), \tag{7}$$

where $k=2.5 \cdot 10^{32}$ cm^6/sec in Xe and $k=0.7 \cdot 10^{32}$cm^6/sec in Ar [8] is the constant of the rate of the reaction

$$N^* + 2N \Rightarrow N_2^* + N;$$

$\Gamma' = \Gamma + \tau_2^{-1}$, τ_2 for the "long-living" state $^3\Sigma_n^+$ is [7], $\tau_2(Ar) = 3.7$ μsec, $\tau_2(Xe)=100$ ns ; Γ stands for the destruction of excimers due to the IR excitation

$$^{1,3}\Sigma_n^+ \Rightarrow {}^3\Pi_g, \tag{8}$$

i.e., the very process observed experimentally, Γ^{ex}, and the ionization, Γ^{ion}. Simple estimations give: $\Gamma^{ex} \approx (10^6 \div 10^7)$sec^{-1}, $\Gamma^{ion} \approx 10^{11}$sec^{-1}.

The following limitations stem from Eq.(7).

1) The number density of excimers for $kn^2 \geqslant \Gamma'$ saturates as a function of the initial pressure P_0; if a P_0 further increases, n_2^* does not grow, but the width of the layer containing excimers decreases

abruptly ($\propto n^{-2}$). According to the estimates based on the data taken from refs [7,8], in xenon the saturation takes place at $n \approx 2 \cdot 10^{21}$ cm^{-3}, i.e. at pressure $P_0 \approx 5.0$ MPa. This prediction has been verified experimentally.

ii) On the other hand, n_2^* does not decrease as long as $\tau_0^{-1} \gg \Gamma_{max} = \max(kn^2/\tau_2\Gamma_2, \ \tau_1^{-1})$ and $d \ll 1$ cm. In xenon $n_e \leqslant 10^{17}$cm^{-3}, and $\tau_1^{-1} \simeq k_1 n_e \leqslant 10^7$ sec^{-1} ($k_1 \simeq (10^{-9} \div 10^{-10})$cm^3/sec at $T_e \simeq 300$K [8]), and one can put $\Gamma_{max} \approx 10^7$sec^{-1}, thus for the "critical density $n_{cr} \approx \Gamma_{max}/k_0$ one has $n_{cr}(^1\Sigma_n^+) \simeq 1.6 \cdot 10^{20}$cm^{-3}, $n_{cr}(^3\Sigma_n^+) \approx 5 \cdot 10^{20}$cm^{-3}.
The former value of the number density of atoms corresponds to the initial pressure of xenon $P_0 \approx 0.6$ MPa, below which the absorption weakens significantly. This fact evidently points out that in xenon the process of excitation of $^1\Sigma_n^+$ excimers is more productive, and the photons required to excite $^3\Sigma_n^+$ do not penetrate through the thin layer d' of the buffer plasma with excimers in the state $^1\Sigma_n^+$.
The results of calculation of n_2^* and the thickness of the buffer plasma layer are given in Table 2 vs the initial pressure of the unperturbed gas.

Table 2

	Xe						Ar
P_0 (10^{5}Pa)	1	3	6	16	30	50	50
n_2^*(10^{12}cm^{-3})	1.3	25	1500	$9 \cdot 10^3$	$2 \cdot 10^4$	$5 \cdot 10^4$	400
d, cm	2.2	0.24	0.1	0.02	$6 \cdot 10^{-3}$	$3 \cdot 10^{-4}$	1

The experimental data for argon ($P_0 = 5.0$ MPa, $d \approx 1$cm) point out that $\tau_0(Ar) \gg \tau_0(Xe)$. The reason for this distinction leading to the formation of excimers Ar$_2^*$ in a thicker layer might be relatively wider windows of transparence in the absorption lines connected with the reaction

$$^1\Sigma_g^+ + \gamma_1^{(1)} \Rightarrow {}^1\Sigma_n^+ \tag{9}$$

(since the mass of the Ar atom is smaller the oscillation level spacing in Ar$_2^*$(≈ 200K) is about two times bigger than that in Xe$_2^*$), and the process

$$^1\Sigma_g^+ + \gamma_1^{(2)} \Rightarrow {}^3\Sigma_n^+ \tag{10}$$

in argon is more productive. Certainly, the resonance excitation Eq.(9) (compare to Eq.(1)) goes faster, but (due to the greater value of k_0, Table 1) in a thinner layer in comparison to the transition of

Eq.(10). Indeed, the life time $\tau_2 \simeq k_0^{-1}$ of ${}^1\Sigma_n^+$ is 5.5 ns (in Xe) and 4.2 ns (in Ar), and that of ${}^3\Sigma_n^+$ is 100 ns and 3.5 μs, respectively [7]. A simple estimate for $k_0(Ar)=u/dn^2 \simeq 5 \cdot 10^{-37} cm^6/s$ with d=1 cm (compare to Table 2) also means that the process Eq.(10) is more productive in argon.

Future calculations like those carried out for xenon show that
1) if $T_e \simeq 300K$ in argon, the initial pressure $P_0 \simeq 5.0$ MPa is critical, like in xenon, i.e., the absorption on excimers will be weaker both for $P_0 \gg 5$ MPa and $P_0 \ll 5$ MPa;
ii) the observation of absorption at $P_0 \gg 5$ MPa might indicate a higher temperature of buffer electrons.
The normalized intensity of the IR absorption was calculated according to the formula

$$I/I_0 = \exp (-\tau^{ex} n_2^* d) \qquad (11)$$

where $d \simeq 1 cm$ and $(I/I_0) \simeq 0.14$ for argon (the experimental value $(I/I_0)^{exp} \simeq 0.10$); and d=d' for xenon (see Fig.5, where squares and rhombuses represent experimental points and the curve calculated). The cross section of the absorption on excimers τ^{ex} was evaluated as $10^{-14} cm^2$ in xenon and $0.5 \cdot 10^{-14} cm^2$ in argon. This estimate of τ^{ex} can be obtained if one views the dipole transition Eq.(8) as a process inverse to the resonance fluorescence with the characteristic value of the transition dipole moment divided by the elementary charge taken to be equal to about 0.4 nm. The latter value is in an agreement with the data on characteristic dimensions of noble gas excimers [7].
In conclusion, though the above kinetic model is compared well with the experimental data, the evaluation of τ^{ex} remains to be improved.

REFERENCES

[1] V.E. Fortov, J.T. Iakubov, Physics of nonideal plasma, (USA, New York, 1989)

[2] V.B.Mintsev, Yu.B.Zaporoghets, Contr.Plasma Phys, **29**, 493 (1989)

[3] V.E.Fortov , V.E.Bespalov , M.I.Kulish , S.I.Kuz, in Strongly Coupled Plasma Physics (Yamada Science Foundation, 1990) p.571

[4] M.Peres, J.Costeraste. Shock Waves in Condensed Matter. (1987) p.703

[5] S.Arai, T.Oka, M.Kogoma, M.Imamura, J.Chem.Phys. **68**, 4595 (1978)

[6] B.L.Borovich, V.S.Zuev. Zh.Eksp.Theor.Phys. **58**, 1794 (1970)

[7] M.V.McCuscer, in Excimer Lasers (Springer Verlag, 1979)

[8] B.M.Smirnov, Ions and Exited Atoms in Plasmas (Atomizdat, Moscow, 1974)

LIGHT FROM DENSE PLASMAS

K. Günther

OSRAM GmbH

Nonnendammallee 41–61, D–1000 Berlin 1

Abstract

Nonideal plasmas are attractive subjects not only for the research on strongly coupled particles in ionized gases but also for many technical applications. Strong interaction forces are directly connected with a high power dissipation of the plasma which is important for switch gears, high power light sources, welding etc. This paper will explain the demands and necessities connected with a modern high intensity light source, the state of the art, and the future possibilities of dense plasmas as the working medium of discharge lamps.

Introduction

More than 90 per cent of the light which is produced in all kinds of light sources is emitted by plasmas. Two basic principles can be distinguished to transfer the electric energy fed into the

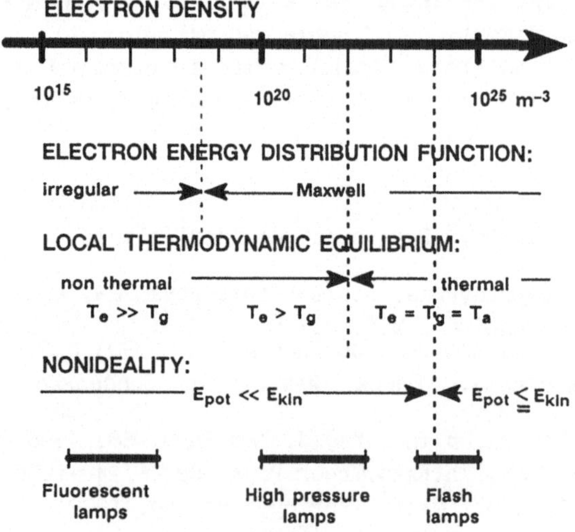

Fig. 1. Plasmas of discharge lamps
    T_e, T_g, T_a: Temperature of electrons, of the gas, and of excitation

lamp in radiation of a wanted spectral distribution. In the first case a low pressure plasma far from thermodynamic equilibrium is optimized for a preferred excitation of resonance levels and a highly efficient monochromatic uv radiation which can be transformed by means of suited phosphors into the visible spectrum. The second possibility is to optimize the chemical composition, the geometry, the pressure, and the spatial temperature distribution of a high pressure plasma in thermodynamic equlibrium in such a way that the essential radiative transitions are located in the wanted spectral range. Considering the range of temperature and pressure, nonideal effects must be expected with respect to the ionization equlibrium, the transport processes, and the radiation (see Fig. 1).

Application demands

The following properties should be optimised or kept within definite limits that a light source can solve a special illumination problem.
Efficacy is the most important feature of a source which describes the fraction of energy spent into physiological effective radiation

$$\eta = 680 \ \text{lm/W} \ \frac{\int \Phi_\lambda \cdot V_\lambda \, d\lambda}{P_{el}}$$

where Φ_λ is the spectral radiation flux, V_λ the eye efficiency curve, and P_{el} the electric power of the lamp.
The colour properties can be characterized by the correlated colour temperature CCT which is a rough measure of the energy distribution over the visible spectral range, and the colour rendering index CRI which describes the quality of colour rendering of the source compared with that of a black body at a temperature equal to the CCT. There are sophisticated procedures to calculate CCT and CRI from a given spectral distribution [1]. The demands to CRI depend on the application field and are very high for indoor and especially for residence lighting. The optimum CCT depends on the illuminance and is given by the well-known KRUITHOFF cosiness curve.
Luminance and flickerung must be considered for the construction of the lighting luminaire and to avoid stroboscopic effects in case of moving objects.
Furthermore, the full radiation flux can be retarded after switching on if the active medium in the lamps must be evaporated by means of a starting discharge. A warm lamp can need extremely high ignition voltages to restart.
The costs of an illumination system include power consumption, manufacturing effort of the lamps and luminaires, installation and power supply. They should be minimised considering the necessary illuminance level and quality. There are two principal ways to minimize the costs of illumination:
  – to improve the efficacy of the system
  – to improve the colour rendering properties of efficient lamps to substitute inefficient lamps
     having high colour rendering in indoor applications.
In this sense, research for the improvement of the efficacy and of the colour rendering properties must be seen in a close interdependence.

Efficacy and colour rendering

An absolute maximum radiation efficacy of 680 lm/W is established for a radiation which is concentrated on a wavelenght of 555 nm, an evaluation of colours is not possible of course. The combination of best colour rendering index CRI = 100 with the theoretical optimum for η yields 250 lm/W in case that the total radiation is distributed like that of a blackbody between 400 and 700 nm. An interesting compromise is the concentration of the radiation to three spectral lines at 470, 550 and 610 nm. OPSTELTEN et al. [2] calculated maximum efficacies of 375 lm/W at CRI = 82. It should be taken into consideration that the figures given before do not include the thermal and electrical losses within the light source.

The general problem is to transfer the electric input energy with minimum losses into the spectral range between 400 and 700 nm using appropriate elementary processes and/or setting favorable thermodynamic conditions.

Dense plasma radiation systems

Among the high pressure lamps a large group be summarized which essentially works on the basis of the optimized and well understood mercury discharge with efficacies up to 60 lm/W and CRI = 60.

The most effort to improve this system was spent in finding additives which fill up the mercury spectrum and increase η, CRI, and CCT. The aspects for their selection are
- favourable optical transitions
- energy of upper level < Hg levels
- sufficient vapour pressures at attainable temperatures
- chemical and physical compatibility with wall and electrode materials
- technologic problems on manufacturing

Approved compositons are Na, Sc and Na, Tl, In for medium CRI and high efficacy and mictures of rare earth halides for high CRI.

The maximum figures are determined by the limited wall loadings of the discharge tubes, they are expected to raise remarkably by the development of new ceramics and sealing technologies, and by the introduction of new complex compounds as radiating additive (see, e.g. [3]).

A second basic system uses the typical extreme high radiation efficiency of the resonance line of one–electron elements (in hydrogen, e.g., the L_α represents more than 50 per cents of the total radiation of the plasma under LTI conditions). The only resonance transition of such an element near the maximum of the eye efficiency curve belongs to sodium, and the development of alkali resistive Al_2O_3 ceramics together with stable sealing technologies allowed the use of rather high sodium and mercury pressures of about 0,1 and 1 bar, respectively. Pressures of this order of magnitude are necessary to broaden the resonance lines so that the efficiency losses due to self absorption are tolerable. High pressure sodium lamps show typical high efficacies but rather poor CRI and CCT data of 20 and 2000 K, respectively [4,5]. High pressure sodium lamps exhibit an excellent operation performance as long life, stability of colour temperature and light output because of their simple radiation system consisting of Na and Hg only. Therefore, many attempts have been made to improve the colour rendering properties, especially by increasing the radiation intensity in the blue and green spectral range.

A promosing approach to improve high pressure sodium lamps is the pulsed mode operation using medium or high pulse duty factors where the average power of the lamp is adjusted to be the same as at conventional operation. An instantaneous power input far above the allowed stationary thermal loading of the lamp heats the discharge plasma depending on its composition to temperatures which establish a new energy balance between heating and dissipation. The following effects are expected:

- Because the heat flux from the discharge is proportional to T and the radiation flux raises with exp(-E/kT) a more than linear increase of the efficacy should be expected (E: excitation energy of the radiation).
- For pulse lengths of 10^{-4} s or less the thermalization of plasma and wall proceeds so slowly that the heat flux from the plasma is hindered.
- The rise of temperature excites transitions from higher energy energy levels and fills the blue part of the spectrum.

Recent measurements of DAKIN and RAUTENBERG [6], BRATES and WYNER [7], and GÜNTHER et al. [8] yielded CCT > 3000 K and CRI > 80 at rather high efficacies. Nevertheless, there are remaining problems in understanding the dynamic processes during the rise and decay of the power pulses, in maintaining a minimum keep alive power, and in suppressing plasma resonances. The theoretical understanding of such discharge modes is limited until now because of the interference of the dynamic effects during the heating and decay periods with nonideal effects due to the high particle densities at low temperatures. At ratios of the mean potential and kinetic energies of the charge carriers γ and charged particle numbers in the Debye sphere N_D

$$\gamma = \frac{E_{pot}}{E_{kin}} = \frac{e^2(2n_e)^{1/3}}{kT} \qquad N_D = 2n_e \cdot \frac{4}{3}\pi D^3$$

$$\gamma < 0.1 \qquad \text{ideal}$$
$$0.1 < \gamma < 0.5 \qquad \text{weakly nonideal}$$
$$\gamma < 0.5 \qquad \text{nonideal}$$

on finds for the different high pressure lamp types the following nonideality parameters:

	n_e cm^{-3}	T K	γ	N_D
high pressure Hg	$5*10^{15}$	6000	0.06	6
high pressure Na	10^{16}	4000	0.1	3
high pressure Na (pulsed)	10^{18}	5000	0.4	0.3

Thus, the ionization equilibrium should be calculated using quantum statistical methods (see, e.g. KRAEFT et al. [9], the transport properties should include nonideal corrections according to GUENTHER and RADTKE [10], and the spectral distribution of the radiation must consider the effects of atomic interactions on the potential energy systems and the corresponding wave functions (see, e.g. de GROOT et al [11]).

A further interesting approach to create new light sources is the use of clusters in metal vapours as a source of intense visible radiation. In closed light sources these clusters can be generated using a dissociation partial pressure of the metal component of metal halide vapours which can be positioned above the saturation vapour of the metal in a special region of the radial temperature profile. In this region a continuous generation of clusters takes place as a result of the condensation of the supersaturated metal vapour. The radiation of such cluster bodies can be described as proportional to T^5 in contrast to the STEFAN-BOLTZMANN dependence of T^4 and a shift of the continuous spectral distribution towards shorter wavelenghts. The reason can be found in the dipole radiation character and in the scattering properties of small size particles [12]. Open questions are the structure of discrete energy level systems of chusters consisting of only few atoms and the possibilities to optimize their spectral emission characteristics to construct intense light sources of high efficacy at good colour rendition.

References.

[1] CIE-Publ. 13.2: Method of measuring and specifying colour rendering of light sources; 1974

[2] Opstelten, J.J.; Radiclovic, D; Verstegen, J.M. P.J.: Philips Techn. Rdsch. 35 (1975/76) 385

[3] Fischer, E.; E. Schnedler: Proc. 5th Int. Conf. on Sci. and Techn. of Light Sources, York 1989

[4] De Groot, J.;   J. van Vliet: The High Pressure Sodium Lamp; Philips Techn. Library; Deventer 1986

[5] Van Vliet, J.: Proc. XXth ICPIG, Il Ciocco 1991; Invited lecture

[6] Dakin, J.T.; T.H. Rautenberg: J. Appl. Phys. 56 (1984) 118

[7] Brates, N.: Wyner, E.F.: J. Illum. Eng. Soc. (1987) 50

[8] Günther, K.: Kloss, H.G.; Lehmann, T.; Radtke, R.; Serick, F.: Proc. 5th Int. Conf. Sci. and Techn. of Light Sources, York (1989)

[9] Kraeft, W.D.; Kremp, D.; Ebeling, W.; Roepke, G.: Quantum Statistics of Charged Particle Systems; Akademie-Verlag Berlin 1986

[10] Günther, K.; Radtke, R.: Electric properties of weaky nonideal plasmas; Akademie-Verlag Berlin

[11] De Groot, J.J.; Schleyen, J.; Woerdman, J.P.: Proc. 39th Gaseous Electronics Conference, Madison 1986

[12] Weber, B.: Die mikrowellenangeregte Cluster-Entladung: eine neuartige Lichtquelle. Dissertation Universität Karslruhe 1990

COLLISION FREQUENCY OF NON-IDEAL PLASMAS: INFLUENCE OF PLASMA OSCILLATIONS

M. SKOWRONEK

Laboratoire des Plasmas Denses. Université P. & M. Curie, Tour 12 E5,
4 pl. Jussieu, F-75252 Paris Cedex 05 (France).

Abstract

From the spectral results obtained in flashtube produced plasmas, it has been concluded that plasma oscillations may have the dominant role on the conduction electron scattering. The conductivity of different strongly coupled plasmas in a wide range of plasma parameters is studied, from which the electron-ion collision frequency is calculated. An approximation formula is proposed: $<v_{ei}> \approx f_p/n_D$ which describes well all the measurements. A relation may allow the selection of the efficient oscillations.

Introduction.

The electric conductivity is one of the most essential plasma characteristics that determines its dissipative heating and the interaction with the electromagnetic field. In ideal plasmas, the electron conduction can be described by Spitzer's theory[1]. In pulsed experiments, the plasma may reach a state in which the interaction potential energy between particles is of the same order as their kinetic energy. The plasma is then strongly coupled. Strong correlations in plasmas may be due, either to a small number of particles in Debye sphere (at low temperature and high electron density) or to the presence of turbulences due to drift currents[2-4]. Different theoretical models have been proposed to describe the physics of strongly coupled plasmas[5-10]. In order to check the validity of these models, reliable measurements from different plasmas covering a large range of electron densities and temperatures are necessary.

Experiments have been carried out with an explosive plasma generator working in air, neon, argon and xenon by Yu. V. Ivanov et al.[11]. A series of measurements in strong discharges under high caesium vapour densities have been reported by P. P. Kulik et al.[12]. Measurements on very intense capillary discharges in plastic tubes have been recently reported by R.L. Shepherd et al.[13]. But, in all these cases, the diagnostics seem to be not suffficiently precise.

Experiments performed in high pressure arcs [14], have been designed to meet the requirements of precise mesurements and reliable diagnostics. Typically, the plasma parameters are: $n_e \approx 10^{18} cm^{-3}$,

$T \approx 1$ eV, $n_D < 2$. The mean interaction parameter is $\Gamma = U/E$, where U is the potential energy density and E the kinetic energy density. Taking: $\Gamma = e^2/(6\pi\varepsilon_0 d_e kT)$ where d_e is the mean distance between particles defined by: $(4\pi/3)d_e^3(n_e+n_i) = 1$. As $n_e = n_i = n$, then:

$$\Gamma = [(\pi/3)^{1/3}e^2/3\pi\varepsilon_0 k]n^{1/3}/T = 2.26 \times 10^{-5} n^{1/3}/T \quad \text{MKS.} \quad (1)$$

In the typical case: $n = 10^{24} m^{-3}$, $T = 12,000$ K, $\Gamma = 0.2$.

Since the ionisation rate often exceed 0.8 and since they are well reproducible, flashtube produced plasmas are very convenient for the study of non-ideality effects.

We present, in the following, the main results obtained by optical and electrical measurements in flashtubes. The methods are only shortly summarized. For more details, one is asked to refer to preceding papers[14-19].

First, the methods allowing the determination of the radial profiles $n_e(r)$ and $T(r)$ are briefly described. The linewidths and shifts are affected by the non-ideality. This may give some insight on its effect on both types of collisions: long range collisions and near collisions. From conductivity measurements, the election-ion collision frequency is deduced. The results are discussed in the frame of a physical interpretation based on the possibility of a build-up of suprathermal plasma oscillations [20].

Experimental set-up and plasma diagnostics

The plasma are produced in fused quartz linear flashtubes having an inner diameter $\phi = 6$mm and a distance between the electrodes l =150 mm. The tubes are filled with Ar, Kr or Xe at initial pressures in the range 50 Torr to 1 atmosphere. After the breakdown, a simmer is produced in the lamp during 120ms with a current intensity less than 1A. This procedure ensures the centering of the discharge on the tube axis and avoids plasma-wall perturbative phenomena[15-16]. The main discharge is triggered. The intensity is set in the range 1 kA to 3 kA. It is constant within 1% during about 1ms. The voltage drop across the tube is measured trough a voltage divider with a precision of ±1%. The current intensity is measured by a coaxial shunt having a resistance: $R = 0{,}25$ mΩ ($\pm 0{,}2\%$). The signals are recorded on a 12-bit memory oscilloscope. An electronic camera is used to follow the filling of the tube by the plasma. A gated intensified photodiode array records each spectrum during 1µs at the time of best plasma filling.

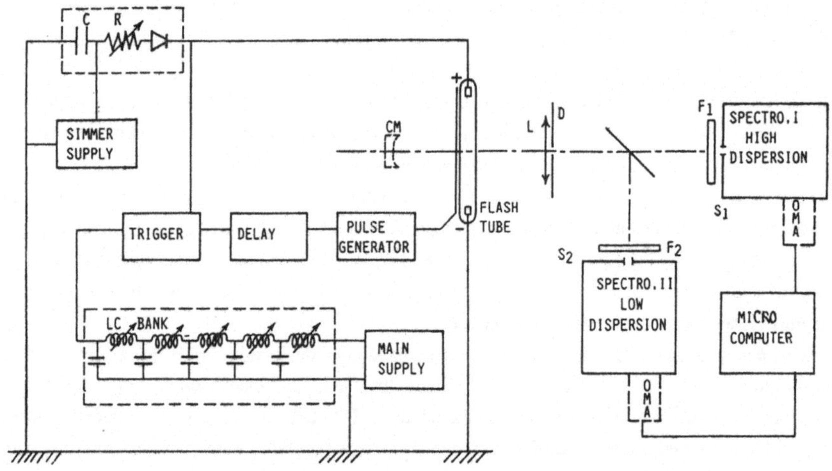

Fig. 1. Experimental set-up: L : focusing lens; D : diaphragm; F1 and F2 : interference filters; CM : cylindrical mirror for backlighting.

The electron density n_e is deduced from continuum measurements in the spectral range: 360 nm - 385nm where the plasma is optically thin and the spectrum is free of lines. The profil $n_e(r)$ is obtained trough Abel inversion. The temperature on axis T(0) is obtained from optically thick lines in the infrared. T(r) is obtained by means of special tubes with end-on view and a computing way assuming a constant pressure and LTE. T(r) and $n_e(r)$ are found to be flat over 2/3 of the tube diameter. $n_e(r)$ is also verified by laser interferometry. Another method with introduction of less than 3% H_2, using H_α line width has given a good agreement. As we have used independant methods we estimate a 5% precision on the plasma parameters.

Conclusions about the collisions in NIP.

The following conclusions are obtained from the spectral study:

a) the intensity of the continuum is in good agreement with Hofsaess's theory[21];

b) the measured linewidths are significantly smaller than the theoretical one; their variation with n_e is non-linear;

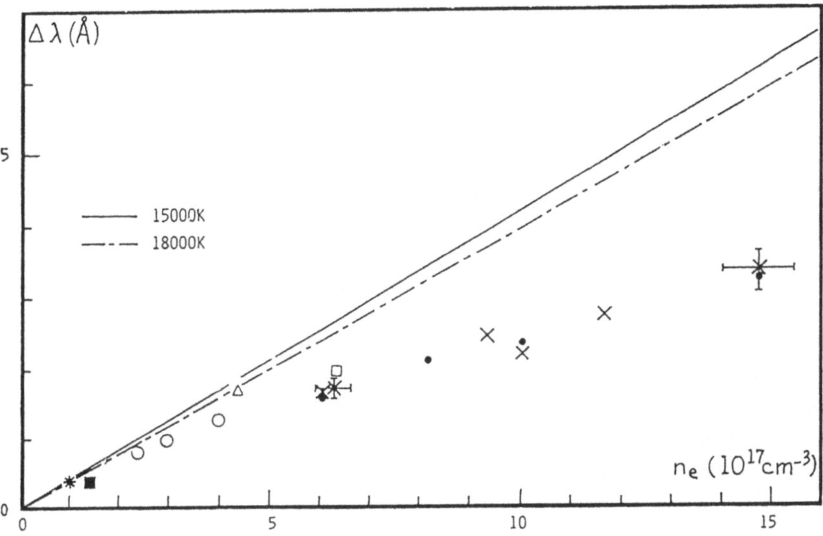

Fig. 2. Measured linewidth of ArII 4806 Å and 4847Å versus the electron density. ×4806 Å, •4847 Å: present work; * Chapelle et al.[22]; OKonjevic et al.[23]; □Labat et al.[24]; ΔRoberts[25]; ■Nick and Helbig[26].

c) the measured shifts of ionic lines are small; their variation with n_e is non-linear;

d) the measured H_α linewidths are in agreement with Griem's theory.

It is convenient to divide the collisions into two categories:

a) close-collisions are quite insensitive to non-ideality nor to collective waves; these collisions provide the main contribution to the recombination continuum. Also, these collisions are responsible of the H_α broadening.

b) far-collisions are strongly influenced by non-ideal effects; these collisions play the main role on the shift of noble gas lines and partially to their linewidth.

From the above mentionned observations, it may be deduced that, in non-ideal plasmas, far-collisions are strongly screened or diminished. As the plasma becomes less collisional, it seems that plasma waves can propagate with less damping. The possibility of a build-up of such plasma oscillations in a non-ideal plasma has been suggested and discussed[20]. An indication of the existence of such oscillations may be deduced from the profiles of different lines [27]. One example is displayed on figure 3.

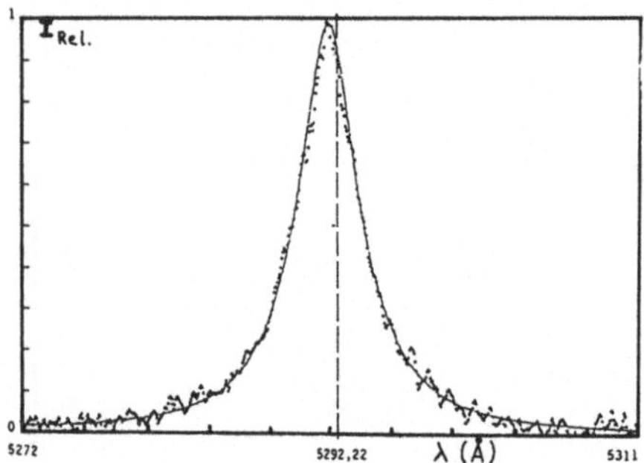

Fig.3. Experimental profile of Xe II 5292.2Å line. $N_e = 1.23\ 10^{24} m^{-3}$; T= 13,900K
FWHM = 4.8 Å; d = -0.6Å.
Modulation of the line is seen on the wing and also on the top of the line.

Conductivity values and theoretical background.

The conductivity measurements previously reported[19] agree well with those of other authors[28-29] in the same range of lowest electron density. We have deduced the electron-ion collision frequency ν_{e-i} from the conductivity σ through the formula: $\sigma = ne^2/m\langle \nu \rangle$ where $\langle \nu \rangle = \nu_{e-i} + \nu_{e-a} + \nu_{e-a*}$ and ν_{e-a} is the electron-atom (in the fundamental state) collision frequency and ν_{e-a*} is the electron-atom (in the excited state) collision frequency. On figures **4, 5 and 6** , are displayed the values ν_{e-i} versus $\sqrt{n}$, respectively in the case of Ar, Kr and Xe, after correction of the e-a and e-a* collision frequency, deduced from electric conductivity measurements.

Two types of conclusions may immediately appear:

a) the experimental values are significantly higher than those given by a crude extension of Spitzer's law;

b) for each tube (each initial pressure), n_{e-i} varies linearly with $\sqrt{n}$, proportionally to the plasma frequency.

Another mechanism has been invoked that gives an anomalous collision rate due to low frequency ion-sound turbulence, through oscillating microfields.

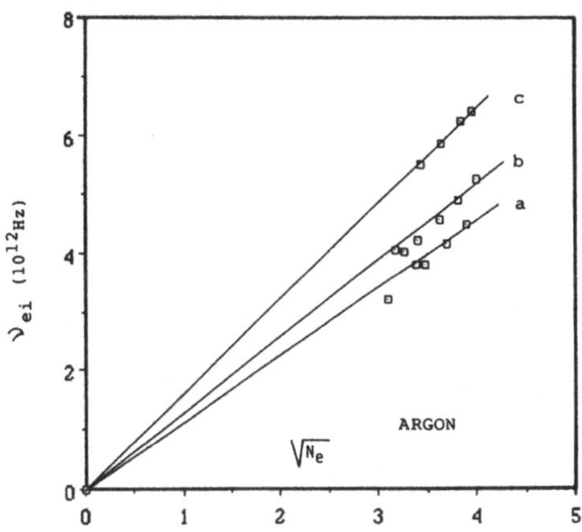

Fig. 4. v_{e-i} (in 10^{12}Hz) in Ar vs $\sqrt{n}$ ($\times 10^{12}$). Initial pressures : 200, 400, 600 Torrs.

The collision frequency would be given by the formula: $v_{e-i} \approx (W/nkT)\, \omega_{pe}$.
Here, ω_{pe} is the electron plasma frequency and W is the total energy in waves having the value [25-26] :

$$W/nkT = (M_i \varepsilon_0 E^2/2m_e nkT)^{1/4}. \quad (2)$$

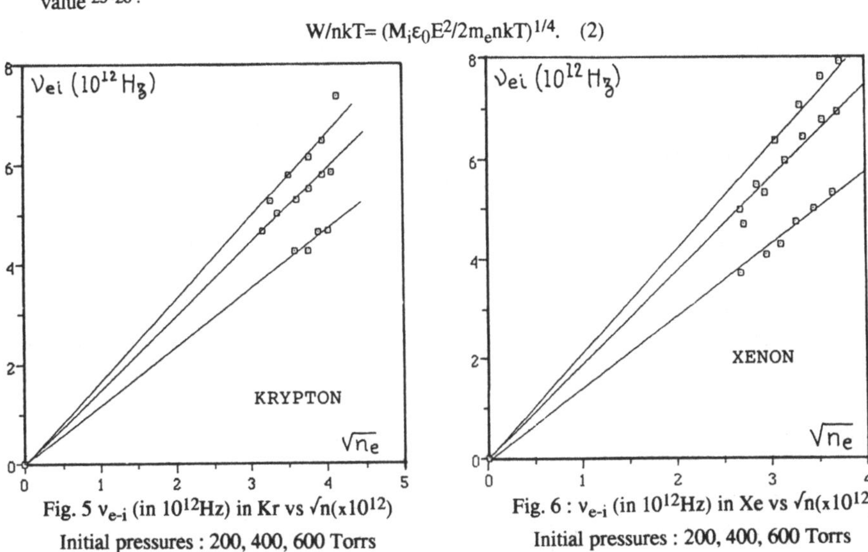

Fig. 5 v_{e-i} (in 10^{12}Hz) in Kr vs $\sqrt{n}(\times 10^{12})$
Initial pressures : 200, 400, 600 Torrs

Fig. 6 : v_{e-i} (in 10^{12}Hz) in Xe vs $\sqrt{n}(\times 10^{12})$
Initial pressures : 200, 400, 600 Torrs

With our plasma parameters, $W/nkT \approx 0.1$ when calculated from the above formula. It is in the same order of magnitude as the experimental one. An objection against this mechanism is that the dependance with the ion mass has not been verified.

In addition to accelerating a small number of particles to high energies and producing a change in the distribution function, the plasma turbulence acts to scatter the particles. Thus, the effective collision frequency depends on the level of turbulence. From a strongly turbulent plasma the average collision frequency $< v_{e-i}>$ is given by the following expression[31]:

$$< v_{e-i}> = \omega_{pe} f(n,T).$$

It is known that the average collision frequency is enhanced by a factor g over the value taken in the quiescent plasma case : $g = 1/n_D$. In a first attempt[19], we have tried to introduce the scattering of the conduction electrons by the plasma waves in a partial manner following Kurilenkov and Valuev[24] but the result is not convincing.

In this paper, the e-i collision frequency is analyzed starting from the expression:

$< v_{e-i}> = f_p/ n_D$ with : $f_p = 8.985 \sqrt{n}$ (MKS) and $n_D = 1.375 \times 10^6 \, (T^{3/2}/n^{1/2})$ (MKS).

Recently, Mohanti and Gilligan[34] have calculated a corrected screening distance r_s for the replacement of the Debye length λ_D in non-ideal plasmas, when the interaction parameter Γ varies in the range 0-1. They have shown that the ratio r_s / λ_D may be approximated by a simple function (for instance linear) and varies between 1 to 2 in the variation range of Γ. If one replaces λ_D by r_s in order to calculate a new value for n_D, the improvement brought by this new value is not significant.

Due to the dispersion relations of the plasma oscillations and the unknown relation between the wavelenth of the efficient oscillations and the mean electron energy, the plasma frequency has been approximated.

Approximation result.

The following approximation describes well the collision frequency in the entire range. Using the following variable change, with more convenient units:

$$n = 10^{24} N, \quad T = 10^4 T' \text{ and } < v_{e-i}> = 10^{12} nu \text{ th, the approximation becomes:}$$

$$nu \text{ th} = 4.783 * F_p/(N_D^\wedge(0.5 + 2.5 * N_D)$$

where $F_p = f_p * [1 + 1.5 * T'/(N*M/40)]$ and $N_D = n_D/(1+Z)^\wedge 1.5$ taking into account a dispersion relation for the plasma waves and the charge correction

The table **I** displays successively, the nature of the gas (the reference of the author), the temperature in unit 10^4 K, The electron density N_e in unit $10^{24} m^{-3}$, the conductivity in Siemens, the electron-ion collision frequency deduced from the conductivity nu ei in unit 10^{12} Hz , the charge of the ion, the value of the interaction parameter G, the number of particles in Debye sphere, the plasma frequency, and finally the ratio of the empirical theoretical value to the e-i experimental collision frequency. The agreement is good: for Vitel's experiments the standard deviation is about 5%. For all the experiments reported the stantard deviation is about 28% this is related to the poor precision of the plasma parameters and the conductivity measurements. The agreement is good on more than 5 orders of magnitude, despite the simplicity of the formula.

Conclusion

Different experimentalists have measured the conductivity of strongly coupled plasmas in a very wide range of plasma parameters. In the case of flashtube experiments a great care was taken to measure the plasma parameters with a great precision about 5%. In this case, nevertheless, the variation range of the plasma parameters is restricted. We have deduced from the spectral study (linewidth and lineshift) that, in these plasma having 1 to 2 particles in Debye sphere, that he most important electron-ion interaction which takes place at long distances (on the order of the Debye length l_D) is strongly screened.

As the plasma becomes less collisional, the proposed hypothesis that Langmuir plasma oscillations can exist in non-ideal plasmas[30] seems to be fulfilled. We have found that the collision frequency is proportional to:

$$F_p/(N_D{}^{(0.5+2.5*N_D)}) \qquad (3)$$

$$\text{with } F_p = f_p*[1+1.5*T'/(N*M/40)] \text{ and } N_D = n_D/(1+Z)^{1.5}$$

The approximation formula we have proposed is valid for the entire range of experiments. This would allow the selection of the most efficient oscillations.

Table1.

Gas	T1E4K	NeE24m-3	Cond Sm-1	Nuei1E12Hz	Z	gamma	nD	fp	Nuth/nuei
Ar	1,62	0,96	7720	3,20	1	0,14	2,89	8,80	1,008
[15]	1,69	1,14	8070	3,80	1	0,14	2,83	9,59	0,947
	1,74	1,20	8420	3,82	1	0,14	2,88	9,84	0,908
	1,80	1,36	8710	4,17	1	0,14	2,85	10,48	0,894
	1,81	1,51	8950	4,48	1	0,14	2,72	11,04	0,967
	1,57	1,01	6820	4,05	1	0,14	2,69	9,03	0,982
	1,60	1,15	7250	4,22	1	0,15	2,60	9,64	1,057
	1,64	1,31	7600	4,56	1	0,15	2,52	10,28	1,093
	1,68	1,45	7840	4,89	1	0,15	2,49	10,82	1,084
	1,70	1,60	8030	5,24	1	0,16	2,41	11,37	1,113
	1,56	1,17	5680	5,51	1	0,15	2,48	9,72	0,913
	1,59	1,32	6020	5,84	1	0,16	2,40	10,32	0,953
	1,61	1,47	6270	6,24	1	0,16	2,32	10,89	0,987
	1,65	1,56	6470	6,40	1	0,16	2,33	11,22	0,965
	1,58	1,07	6490	4,04	1	0,15	2,64	9,29	1,048
Kr	1,62	1,28	7150	4,25	1	0,15	2,51	10,17	0,964
[15]	1,64	1,41	7640	4,26	1	0,16	2,43	10,67	1,063
	1,69	1,52	7540	4,64	1	0,15	2,45	11,08	0,987
	1,74	1,62	7860	4,66	1,01	0,15	2,48	11,44	0,984
	1,43	1,00	5280	4,65	1	0,16	2,35	8,99	0,925
	1,46	1,13	5410	5,02	1	0,16	2,28	9,55	0,947
	1,50	1,30	5850	5,29	1	0,16	2,22	10,24	0,997
	1,53	1,43	6080	5,52	1	0,17	2,18	10,74	1,020
	1,56	1,56	6250	5,80	1	0,17	2,15	11,22	1,027
	1,57	1,66	6480	5,86	1	0,17	2,10	11,58	1,079
	1,39	1,07	4860	5,28	1	0,17	2,18	9,29	0,963
	1,42	1,23	5070	5,80	1	0,17	2,10	9,96	0,980
	1,45	1,43	5320	6,16	1	0,18	2,01	10,74	1,044
	1,48	1,56	5510	6,49	1	0,18	1,98	11,22	1,043
	1,49	1,72	5750	7,40	1	0,18	1,91	11,78	1,003

Xe	1,26	0,72	4800	3,72	1	0,16	2,29	7,62	0,978
[15]	1,30	0,88	5230	4,09	1	0,17	2,17	8,43	1,056
	1,34	0,97	5430	4,29	1	0,17	2,17	8,85	1,050
	1,35	1,07	5730	4,74	1	0,17	2,09	9,29	1,052
	1,39	1,20	6070	5,00	1	0,17	2,06	9,84	1,065
	1,43	1,34	6310	5,33	1	0,17	2,03	10,40	1,064
	1,14	0,74	3780	4,70	1	0,18	1,95	7,73	1,030
	1,20	0,87	3980	5,32	1	0,18	1,94	8,38	0,971
	1,21	1,00	4190	5,97	1	0,19	1,83	8,99	0,986
	1,24	1,12	4380	6,43	1	0,19	1,79	9,51	0,981
	1,26	1,26	4630	6,77	1	0,19	1,73	10,09	1,018
	1,30	1,38	4930	6,95	1	0,19	1,73	10,55	1,026
	1,10	0,72	3410	4,97	1	0,18	1,87	7,62	1,019
	1,12	0,82	3530	5,48	1	0,19	1,80	8,14	1,018
	1,16	0,94	3710	6,37	1	0,19	1,77	8,71	0,940
	1,21	1,10	3910	7,08	1	0,19	1,74	9,42	0,915
	1,24	1,25	4110	7,63	1	0,20	1,70	10,05	0,921
	1,26	1,40	4350	7,92	1	0,20	1,64	10,63	0,963
Ar	2,22	28,00	18500	39	1,01	0,31	0,86	47,54	1,212
[11]	2,03	55,00	15000	91	1	0,42	0,54	66,63	0,795
	1,93	81,00	16000	113	1	0,51	0,41	80,87	0,805
	1,90	140,00	22500	83	1	0,62	0,30	106,31	1,502
	1,78	170,00	21000	87	1	0,70	0,25	117,15	1,631
Xe	3,01	250,00	44500	143	1,64	1,27	0,45	142,07	1,160
[11]	2,75	590,00	65500	241	1,63	1,83	0,26	218,24	1,178
	2,70	790,00	68500	298	1,58	1,93	0,22	252,54	1,140
	2,61	1400,00	63000	573	1,74	2,93	0,15	336,19	0,886
	2,51	1600,00	67000	565	1,78	3,33	0,14	359,40	1,005
	2,46	2000,00	82000	527	1,87	4,04	0,12	401,82	1,280
	2,27	2000,00	75000	491	1,87	4,38	0,11	401,82	1,432
Ne	1,98	11,00	10000	24	1	0,25	1,16	29,80	1,258
[11]	1,96	19,00	10000	30	1	0,31	0,87	39,16	1,397
air[11]	1,10	1,30	1200	9,3	1	0,22	1,39	10,24	1,690
Cs	0,6	18,00	2600	175	1	0,99	0,15	38,12	0,294
[12]	0,8	32,00	4200	202	1	0,90	0,17	50,83	0,328
	1,2	32,00	7600	96	1	0,60	0,32	50,83	0,615
	1,6	25,00	8900	78	1	0,41	0,56	44,93	0,613
	0,6	48,00	2500	466	1	1,37	0,09	62,25	0,205
	0,8	98,00	4200	625	1	1,30	0,10	88,95	0,213
	1,2	100,00	9200	299	1	0,87	0,18	89,85	0,387
	1,6	80,00	14300	156	1	0,61	0,31	80,36	0,599
	0,6	250,00	8100	711	1	2,37	0,04	142,07	0,407
	0,8	260,00	9200	689	1	1,80	0,06	144,88	0,368
	1,2	240,00	16100	385	1	1,17	0,12	139,20	0,518
	1,6	220,00	19200	312	1	0,85	0,19	133,27	0,545
CH[13]	11,6	60000,00	40000	42000	3	6,84	0,22	2200,87	0,084

References

[1] Lyman Spitzer Jr., Physics of fully Ionized Gases (Interscience, New-York, 1967) 2nd revised edition.

[2] S. M. Hamberger and M. Friedman, Phys. Rev. Lett. **21**, 674 (1968).

[3] T. M. O'Neil, Phys. Rev. Lett. **25**, 995 (1970).

[4] Setsuo Ichimaru, Phys. Rev. A **15**, 744 (1977).

[5] Yu. K. Kurilenkov and A. A. Valuev, Beitr. Plasma Physik, **24**, 161 (1984).

[6] Setsuo Ichimaru and Shinegori Tanaka, Phys. Rev. A, **32**, 1790 (1985).

[7] G. Rinker, Phys. Rev. B, **31**, 4207 (1985).

[8] W. D. Kraeft, D. Kremp, W. Ebeling and G. Röpke, Quantum Statistics of Charged Particles, Plenum Press, New-York and London (1986).

[9] G. Rinker, Phys. Rev. A, **35**, 1284 (1988).

[10] G. Röpke and R. Redmer, Phys. Rev. A, **39**, 907 (1989).

[11] Yu. V. Ivanov, V.B. Mintsev, V.E. Fortov and A. N. Dremin, Sov. Phys. JETP, **44**, 112 (1976).

[12] P. P. Kulik, V. A. Ryabii and I. V. Ermokhin, Non-ideal Plasmas (in russian), Energoatomizdat, Moscou (1984).

[13] R. L. Shepherd, D.R. Kania and L.A. Jones, Phys. Rev. Lett., **61**, 1278 (1988).

[14] K. Günther, M. M. Popovic, S.S. Popovic and R. Radtke, J. Phys. D: Appl. Phys., **9**, 1139 (1976).

[15] Y. Vitel, M. Skowronek, K. Benisty and M. M. Popovic, J. Phys. D: Appl. Phys., **12**, 1125 (1979).

[16] Y. Vitel and M. Skowronek, Revue Phys. Appl., **22**, 193 (1987).

[17] Y. Vitel, J. Phys. B: At. Mol. Phys., **20**, 2327 (1987).

[18] Y. Vitel and M. Skowronek, J. Phys. B: At. Mol. Phys., **20**, 6477 and 6493 (1987).

[19] Y. Vitel, M. Skowronek and A. Mokhtari, J. Phys. B: At. Mol. Phys., **23**, 651 (1990).

[20] V. M. Batenin, M.A. Berkovskii, A. A. Valuev and YU. K. Kurilenkov, High Temperature, **25**, 145 (1987) and **25**, 299 (1987).

[21] D. Hofsaess, J. Quant. Spectrosc. Radiat. Transfer, **19**, 339(1978).

[22] J. Chapelle, A. Sy, F. Cabannes and J. Blandin, J. Quant. Spectroscop. Radiat. transfer, **8**, 1201(1968).

[23] N. Konjevic, J. Labat, Lj. Cirkovic and J. Puric, Z. Physik, **230**, 35(1970).

[24] J. Labat, S. Djenize, Lj. Cirkovic and J. Puric, J. Phys. B: Atom. Molec. Phys., 7,1174, (1974).

[25] D.E. Roberts, J. Phys. B: Proc. Phys. Soc., **1**, 53(1968).

[26] K.P. Nick and V. Helbig, Physica Scripta, **33**, 55(1986).

[27] Y. Vitel, Thèse de doctorat es sciences, décembre 1986, Univ. P. et M. Curie.

[28] M. M. Popovic, VIthSPIG, Split (Yougoslavia), Invited Papers, p. 651, M.V. Kurepa, Editor , Beograd (1972).

[29] S.S. Popovic, PhD Thesis, Univ. of Beograd, Yougoslavia,(1977).

[30] Yu. K. Kurilenkov and A.A. Valuev, Beitr. Plasmaphysik, **24**, 161 and 529 (1984).

[31] B.B Kadomtsev, Phénomènes collectifs dans les plasmas, traduction française, Editions Mir, Moscou(1979).

[32] D.L. Book, NRL Plasma Formulary, revised 1987, NRL Publication 0084-4040.

[33] Setsuo Ichimaru, Basic Principles of Plasma Physics, W. A. Benjamin editor, Reading, Mass. USA (1973).

[34] R. B. Mohanti and J.G. Gilligan, J. Appl. Phys., **68**, 5044 (1990).

Gas-Liner Pinch as a Source for High Density Plasma Spectroscopy

N. I. Uzelac[†], S. Glenzer, and H.-J. Kunze
Institut für Experimentalphysik V, Ruhr-Universität, 4630 Bochum, Germany
[†]Permanent address: Institute of Physics, 11000 Belgrade, Yugoslavia

Abstract

The gas-liner pinch is used for spectroscopic investigations in dense ideal plasmas. Features are discussed that make the device also suitable for observing Stark broadening and shifts of spectral lines in weakly non-ideal plasmas.

1 Introduction

Investigations of Stark broadening are of great interest for understanding the interaction of emitters with charged particles in a plasma, and reliable data on profiles are an important and often the only available tool for plasma diagnostics. Standard impact-theory approximations, which are succesfully used for evaluating Stark broadening in ideal plasmas, are not valid in the non-ideal regime (see e.g. Ref. [1]). Therefore, reliable measurements of line shapes in non-ideal plasmas are of the utmost importance for plasma diagnostics, as well as for establishment of an adequate theory.

An ideal plasma source for investigating Stark broadening should be: (1) homogeneous,(2) stationary, (3) optically thin, (4) capable of achieving relatively high electron temperatures ($T_e > 10000$K) in order to obtain sufficient numbers of ionized atoms, (5) capable of achieving relatively high electron density ($N_e > 10^{17}$ cm^{-3}) to make Stark broadening the predominant and readily measurable line broadening mechanism, and (6) accessible to reliable plasma diagnostic methods.

The value of the non-ideality parameter γ defined as $\gamma = Ze^2(N_e + N_i)^{1/3}/kT_e$, where N_e and N_i are concentrations of charge carriers, T_e is the electron temperature, and Z is the spectroscopic charge number, should be $0.1 \leq \gamma \leq 0.5$ (the weakly non-ideal case), or $\gamma > 0.5$ for the non-ideal case.

2 The Plasma Source

A convenient plasma source for such purposes is the gas-liner pinch shown schematicaly in Fig. 1 developed at the Ruhr-Universität Bochum [2, 3, 4, 5]. It is a large-aspect ratio (ratio of outer shell diameter to distance between electrodes) z-pinch, where the so-called driver gas (usually hydrogen or helium) is injected through a fast acting electromagnetic valve with an annular nozzle, forming initially a hollow gas cylinder near the wall. After preionization (a 50 nF capacitor charged to 20 kV is discharged through 50 annularly mounted needles) a 11.1μF capacitor bank (25 - 35 kV) is discharged through the initial plasma shell resulting in a compressed plasma column 1-2 cm in diameter and 5 cm in length.

Another fast valve independently injects the test gas (containing the element under investigation) along the axis of the discharge tube. If this injection is properly timed, the atoms and ions of the test

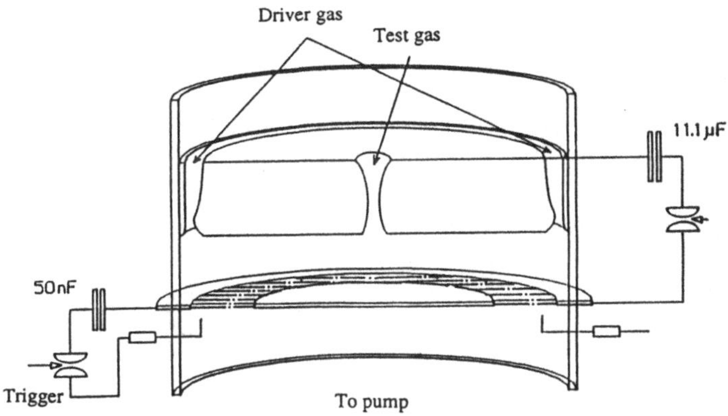

Figure 1: Schematic of the gas-liner pinch

gas remain concentrated in the central part of the discharge where the plasma is rather homogeneous, the colder and less dense boundary layers being located in the envelope formed by the driver gas.

Normaly one arrives at electron densities N_e up to a few times 10^{18} cm^{-3} and temperatures kT_e up to 40 eV. However, with heavier elements (like e.g. Kr or Xe) as test gas one can get to high densities (of the order of 10^{18} cm^{-3}) at the same time keeping the temperature low enough (a few eV) to arrive at the weakly non-ideal or even non-ideal plasma conditions.

Features of the gas-liner pinch Features of the plasma produced in the gas-liner-pinch that make it a suitable source for plasma spectroscopy are:
-Test gas atoms and ions emit radiation only from a homogeneous plasma region, so no Abel inversion is needed although observations are made side-on.
-Radiative transport effects are easily checked by variation of the test gas concentration.
-No distortion of the line by absorption in the cool boundary layers since no test gas atoms or ions are present there.
-By changing times of observation after maximum compression, one may scan the radiation for different plasma parameters.

3   Plasma Diagnostics

Electron densities and ion temperatures at the center of the plasma are determined by 90° Thomson scattering of intense laser radiation by the plasma electrons [6]. At high densities, the collective regime with the scattering parameter $\alpha > 1$ is reached, the spectral width of the scattering spectrum reflecting the thermal ion motion. The evaluation procedure takes into account also effects on the scattering spectra caused by the impurity ions [7], thus yielding ion and electron temperatures (T_i and T_e) and the temperature and concentration of the impurity ions (T_{imp} and N_{imp}). From the shapes of the spectra macroscopic plasma turbulence, instabilities or relative drifts of ions and electrons can be recognized if present.

The accuracy of the measured plasma parameters depends on the gas under investigation, but generally is in the range of 10% to 20% for both electron density and temperature.

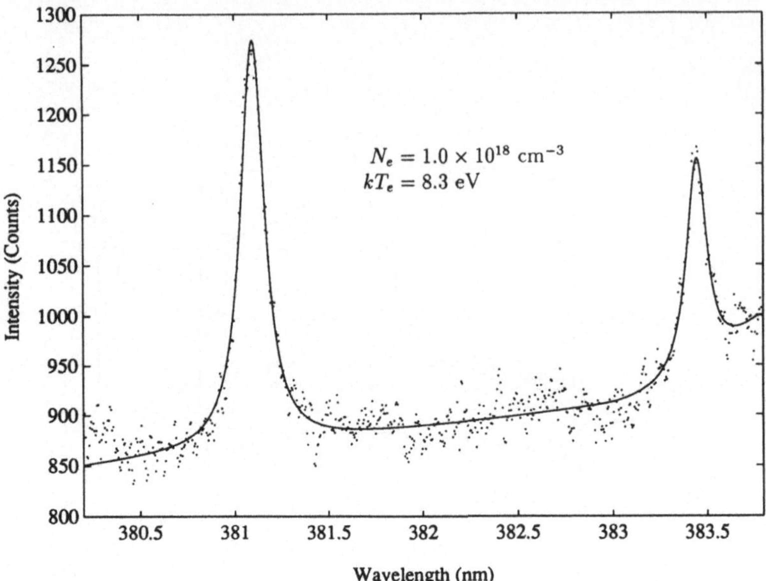

Figure 2: A recorded line spectrum of the $3s\ ^2S - 3p\ ^2P^o$ transitions in Ovi for a value of the non-ideality parameter $\gamma = 0.11$; dotted line, recorded; full, Voigt function best fit.

4   Plasma Spectroscopy

Line profiles can be obtained for different combinations of plasma parameters by recording spectra at different times during the discharge, starting at the point of maximum compression and observing the emission in the decay phase of the plasma as long as it can be considered homogeneous with respect to the test gas.

Spectrocopic investigations of the plasma produced in the gas-liner pinch can be carried out both side- and end-on with two spectrographs, both including gated OMA systems. Line profiles are recorded on a single shot basis in the 200 – 850 nm range with linear reciprocal dispersions of 0.04 – 0.004 nm/pixel and in the VUV in the 33 – 190 nm range with 0.02 – 0.005 nm/pixel resolution.

Fig. 2 shows a spectrum of the 3s-3p doublet in Ovi obtained at weakly non-ideal conditions ($\gamma = 0.11$). Investigations of profiles of those lines have not shown non-ideality effects, but the goal of these experiments had been the investigation of Stark widths along the isoelectronic sequence of Li, so non-ideality was not specifically strived for.

Serious experimental problems usually encountered with line profile measurements in plasmas with higher density are the occurance of self-absorption and inhomogeneity in the plasma.

Self-absorption   As already mentioned one of the features of the gas-liner pinch is the possibility of controlling the optical depth of the test gas plasma. By injecting less gas into the discharge, one can decrease the concentration of the absorbers, while electron density and temperature remain essentialy unchanged (see Fig. 3).

The presence of self-absorption can be easily checked by placing a mirror behind the pinch and observing the increase in the radiation intensity over the line profile.

Homogeneity   The homogeneity with respect to the test gas ions can be checked by recording and comparing continuum emission and line intensities over the cross section of the plasma column.

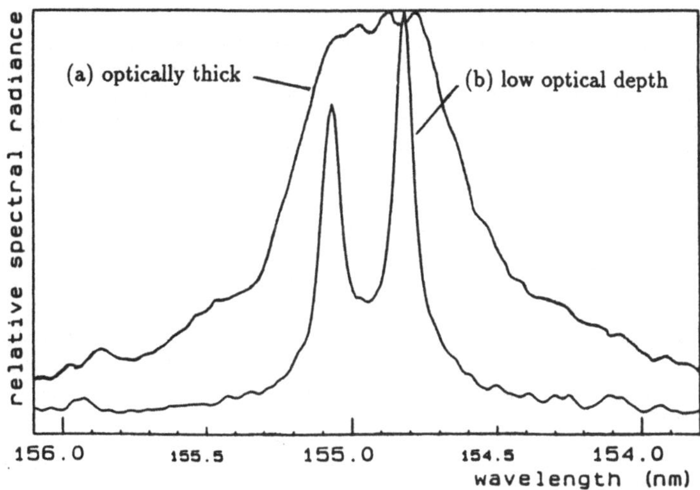

Figure 3: CIV resonance lines 2s-2p with different amounts of methane as test gas [8]. The values of plasma parameters are roughly the same for both (a) and (b), but for achieving the optically thick line (a), ten times more test gas was used than for (b).

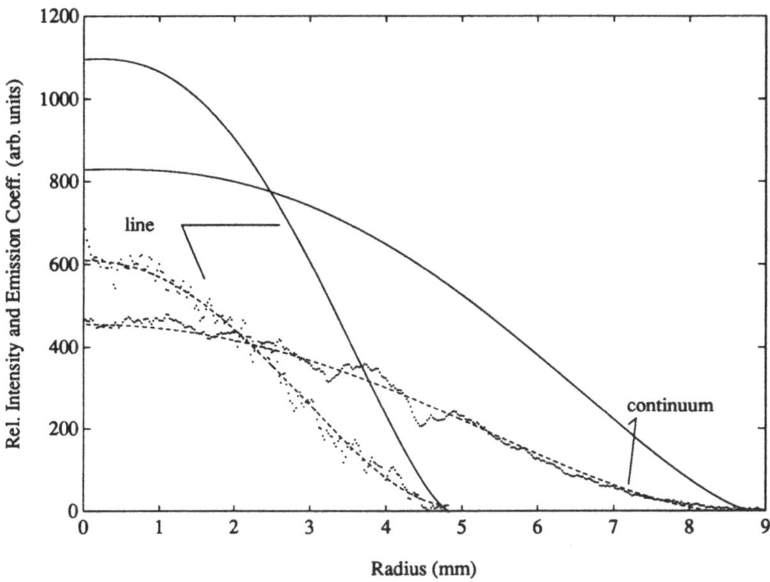

Figure 4: Radial distribution of line and continuum radiation from the gas-liner pinch plasma. Dotted lines are recorded intensities, and full lines are the corresponding radial emission coefficients obtained as Abel inversions of polynomial fits (dashed) to those intensities.

This is achieved by imaging the cross section of the plasma onto the entrance slit of the monochromator, the intensity distribution along the height of the entrance slit thus reflecting directly the radial distribution of the radiance in the respective wavelength interval. The intensity distribution is recorded with an OMA system, the diode array being aligned along the height of the exit slit. Fig. 4 shows the emission of line radiation confined to the central part of the discharge as compared to the continuum radiation which extends to the outer radius of the plasma column. One can deduce from this example that the assumption of line radiation coming from a homogeneous plasma region, contributes to the estimated error of Stark width with less than 3 %.

5   Conclusions

The preceding considerations show that the gas-liner pinch has great advantages in aspects that are critical in the spectroscopy of non-ideal plasmas. Both homogeneity and self-absorption of the plasma region under investigation can be tested, and to a certain extent also avoided. By appropriate choice of the test gas and the operating parameters we expect to approach non-ideality conditions, not losing any of the advantages of the gas-liner pinch plasma mentioned above.

References

[1] N. Konjević and N. I. Uzelac, J. Quant. Spectrosc. Radiat. Transfer **44**, 61 (1990).

[2] H.-J. Kunze, in *Spectral Line Shapes*, Vol. **4**, edited by R. J. Exton (A.Deepak Publ., Hampton, 1987).

[3] K.-H. Finken and U. Ackermann, Phys. Letters **85A**, 278 (1981).

[4] K.-H. Finken and U. Ackermann, J. Phys. D: Appl. Phys. **15**, 615 (1982).

[5] S. Glenzer, J. Musielok, and H.-J. Kunze, Phys. Rev. A **44**, 1266 (1991).

[6] H.-J. Kunze, in *Plasma Diagnostics*, edited by W. Lochte-Holtgreven, (North-Holland, Amsterdam, 1968).

[7] A. W. DeSilva, T. J. Baig, I. Olivares, and H.-J. Kunze, Phys. Fluids [accepted for publication].

[8] F. Böttcher, J. Musielok, and H.-J. Kunze, Phys. Rev. A **36**, 2265 (1987).

NUCLEAR FUSION IN DENSE PLASMAS: SUPERNOVAE TO ULTRAHIGH-PRESSURE LIQUID METALS

Setsuo   Ichimaru

Department of Physics, University of Tokyo

Bunkyo, Tokyo 113, Japan

Abstract

This review begins with classifying nuclear reactions in three elements: *binary processes, few-particle processes,* and *many-particle processes,* and thereby elucidates the special features for the nuclear fusion in *dense plasmas.* These analyses are then applied to estimation of the nuclear reaction rates in specific examples of the dense plasmas, namely, ^{12}C-^{12}C reactions in a white-dwarf progenitor of supernova, p-p reactions in the solar interior, d-d reactions in palladium hydride, and p-d or p-^7Li reactions in pressurized liquid metals. The special role that the many-particle processes play in dense plasmas is remarked and the similarilty between nuclear reactions in supernovae and those projected in the ultrahigh-pressure liquid metals is particularly emphasized.

Dense Plasmas: Examples

Dense plasmas under present consideration include the solar interior (SI), a white-dwarf progenitor of supernova (SN), isotopes of hydrogen in metal hydrides (MH) such as palladium hydride, and pressurized liquid metals (PM).

The Sun has the radius, $R_S \approx 6.96 \times 10^{10}$ cm, and the mass, $M_S \approx 1.99 \times 10^{33}$ g. Its mass density is 1.41 g/cm^3 on average, and $\sim 1.56 \times 10^2$ g/cm^3 near the center. The hydrogen contents (in mass ratio) are 0.36 near the center and 0.73 near the surface. The total luminosity is $L_S \approx 3.85 \times 10^{26}$ W. Hence the average luminosity per unit mass is $\sim 1.93 \times 10^{-7}$ W/g, which is to be accounted for by the rate of p-p reactions.

For a progenitor of supernova, we consider a carbon-oxygen white dwarf with a central mass-density of $10^7 \sim 10^{10} \mathrm{g/cm^3}$ and temperature of $10^7 \sim 10^9 \mathrm{K}$. Thermonuclear runaway leading to supernova explosion is expected to take place when the thermal output due to nuclear reactions exceeds the rate of energy losses. Assuming that neutrino losses are major effects in the latter, one estimates (*e.g.*, Arnett and Truran, 1969; Nomoto, 1982; Itoh *et al.*, 1989) that a nuclear runaway will take place when the nuclear power generated exceeds $10^{-9} \sim 10^{-8}$ W/g.

Nuclear reactions between hydrogen isotopes trapped in a metal hydride, such as palladium deuteride (PdD), offer a unique opportunity of studying reaction processes in *microscopically inhomogeneous* metallic environment of regular or irregular (*e.g.*, due to defects) lattice fields produced by the metal atoms. (For earlier experiments, see *e.g.*, Jones *et al.*, 1989; Ziegler *et al.*, 1989; Gai *et al.*, 1989.) The key to achieve an observable reaction rate ($>10^{-24} \mathrm{s}^{-1}$, say) in such a metal hydride is said to depend on the following factors which appear mutually contradictory (Ichimaru, Ogata, and Nakano, 1990): Reacting nuclei 1) are in *non-equilibrium* "fluidlike" or itinerant states, avoiding lattice trapping, and 2) have *lower* "effective temperatures" to be able to utilize the "many-particle processes" considered below.

Pressurized liquid metals offer another intersting environment in which to study nuclear reactions (Ichimaru, 1991). It will be shown that $d(p, \gamma)^3\mathrm{He}$ and $^7\mathrm{Li}(p, \alpha)^4\mathrm{He}$ reactions can take place at a power-producing level on the order of a few $\mathrm{kW/cm^3}$ if such a material is brought to a liquid-metallic state under an ultrahigh pressure on the order of 10 Mbar at a mass density of $3\text{-}7\mathrm{g/cm^3}$ and a temperature of 500-700K, slightly above the estimated melting conditions for hydrogen. Such a range of physical conditions may be accessible through an extention of ultrahigh-pressure metal technology (*e.g.*, Nellis *et al.*, 1988; Mao, Hemley, and Hanfland, 1990).

Plasma Parameters

We consider reaction rates between nuclei of species, i and j (which may be the same), in a dense matter at mass density ρ_m, temperature T and electron density n_e. The *ion-sphere radius* (*e.g.*, Ichimaru, 1982) for the i-species with charge number Z_i is given by

$$a_i = \left(\frac{3Z_i}{4\pi n_e} \right)^{1/3},$$ (1)

and an average internuclear spacing,

$$a_{ij} = \frac{a_i + a_j}{2}.$$ (2)

Here the condition for charge neutrality, $\sum_i Z_i n_i = n_e$, has been assumed.

The (unscreened) Coulomb coupling parameters are defined as

$$\Gamma_{ij} \;=\; \frac{Z_i\,Z_j\,e^2}{a_{ij}\,k_B\,T} \tag{3}$$

where k_B is the Boltzmann constant. The ratio between a thermal de Broglie wavelength and an average internuclear spacing is

$$\Lambda_{ij} \;=\; \frac{\hbar}{a_{ij}\,\sqrt{2\,\mu_{ij}\,T}} \tag{4}$$

with μ_{ij} denoting the reduced mass between i and j. The system of nuclei may be regarded as classical when $\Lambda_{ij} < 1$.

Light particles such as electrons and muons in the system may act effectively to modify (*i.e.*, to screen) the internuclear potentials. Let such a screening function be denoted by $S(r)$. Since $S(r{=}0){=}1$, an expansion

$$S(r) \;=\; 1 - r/D_S \;\ldots \tag{5}$$

defines the short-range screening length D_S. A useful example for the screening

TABLE I   PLASMA PARAMETERS

Assumed or calculated quantities	SN	SI	MH	PM	PM
Matter	C	H	Pd-D	D-H	Li-H
$\rho_m(g/cm^3)$	5×10^9	1×10^2	12.4	3.9	6.8
T (K)	1×10^8	1.5×10^7	300	600	550
n_i (cm^{-3})	2.5×10^{32}	6.0×10^{25}	6.2×10^{22}	7.8×10^{23}	5.1×10^{23}
a_{ij} (10^{-8} cm)	1×10^{-3}	0.16	1.56	0.54	0.62
Γ_{ij}	61	0.07	285	520	492
Λ_{ij}	0.203	0.011	0.235	0.458	0.362
Θ	1.8×10^{-4}	2.3	9.1×10^{-4}	1.1×10^{-3}	1.3×10^{-3}
D_S (10^{-8} cm)	3.2×10^{-3}	0.36	0.20	0.36	8.7×10^{-2}
Γ^S_{ij}	45	0.04	25.5	104	82

function has been provided (Ichimaru, 1991) as

$$S(r) = A \exp(-K_s r) + (B + C K_b r) \exp(-K_b r) .\qquad (6)$$

Here K_S represents a screening parameter of the "free" electrons, which depends on their degeneracy, $\Theta = k_B T / E_F$, with $E_F = \hbar^2 (3\pi^2 n_e)^{2/3} / 2m$, the Fermi energy. The last term of Eq. (6) describes the screening action of the "1s-bound" electrons when such need to be taken into account; $2/K_b$ then represents a corresponding "Bohr radius." In the cases where those screening effects of light particles are significant (Tanaka and Ichimaru, 1984; Ichimaru *et al.*, 1990; Ichimaru and Ogata, 1991), one introcuces *screened* Coulomb-coupling parameters as

$$\Gamma^S_{ij} = \Gamma_{ij} S(r=a_{ij}) .\qquad (7)$$

Table 1 lists the assumed or calculated values for the plasma parameters in those examples of dense plasmas mentioned earlier.

Elements of Nuclear Reactions

In analyzing the nuclear reactions in dense plasmas, we single out reacting (R) nuclei, i and j, and name all others as "spectator (S)" nuclei. In addition to those "R" and "S" nuclei, the system contains electrons (and/or muons). The S-wave scattering between "R" nuclei with relative kinetic energy E is described by the Schrödinger equation

$$\left[-\frac{\hbar^2}{2\mu_{ij}} \frac{d^2}{dr^2} + W_{ij}(r) - E \right] r\,\psi_{ij}(r) = 0 ,\qquad (8)$$

where $W_{ij}(r)$ is the effective potential of scattering and μ_{ij} denotes the reduced mass. Nuclear reaction rate at E is given by (Salpeter and Van Horn, 1969; Ogata, Iyetomi, and Ichimaru, 1991)

$$R_{ij}(E) = \frac{8 S_{ij} r_{ij}^*}{(1 + \delta_{ij})\hbar} \frac{n_i n_j}{n_i + n_j} |\psi_{ij}(0)|^2 ,\qquad (9)$$

where $r_{ij}^* = \hbar^2 / 2\mu_{ij} Z_i Z_j e^2$ are the nuclear "Bohr radii". The reaction rates at temperature T are finally obtained through an average:

$$R_{ij}(T) = \langle R_{ij}(E) \rangle_R \qquad (10)$$

over the states of "R" nuclei.

BINARY PROCESSES   When

$$W_{ij}(r) = W_0(r) = Z_i Z_j e^2 / r \tag{11}$$

is assumed in (8), the cross sections of nuclear reactions are expressed in a form,

$$\sigma_{ij} = \frac{S_{ij}}{E} \exp\left[-\pi \left(\frac{E_G}{E}\right)^{1/2} \right], \tag{12}$$

where $E_G = Z_i Z_j e^2 / r_{ij}^*$. Equation (10) then gives the *Gamow reaction rate*,

$$R_G = \frac{32 S_{ij} r_{ij}^* n_j \tau_{ij}}{3^{3/2} (1 + \delta_{ij}) \hbar} \exp(-\tau_{ij}), \tag{13}$$

with $\tau_{ij} = 3(\pi/2)^{2/3}(E_G/T)^{1/3}$. $E_G \gg T$ has been assumed in the derivation of (13).

In Table 2, values of the nuclear reaction parameters are listed for those dense plasmas exemplified in Table 1. We note that the calculated nuclear-power output P_G for SI can account for the solar luminosity, while that for SN is far below the condition for a nuclear runaway by a factor of some 10^{-25}.

FEW-PARTICLE PROCESSES   The presence of electrons or other leptons acts to screen the Coulombic repulsion from $W_0(r)$ to $W_S(r) = W_0(r)S(r)$. Hence, we may take

$$W_{ij}(r) = W_S(r) = W_0(r) - E_S \ldots \tag{14}$$

with $E_S = Z_i Z_j e^2 / D_S$. The exponential factor for penetration in Eq. (12) now takes a form, $\exp\{-\pi[E_G/(E+E_S)]^{1/2}\}$. When $E_S < k_B T$, the leptonic screening introduces only a weak perturbation to R_G (Tanaka and Ichimaru, 1984; Ichimaru and Ogata, 1991).

If, however, a cold-fusion condition

$$E_S \gg k_B T \tag{15}$$

is satisfied as in the present MH and PM cases, we find that Eq. (13) is replaced by (Ichimaru, Ogata, and Nakano, 1990)

$$R_S = \frac{8 S_{ij} r_{ij}^* n_j}{(1 + \delta_{ij}) \hbar} \exp\left[-\pi \left(\frac{D_S}{r_{ij}^*}\right)^{1/2} \right], \tag{16}$$

a rate *independent* of *T*, in line with the original idea (Cameron, 1959) for the *pycnonuclear reactions*. The analysis is applicable irrespective of whether the "R" nuclei are in a fluid, molecular, or crystalline state. These are the extent to which the leptons may affect the nuclear reactions.

MANY-PARTICLE PROCESSES The most significant effects in dense plasmas are those of the screening potentials (*e.g.*, Ichimaru, 1982)

$$H_{ij}(r) = W_S(r) + k_B T \ln[g_{ij}(r)], \tag{17}$$

produced collectively through averages over the "S" nuclei, where $g_{ij}(r)$ denote the resultant joint probability densities for the "R" pairs. When "R" and "S" nuclei are in classical fluid states with $\Lambda_{ij} < 1$, one can determine $H_{ij}(r)$ accurately through analyses (Ogata, Iyetomi, and Ichimaru, 1991) combining between Monte Carlo (MC) sampling at intermediate distances ($0.4 < r/a_{ij} < 2$) and the short-range Widom (1963) expansion, as

$$\frac{H_{ij}(r)}{k_B T \, \Gamma_{ij}} = \begin{cases} A - \dfrac{B^2}{4h_1} - h_1 x^2 & \text{for } x < B/2h_1 \\[2mm] A - Bx + \dfrac{1}{x}\exp(C\sqrt{x} - D) & \text{for } B/2h_1 \leq x < 2 \end{cases} \tag{18}$$

where $x = r/a_{ij}$, and

$$h_1 = \frac{\left(Z_i^{1/3} + Z_j^{1/3}\right)^3}{16(Z_i + Z_j)},$$

$$A = 1.356 - 0.0213 \ln \Gamma_{ij}, \qquad\qquad B = 0.469 - 0.0130 \ln \Gamma_{ij},$$

$$C = 9.29 + 0.79 \ln \Gamma_{ij}, \qquad\qquad D = 14.83 + 1.31 \ln \Gamma_{ij}. \tag{19}$$

Enhancement factors for the reaction rates over Eq. (13) or (16) are then calculated as (Alastuey and Jancovici, 1978)

$$A_{ij} = \exp\{\langle H_{ij}(r)\rangle/k_B T\} \tag{20}$$

where $\langle\cdots\rangle$ means a path-integral "R" average with respect to the penetrating wave functions from $r=0$ to the classical turning radii r_t and back. This average can be evaluated through an exact solution (Ogata, Iyetomi, and Ichimaru, 1991) to Eq. (8) in which $W_{ij}(r) = W_S(r) - H_{ij}(r)$.

Equation (20) implies that when $r_t \ll a_{ij}$, the enhancement factor is evaluated as

$$A_{ij} = \exp\{H_{ij}(0)/k_B T\}. \tag{21}$$

It should be remarked that $H_{ij}(0)$ correspond to the increments in the Coulombic chemical potentials for the "R" pair before and after the reactions. (Widom, 1963; DeWitt, Graboske, and Cooper, 1973). Approximately, one finds (Salpeter and Van Horn, 1969; Ichimaru *et al.*, 1990; Ichimaru and Ogata, 1991)

$$H_{ij}(0) \approx 1.06 \times \Gamma_{ij}^S. \tag{22}$$

TABLE 2   NUCLEAR REACTIONS AND ENHANCEMENT

Assumed or calculated quantities	SN	SI	MH	PM	PM
Matter	C	H	Pd-D	D-H	Li-H
ρ_m (g/cm^3)	5×10^9	1×10^2	12.4	3.9	6.8
T (K)	1×10^8	1.5×10^7	300	600	550
Reactions	^{12}C-^{12}C	p-p	d-d	p-d	p-^{7}Li
S_{ij} (keV barn)	8.8×10^{19}	3×10^{-22}	106	2.5×10^{-4}	10^2
E_G (keV)	7.78×10^5	50	80	67	787
$r_{ij}*$ (10^{-13}cm)	6.7×10^{-2}	29	18	22	5.5
R_G (s^{-1})	1.3×10^{-44}	3.5×10^{-18}	- -	- -	- -
P_G (W/g)	4×10^{-34}	$1.\times10^{-4}$	- -	- -	- -
R_S (s^{-1})	- -	- -	1.6×10^{-36}	3.6×10^{-53}	1.2×10^{-46}
A_{ij}	5.0×10^{24}	1	2.9×10^{12}	3.9×10^{44}	4.1×10^{37}
R_{ij} (s^{-1})	6.8×10^{-20}	3.5×10^{-18}	4.6×10^{-24}	1.4×10^{-8}	4.5×10^{-9}
POWER (W/cm^3)	12	9.5×10^{-4}	3.7×10^{-13}	6.9×10^3	6.3×10^3

Reaction Rates

In the cases of SN, the basic reaction rates are R_G. SN mechanisms (Cameron, 1959; Salpeter and Van Horn, 1969; Nomoto, 1982; Ogata, Iyetomi, and Ichimaru, 1991) depend strongly on enhancement due to the many-particle processes. The many-particle processes are capable of producing an enhancement factor on the order of 10^{25} as the example in Table 2 illustrates.

In SI, the basic reaction rates are R_G. No significant enhancement is expected either from electron screening or from the many-particle processes. Such is the case (Tanaka and Ichimaru, 1984) also with the inertial-confinement fusion plasmas.

In MH, the reacting nuclei (deuterons, in the present example) are under a strong influence of the inhomogenious lattice field produced by the metal (palladium) atoms; due care must therefore be exercised in the MC sampling of the screening potentials via Eq. (17) (Ichimaru, Ogata, and Nakano, 1990). MH may employ the enhancement (20) and thereby raise the reaction rate to a barely observable level ($\approx 10^{-24}\mathrm{s}^{-1}$), provided that "R" nuclei may be found in an itinerant state; in equilibrium, however, such is not possible due to the trapping in the metallic microfields. It has been remarked (Ichimaru, Ogata, and Nakano, 1990) that key factors in realizing "observable" fusion rates in MH should be sought in feasibility of achieving non-equilibrium "fluidlike" situations without effectively raising "temperatures" for the reacting nuclei.

The nuclear reactions in the PM cases under present study differ in an essential way from those in the MH cases, in that the reacting nuclei are in *metallized, fluid* states where a substantial enhancement of the reaction rate due to the many-particle processes is expected as in the SN cases (Ichimaru 1991). Combination of Eq. (16) with the estimates of Eq. (20) shows in Table 2 that $d(p,\gamma)^3$He and ^{7}Li$(p,\alpha)^4$He reactions can take place at a power-producing level on the order of a few kW/cm^3 if such a material is brought into a liquid-metallic state under an ultrahigh pressure on the order of 10Mbar at a mass density of 3-7g/cm^3 and a temperature of 500-700K, slightly above an estimated Lindemann-melting temperature (Ichimaru, 1991)

$$T_\mathrm{m} \approx \frac{Z_i Z_j e^2}{180\, a_{ij} k_\mathrm{B}}\, F\!\left(k=\frac{4.3}{a_{ij}}\right) \tag{23}$$

for hydrogen. In Eq. (23), $F(k)$ is a screening factor related with the Fourier transform of $S(r)$.

If the mass density and pressure are lowered to 2.4g/cm^3 (4.0g/cm^3) and 6.8Mbar (2.3Mbar) for the D-H (Li-H) material of Table 2, the estimated reaction rate will take on a barely detectable value of $2.3\times10^{-24}\mathrm{s}^{-1}$ ($9.7\times10^{-24}\mathrm{s}^{-1}$). A detection of such a nuclear reaction will make the first laboratory demonstration for the elementary processes in supernova mechanisms and may lead to an examination on the validity of extrapolating cross sections, such as Eq. (12), into regimes of extremely low energies on the order of 0.1eV.

These ranges of the physical conditions may be accessible through extensions of the current ultrahigh-pressure metal technologies (*e.g.*, Nellis et al., 1988; Mao, Hemley, and Hanfland, 1990).

Acknowledgments

The author wishes to thank H. Iyetomi, S. Ogata, and H.M. Van Horn for useful discussions and collaboration on these and related subjects through Japan-U.S. Cooperative Science Program: *Phase Transitions in Dense Astrophysical Plasmas* supported jointly by the Japan Society for the Promotion of Science and the U. S. National Science Foundation.

References

Alastuey, A. and B. Jancovici, 1978, Astrophys. J. **226**, 1034.

Arnett, W.D., J.W. Truran, 1969, Astrophys. J. **157**, 339.

Cameron, A.G.W., 1959, Astrophys. J. **130**, 916.

DeWitt, H.E., H.C. Graboske, and M.C. Cooper, 1973, Astrophys. J. **181**, 439

Gai, S.M., S.L. Rugari, R.H. France, B.J. Lund, Z. Zhan, A.J. Davanport, H.S. Isaacs, and K.J. Lynn, 1989, Nature **340**, 29.

Ichimaru, S., 1982, Rev. Mod. Phys. **54**, 1017.

Ichimaru, S., 1991, J. Phys. Soc. Jpn **60**, 1437.

Ichimaru, S., A. Nakano, S. Ogata, S. Tanaka, H. Iyetomi, and T. Tajima, 1990, J. Phys. Soc. Jpn **59**, 1333.

Ichimaru, S. and S. Ogata, 1991, Astrophys. J. **374**, 647.

Ichimaru, S., S. Ogata, and A. Nakano, 1990, J. Phys. Soc. Jpn **59**, 3904.

Itoh, N., T. Adachi, M. Nakagawa, Y. Kohyama, and H. Munakata, 1989, Astrophys. J. **339**, 354.

Jones, E.S., E.P. Palmer, J.B. Czirr, D.L. Decker, G.L. Jensen, J.M. Thorne, S.F. Taylor, and J. Rafelski, 1989, Nature **338**, 737.

Mao, H.K., R.J. Hemley, and M. Hanfland, 1990, Phys. Rev. Lett. **65**, 464.

Nellis, W.J., J.A. Moriarty, A.C. Mitchell, M. Ross, R.G. Dandrea, N.W. Ashcroft, N.C. Holmes, and G.R. Gathers, 1988, Phys. Rev. Lett. **60**, 1414.

Nomoto, K., 1982, Astrophys. J. **253**, 798.

Ogata, S., H. Iyetomi, and S. Ichimaru, 1991, Astrophys. J. **372**, 259.

Salpeter, E.E. and H.M. Van Horn, 1969, Astrophys. J. **155**, 183.

Tanaka, S. and S. Ichimaru, 1984, J. Phys. Soc. Jpn **53**, 2039.

Widom, B., 1963, J. Chem. Phys. **39**, 2808.

Ziegler, J.F., T.H. Zabel, J.J. Cuomo, V.A. Brusic, G.S. Cargill, III, E.J. O'Sullivan, and A.D. Marwick, 1898, Phys. Rev. Lett. **62**, 2929.

THE EQUATION OF STATE FOR STELLAR ENVELOPES

Werner Däppen

Department of Physics and Astronomy, University of Southern California, Los Angeles,
CA 90089-1342,U.S.A.
and
Institut für Astronomie, Universität Wien, 1180 Vienna, Austria

Abstract: Although the plasma of envelopes of normal stars is only weakly non-ideal, the inadequacy of simple equations of state has been clearly demonstrated by helioseismology. Solar and stellar applications put high demands on the precision and consistency of the realization of any physical formalism. The three principal non-ideal effects are due to internal partition functions of bound systems, to pressure ionization, and to collective interactions of the charged particles. Comparisons between different formalisms help to find the astrophysical observables that can discriminate between equations of state that are based on different treatments of non-ideal effects. A first set of astrophysical observables is provided by helioseismology, which is becoming sufficiently accurate to put constraints on the equation of state.

Introduction

The plasma of envelopes of normal stars is only weakly non-ideal. One would therefore think that finding a good equation of state is not too difficult. Indeed, simple models of the equation of state have been highly successful in many calculations of stellar structure and evolution. However, the inadequacy of simple equations of state has been clearly demonstrated by the still relatively young field of helioseismology, which is the diagnosis of the Sun's internal structure from a set of precisely observed oscillation frequencies. Physically, an oscillation mode is a standing acoustic wave, and its basic property is the connection between local sound speed and the observed oscillation frequency. For extensive reviews on helioseismology, each with many further references, see *e.g.* Deubner and Gough (1984), Bahcall and Ulrich (1988), Christensen–Dalsgaard and Berthomieu (1991), Gough and Toomre (1991), Turck-Chièze *et al.* (1992).

Thanks to the spatial resolution of the solar disk, oscillation modes, classified in terms of spherical harmonics, have been detected for a large range of angular degrees (between $l = 0$ and a few thousands), and their frequencies have been determined with high accuracy (with typical relative errors of 10^{-4}). Because of this unusual observational accuracy, there is so far no solar model that could even nearly predict all observed frequencies. By the very nature of solar oscillations, the basic physical quantity is sound speed. In addititon, the oscillations are largely adiabatic (except very near the surface), and therefore it is the local adiabatic sound speed, a thermodynamic quantity, which is most directly linked to the observed frequencies. This means that a good knowledge of the equation of state is very important (Christensen–Dalsgaard and Däppen, 1992).

Improvements in the equation of state beyond the model of a mixture of ideal gases are difficult. This has both technical and conceptual reasons. As a fundamental conceptual reason I mention the fact that in a plasma environment, already the idea of isolated atoms (and compound ions) has to be abandoned: clearly, the interactions between atoms and environment cannot be neglected, so that, strictly speaking, one has always to deal with complicated many-body states. Of course, many formalisms for this situation of coupled plasmas have been developed. However, usually the motivation of these formalisms is the understanding of qualitative phenomena, and they are neither precise enough nor adequate to describe realistic astrophysical mixtures. As a consequence, they are normally not suited for the purposes of stellar models.

The weakly non-ideal plasma of the solar interior is very complicated if high precision and accuracy is demanded. Given the virtual impossibility of a rigorous formalism, it is no wonder that a large number of more-or-less phenomenological theories have been developed, mainly built around perturbational treatments of ideal gases. The three principal non-ideal effects are due to internal partition functions of bound systems, to pressure ionization, and to collective interactions of the charged particles. Somewhat more specifically, the three issues are the following. First, the internal partition functions contain the difficult problem of excited states, that is, where and how they are to be cut off. Their detailed treatment is an important element in determining the ionization balances. Second, pressure ionization has to be provided by non-ideal interaction terms, because the ideal gases would spuriously recombine toward the central regions of the Sun. And third, the collective interactions between the charged particles mainly lead to a (negative) Coulomb pressure contribution, which is typically of the order of one percent in the solar interior.

Although the present article discusses the equation of state for stellar *envelopes*, specific models can often be easily extrapolated to the central regions of a star. This extension of the domain of validity is favoured by two reasons: First, all atoms become more and more ionized at higher densities and temperatures, leaving only a small number of bound systems with little influence on the overall equation of state (though not on opacity). Second, in the resulting nearly fully ionized plasma, the main effects beyond the classical ideal gas come from partially degenerate electrons and the Coulomb interaction. All equations of state that contain these two effects are thus expected to work well down to the center of the star.

Demands on a stellar equation of state

It is important to realize that a stellar equation of state has to be *formally* precise and consistent, even before the question of the accuracy of the physical description is asked. It has to satisfy four conditions: *i*) a large domain of applicability (in ρ, T), *ii*) a high precision of its numerical realization, *iii*) consistency between the thermodynamic quantities, and *iv*) the possibility to take into account relatively complex mixtures with at least several of the more abundant chemical elements.

More specifically, the first condition demands that the formalism can be used from the stellar surface (the photosphere), where T is typically a few 10^3 K and ρ some 10^{-7} g/cm^3, to the center of a star where T is, again typically, about 10^7 K and ρ some 10^2 g/cm^3. The second condition demands that a given formalism can be cast in an algorithm that converges without ambiguity and with sufficient precision, so that all required thermodynamic derivatives (such as adiabatic gradients) can be computed. Note, that for this only *formal* precision is required: reality of the physical description is a different issue. The third condition, consistency, states that all thermodynamic quantities stem from a single thermodynamic potential. This condition is often violated in two- or more-zone formalisms, which contains a different physical theory in different parts of a star. An example is the *ad hoc* imposition of full ionization in the central region, in order to mimic a pressure-ionization device, in combination with a conventional Saha equation in the envelope of the star. Such a formalism leads to a discontinuous thermodynamic potential and a violation of thermodynamic identities. Thermodynamic identities are, however, often used in calculations of stellar structure and oscillations. As an example of the use of thermodynamical identities, I mention the transformation of the density fluctuation to the pressure fluctuation in linear adiabatic pulsation calculations. The connection between density and pressure changes is given by the adiabatic gradient Γ_1 , and it is therefore imperative that this gradient is consistent with the equation of state and other thermodynamic variables used in the model. This example illustrates the necessity of formal consistency. Finally, the third and last condition, *i.e.* the possibility to describe rather realistic chemical compositions, is less important for the equation of state itself. However, for opacity heavy elements are important, and a good equation of state plays an important role in any opacity calculation.

It is in view of these specific requirements that astrophysicists dare to develop their own formalisms, which are often built around time-honoured intuitive ideas, and therefore lack a rigorous foundation. One of these "home-grown" astrophysical equations of state has been developed as part of the international "Opacity Project" (OP, see Seaton, 1987) by Mihalas, Hummer, and Däppen (Hummer and Mihalas, 1988; Mihalas, Däppen and Hummer, 1988; Däppen *et al.* , 1988; hereinafter MHD). The MHD equation of state is written in the *chemical picture* (Krasnikov, 1977), where bound configurations (atoms, ions and molecules) are introduced and treated as new and independent species. Plasma interactions are treated with modifications of atomic states, *i.e.* the quantum mechanical problem is solved before statistical mechanics is applied.

Clearly, compared to such an intuitive formalism, many of the physical theories for non-ideal plasmas presented in these proceedings are much more sophisticated. However, equally clearly, these physical theories were not developed in view of stellar applications, and would therefore not satisfy the four conditions mentioned above. A first attempt to develop a *stellar* equation of state with a better statistical mechanical foundation was made as part of an opacity project at Livermore (OPAL) (Rogers, 1986; Iglesias *et al.* , 1987). This equation of state (hereinafter Livermore equation of state) is based on the *physical picture*. In the physical picture, only fundamental particles (nuclei and electrons) are explicitly introduced. Therefore, there is no need for a minimax principle; the question of bound states is dealt with implicitly, through the Hamiltonian describing the interaction between the fundamental particles, and the problems of quantum mechanics and statistical mechanics are tackled simultaneously.

Two examples of equations of state

The MHD equation of state (chemical picture)

In the chemical picture, perturbed atoms must be introduced on a more-or-less *ad-hoc* basis to avoid the familiar divergence of internal partition functions (see *e.g.* Ebeling *et al.* 1976). In other words, the approximation of unperturbed atoms precludes the application of standard statistical mechanics, *i.e.* the attribution of a Boltzmann-factor to each atomic state. The conventional remedy of the chemical picture against this is a modification of the atomic states, *e.g.* by cutting off the highly excited states in function of density and temperature of the plasma. Such cut-offs, however, have in general dire consequences due to the discrete nature of the atomic spectrum, *i.e.* jumps in the number of excited states (and thus in the partition functions and in the free energy) despite smoothly varying external parameters (temperature and density).

The MHD equation of state avoids these discontinuities (in the free energy) by introducing "soft" cut-offs in the form of occupational probabilities. These occupation probabilities have the same function as the "hard" cut-offs mentioned above. The occupational probabilities of a state simulate a result from quantum mechanics, denoting the fraction of atoms where the state can exist. Only then, these "available" states are populated according to statistical mechanics. It is clear that such an approach is largely intuitive. However, its advantage is that complicated plasmas can be modelled, with *detailed* internal partition functions of a large number of atomic, ionic, and molecular species. Also, full *thermodynamic* consistency is assured by analytical expressions of the free energy and its first- and second-order derivatives. This not only allows an efficient Newton-Raphson minimization, but, in addition, the ensuing thermodynamic quantities are of analytical precision and can therefore be differentiated once more, this time numerically. Reliable third-order thermodynamic quantities are thus calculated.

In the MHD occupation probabilities, perturbations by charged and neutral particles are taken into account. Correlations between the two effects are neglected (for lack of knowing how to describe them); thus the occupation probabilities due to charged and neutral perturbers are simply multiplied. The resulting weighted internal partition functions Z_s^{internal} of species s are (with is labelling the state i of species s)

$$Z_s^{\text{internal}} = \sum_i w_{is} g_{is} \exp\left[-\frac{E_{is} - E_{1s}}{kT}\right] . \tag{1}$$

The coefficients w_{is} take into account charged and neutral surrounding particles. In physical terms, w_{is} gives the fraction of all particles of species s that can exist in state i with an electron bound to the atom or ion, and $1 - w_{is}$ gives the fraction of those that are so heavily perturbed by nearby neighbours that the state is effectively destroyed. Perturbations by neutral particles are based on an excluded volume treatment and perturbations by charges are calculated from a fit to a quantum-mechanical Stark-ionization theory. Hummer and Mihalas's (1988) choice has been

$$\ln w_{is} = -\left(\frac{4\pi}{3V}\right)\left\{\sum_\nu N_\nu(r_{is} + r_{1\nu})^3 + 16\left[\frac{(Z_s + 1)e^2}{\chi_{is} k_{is}^{1/2}}\right]^3 \sum_{\alpha \neq e} N_\alpha Z_\alpha^{3/2}\right\} . \tag{2}$$

Here, the index ν runs over neutral particles, the index α runs over charged ions (except electrons), r_{is} is the radius assigned to a particle in state i of species s, χ_{is} is the (positive) binding energy of such a particle, k_{is} is a quantum-mechanical correction, and Z_s is the net charge of a particle of species s. Note that $\ln w_{is} \propto -n^6$ for large principal quantum numbers n (of state i), and hence provides a smooth (density-dependent) cutoff for Z_s^{internal}. As far as the energy levels are concerned, no shifts due to plasma effects are assumed in the MHD equation of state. This assumption is based on experimental and theoretical arguments (see Hummer and Mihalas, 1988). Finally, the MHD equation of state also includes a Debye-Hückel term for the Coulomb-pressure correction, partially degenerate electrons, and radiation pressure.

The Livermore equation of state (physical picture)

There is an impressive body of literature on the physical picture. Important sources of information with many references are the books by Ebeling *et al.* (1976), Kraeft *et al.* (1986), Ebeling *et al.* (1991). However, the majority of work on the physical picture was not dedicated to the problem of obtaining a high precision equation of state for stellar interiors. Such an attempt was made for the first time by a group at Livermore as part of an opacity project (Rogers, 1986; Iglesias *et al.* , 1987).

It is clear from the preceding subsection that the advantage of the chemical picture lies in the possibility to model complicated plasmas, and to obtain numerically smooth and consistent thermodynamical quantities. Nevertheless, the heuristic method of the separation of the atomic-physics problem from that of statistical mechanics is not satisfactory, and attempts have been made to avoid the concept of a perturbed atom in a plasma altogether. This has suggested an alternative description, the physical picture, in which only fundamental particles (electrons and nuclei) explicitly appear. In such an approach one expects that no assumptions about energy-level shifts or the convergence of internal partition functions will have to be made. In the contrary, properties of energy levels and the partition functions will come out from the formalism.

To explain the advantages of this approach for partially ionized plasmas, it is instructive to discuss the activity expansion for gaseous hydrogen. The interactions in this case are all short ranged and the pressure is determined from a self-consistent solution of the equations (Hill, 1960)

$$\frac{p}{kT} = z + z^2 b_2 + z^3 b_3 + ... \tag{3}$$

$$\rho = \frac{z}{kT}\left(\frac{\partial p}{\partial z}\right) \tag{4}$$

where $z = \lambda^{-3}\exp(\mu/kT)$ is the activity, $\lambda \equiv h/\sqrt{2\pi m_e kT}$ is the thermal (de Broglie) wavelength of electrons, μ is the chemical potential and T is the temperature. The b_n are cluster coefficients such that b_2 includes all two particle states, b_3 includes all three particle states, etc. The second cluster coefficient for hydrogen includes the formation of H_2 molecules as well as scattering states in the $^1\Sigma_g$ potential. It also includes scattering states in the $^3\Sigma_u$ potential and all excited electronic-state potentials. The third cluster coefficient includes H_3 bound states, $H - H_2$ and $H - H - H$ scattering states. Equation (3) demonstrates that the equation of state for associating gases can be obtained without an explicit knowledge of the occupation numbers of associate pairs. Further details can be found in Rogers (1981, 1986) and also in the review by Däppen, Keady and Rogers (1991).

To illustrate how the physical picture allows avoiding the divergences that plague the chemical picture, I note that b_2 is convergent, because the bound state part of b_2 is divergent but the scattering state part, which is normally omitted in the chemical approach (*e.g.* in MHD), has a compensating divergence. Consequently the total b_2 does not contain a divergence of this type (Ebeling, Kraeft, and Kremp, 1976, Rogers, 1977). A major advantage of the physical picture is that it incorporates this compensation at the outset. As a result, the Boltzmann sum appearing in the atomic (ionic) free energy is replaced with the so called Planck-Larkin partition function (PLPF), given by (Ebeling, Kraeft, and Kremp, 1976)

$$\text{PLPF} = \sum_{nl}(2l + 1)\left[\exp(-\frac{E_{nl}}{kT}) - 1 + \frac{E_{nl}}{kT}\right] \tag{5}$$

The PLPF is convergent without additional cut-off criteria as are required in the chemical picture.

Equation of state comparisons

So far there are no laboratory experiments that could distinguish between equations of state in the chemical and physical picture. Attempts to use constraints from a high-precision optical emission spectrum (*e.g.* Wiese, Paquette and Kelleher, 1973) have failed, because line-broadening effects were overshadowing the subtle details of statistical mechanics (Däppen, Anderson and Mihalas, 1987; Seaton, 1990). As another type of observational diagnosis, solar oscillations promise to test the equation of state in the near future (Christensen–Dalsgaard, Däppen and Lebreton, 1988; Däppen, Keady and Rogers, 1991; Christensen–Dalsgaard, 1991; Christensen–Dalsgaard and Däppen, 1992). At present, however, it seems that detailed and precise comparisons of theoretical results formalisms are still the best way to learn about the merits of each approach. Such comparisons also allow solar physicists to determine how uncertainties in the equation of state propagate in theoretically predicted oscillation frequencies. In this way, a "map" of the $T - \rho$ plane can be drawn, showing the localized "interesting" regions, where the non-ideal effects of one or another kind are most important.

In the following, I discuss what has so far emerged from such comparisons. An early comparisons showed a striking agreement between the MHD and Livermore equation of state for conditions as found in the hydrogen-helium ionization zones of the Sun (Däppen, Lebreton and Rogers, 1990; Däppen, 1990). For convenience, a representative result from this early comparison is shown in Figure 1, which compares the MHD and Livermore results with that of the simple Eggleton, Faulkner and Flannery (EFF) formalism (which is essentially a consistent ground-state-only Saha equation of state including a – here irrelevant – arbitrary pressure-ionization device). The absolute curves of part *a* of Figure 1 are merely able to show the difference between MHD (or Livermore) and the simple EFF results. To see the difference between the MHD and Livermore results, one needs the magnified part *b*, which shows the *relative* differences between MHD and

EFF, and between Livermore and EFF values, respectively. This relative plot now not only allows one to see the difference between MHD and Livermore results but also their striking similarity.

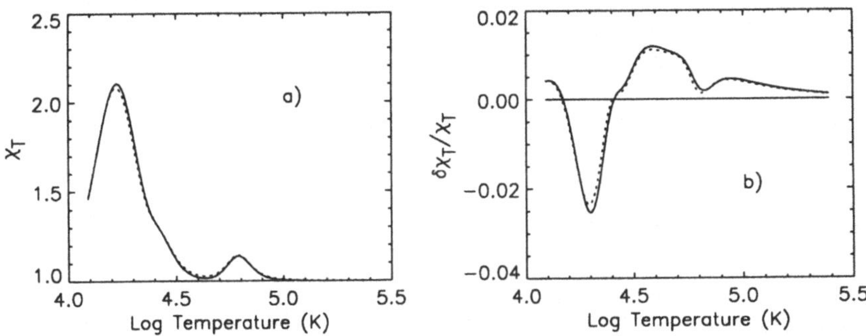

Fig. 1. Comparison of the logarithmic pressure derivative $\chi_T = (\partial \ln p / \partial \ln T)_\rho$ on an isochore with $\rho = 10^{-5.5}$ g cm^{-3}. Part a shows absolute values; the solid line representing EFF, and the dotted line MHD. The chemical composition is hydrogen and helium only, with number abundances of 90% H and 10% He. The Livermore result would lie indistinguishable on the MHD curve. Part b magnifies the effect by showing the relative differences between Livermore and EFF values, $i.e.$ $(\chi_T^{Livermore} - \chi_T^{EFF})/\chi_T^{EFF}$ (solid line) and between MHD and EFF values, $i.e.$ $(\chi_T^{MHD} - \chi_T^{EFF})/\chi_T^{EFF}$ (dotted line). Other thermodynamic quantities essentially show the same behaviour. (From Däppen, Lebreton and Rogers, 1990).

Later, it turned out that this agreement was nearly accidental. The *physical reason* was found by varying the parameters of the MHD equation of state. It followed that on the chosen isochore, all thermodynamical quantities are mainly dominated by the Coulomb pressure correction (Däppen, 1990; Christensen–Dalsgaard, 1991; Christensen–Dalsgaard and Däppen, 1992). The Coulomb correction overshadows the effect of the excited states (which are of course treated differently in the MHD and Livermore approach). However, the Coulomb term acts principally indirectly, at least in the language of the chemical picture, because it is not mainly the free-energy of the Debye-Hückel term itself (it would be 1-2 orders of magnitude too weak), but rather the Coulomb-term induced shift in the ionization equilibrium, which is responsible for the deviation from the unperturbed EFF result.

Of course, solar physicists were happy that two completely different formalisms delivered the same equation of state, but, by the same token, a first attempt to use the Sun as a test was also thwarted. This discovery suggested to upgrade the simple EFF equation of state with the help of the Coulomb interaction term. The resulting equation of state (called CEFF) is a very useful tool for solar physics (Christensen–Dalsgaard, 1991; Christensen–Dalsgaard and Däppen, 1992); at the same time, however, it became also clear that a helioseismic test of the important issue of chemical versus physical picture would be more difficult than first thought.

For reasons not yet fully understood it seems that in the chemical picture, the signature of internal partition functions, such as those employed in the MHD equation of state, is much less visible in the thermodynamic quantities than a naive estimation of the shift in the ionization equilibrium would predict. It is likely that there are accidental cancellations in the derivatives of the free energy. Notice that these cancellation would have nothing to do with those appearing in the physical picture, which lead to the Planck-Larkin partition function. The accidental cancellations of the chemical picture seem to be greatest for the ionization zone of hydrogen and somewhat less for those of helium. For the heavier elements, it appears that the internal partition functions finally lead to the intuitively expected consequences for the thermodynamic quantities.

This last conclusion resulted from a recent comparison involving a representative heavy element, oxygen (Däppen, 1992). Density was chosen as $\rho = 0.005$ g cm^{-3}, a value suggested by a helioseismic study of the solar helium abundance (Christensen-Dalsgaard *et al.* , 1992). For

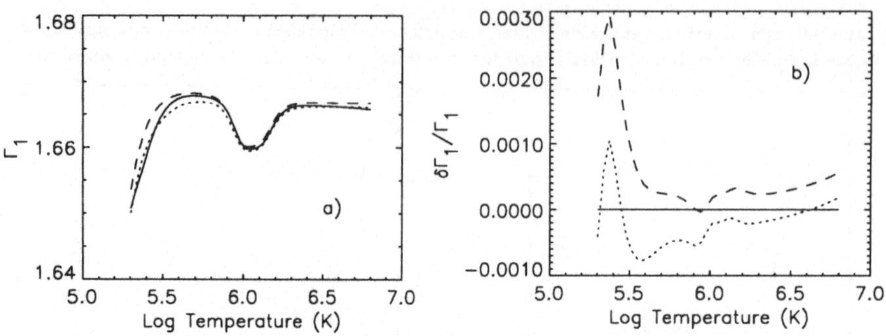

Fig. 2. Γ_1 for $\rho = 5.00 \times 10^{-3}$g cm^{-3}, and a different chemical composition (a representative solar mixture of H, He, and O, with mass abundances of 0.7429, 0.2371, 0.0200, respectively). Part a) shows absolute values, part b) relative differences with respect to the CEFF (EFF plus Coulomb term) equation of state, in a similar fashion as in Figure 1b (dashed: MHD, dotted: Livermore).

convenience, Figure 2 shows the result of this comparison for Γ_1. Here, not only do the large MHD partition functions cause shifts in the ionization balance but these shifts also significantly propagate into the thermodynamic quantities. The effect is large enough that it appears, despite the small relative number of the heavy elements in the mixture, to be within reach of helioseismology (Christensen-Dalsgaard and Däppen, 1992). To examine the MHD ionization fractions, single case was examined ($T = 2.10 \times 10^5 K, \rho = 5.00 \times 10^{-3}$g cm^{-3}), once with the full MHD equation of state, once with a "stripped-down" version of MHD, which does not contain any excited states (but is otherwise identical). The resulting ionization fractions of O^{3+}, O^{4+}, O^{5+} were, respectively, 0.314, 0.248, 0.364 for the stripped-down MHD (without excited states), and 0.304, 0.476, 0.182 for the full MHD. (The result for the stripped-down very closely reflects the ground-state weights of the ions). Not unexpectedly in view of the Planck-Larkin partition function, the Livermore equation of state predicts ionization fractions close to those of the stripped-down MHD equation of state (Rogers, *private communication*).

This comparison for the first time establishes a clear case of disagreement between the MHD and Livermore results. Clearly, the origin of the discrepancy in the ionization degrees is due to the treatment of the excited states. Of course, only some 2 percent of the matter in the Sun consist of elements heavier that H and He, and therefore the signature of the MHD-Livermore discrepancy on thermodynamic quantities (Figure 2) is small (of the order of 10^{-3}). Nevertheless, as has been demonstrated by Christensen-Dalsgaard and Däppen (1992), even the resulting tiny sound-speed differences are within reach of a helioseismological diagnosis.

Conclusion

Even weakly coupled plasmas can pose tough problems if a high accuracy is demanded. Solar oscillations are an example for a case where the present observational material is much better than the theoretical models. Stellar oscillations also promise interesting constraints. Their observations will not be as detailed as those of the Sun: however, a sample of more than one star stars will access a variety of different physical conditions and chemical compositions. Theoretical comparisons between different formalisms will be very important to localize the interesting regions in the T-ρ plane, where the different formalisms lead to distinct thermodynamic quantities. As for the Sun, such comparisons have already resulted in a better (and yet simple) equation of state.

Acknowledgement: I thank W. Ebeling, A. Förster and R. Radtke for organizing a superb meeting. I also thank Forrest Rogers for providing results of the Livermore equation of state.

References

Bahcall, J. N., and Ulrich, R. K., 1988, *Rev. Mod. Phys*, **60**, 297-372.

Christensen-Dalsgaard, J. 1991, in *Challenges to theories of the structure of moderate-mass stars*, eds. D.O. Gough and J. Toomre (Lecture Notes in Physics, 388, Springer, Heidelberg), 11-36.

Christensen-Dalsgaard, J., and Berthomieu, G., 1991, in *Solar Interior and Atmosphere*, eds. A.N. Cox, W.C. Livingston and M. Matthews, (Space Science Series, University of Arizona Press), (in press).

Christensen-Dalsgaard, J., and Däppen, W. 1992, *Astronomy and Astrophysics Review*, (submitted).

Christensen-Dalsgaard, J., Däppen W., and Lebreton Y. 1988, *Nature*, **336**, 634-638.

Christensen-Dalsgaard, J., Däppen W., Dziembowski, W.A., Gough, D.O., Kosovichev, A.G., and Thompson, M.J. 1992, *Mon. Not. R. astr. Soc.*, (submitted).

Däppen, W. 1990, in *Progress of seismology of the sun and stars*, eds. Y. Osaki and H. Shibahashi (Lecture Notes in Physics, 367, Springer, Heidelberg), 33-40.

Däppen, W. 1992, in *Astrophysical Opacities*, eds. C. Mendoza and C. Zeippen (*Revista Mexicana de Astronomia y Astrofisica*), (in press).

Däppen, W., Anderson, L.S. and Mihalas, D. 1987, *Astrophys. J.*, **319**, 195-206.

Däppen, W., Lebreton, Y., and Rogers, F. 1990, *Solar Physics*, **128**, 35-47.

Däppen, W., Keady, J., and Rogers, F. 1991, in *Solar Interior and Atmosphere*, eds. A.N. Cox, W.C. Livingston and M. Matthews, (Space Science Series, University of Arizona Press), (in press).

Däppen, W., Mihalas, D., Hummer, D.G., and Mihalas, B.W. 1988, *Astrophys. J.*, **332**, 261-270.

Deubner, F.-L., and Gough, D. O., 1984, *Ann. Rev. Astron. Astrophys.*, **22**, 593-619.

Ebeling, W., Kraeft, W.D. and Kremp, D. 1976, *Theory of Bound States and Ionization Equilibrium in Plasmas and Solids*, (Berlin, DDR: Akademie Verlag).

Ebeling, W., Förster, A., Fortov, V.E., Gryaznov, V.K., and Polishchuk, A.Ya. 1991, *Thermodynamic Properties of Hot Dense Plasmas*, (Stuttgart, Germany: Teubner).

Eggleton, P.P., Faulkner, J., and Flannery, B.P. 1973, *Astron. Astrophys.*, **23**, 325-330.

Gough, D. O., and Toomre, J. 1991, *Ann. Rev. Astron. Astrophys.*, **29**, 627-684.

Hill, T.L., 1960, *Statistical Thermodynamics*, (Addison-Wesley), Chapt. 15.

Hummer, D.G., Mihalas, D. 1988, *Astrophys. J.*, **331**, 794-814.

Iglesias, C.A., Rogers, F.J., and Wilson, B.G., 1987, *Ap.J.*, **322**, L45.

Krasnikov Yu.G. 1977, *Zh. Eksper. teoret. Fiz.*, **73**, 516 (1978, *Soviet Phys. - JETP*, **46** No. 2, 170; author's name misspelt as "Karsnikov").

Mihalas, D., Däppen W., and Hummer, D.G. 1988, *Astrophys. J.*, **331**, 815-825.

Rogers, F.J., 1977, *Phys. Lett.*, **61A**, 358.

Rogers, F.J., 1981, *Phys. Rev.*, **A24**, 1531.

Rogers, F.J. 1986, *Astrophys. J.*, **310**, 723-728.

Seaton, M. 1987, *J. Phys. B:Atom. Molec. Phys.*, **20**, 6363-6378.

Seaton, M. 1990, *J. Phys. B:Atom. Molec. Phys.*, **23**, 3255-3296.

Turck-Chièze, S., Däppen, W., Fossat, E., Provost, J., Schatzman, E., and Vignaud, D. 1992, *Physics Report*, (submitted).

Wiese, W. L., Kelleher, D. E., and Paquette, D. R. 1972,, *Phys. Rev.*, **A6**, 1132-1153.

THE PLASMA PHASE TRANSITION OF HYDROGEN
IN GIANT PLANETS

D. Saumon

Lunar and Planetary Laboratory, University of Arizona,
Tucson, AZ 85721, USA

G. Chabrier

Ecole Normale Supérieure*
46, Allée d'Italie, 69394 Lyon Cedex 07, France

W. B. Hubbard and J. I. Lunine

Lunar and Planetary Laboratory, University of Arizona,
Tucson, AZ 85721, USA

Abstract

Our knowledge of the interior structure of giant planets is derived by indirect methods
which depend heavily on the equation of state (EOS) assumed in the models. Conversely,
Jupiter and Saturn are the only large reservoirs of metallic hydrogen accessible to observations and constitute the only sites where pressure ionization of hydrogen is known to occur.
They are natural laboratories where we can test theories of hydrogen under high pressure.

A new EOS for He/H mixtures, which includes a description of the plasma phase
transition of hydrogen, is used to compute interior models of Jupiter and Saturn, subject
to the constraints of the measured gravitational harmonics of both planets. We emphasize
the strong dependence of such models and of their astrophysical significance on accurate
theories of dense plasmas and of pressure ionization.

Introduction

The last two decades have seen a very fruitful exchange between the fields of astrophysics and
dense matter physics, which started with the spectacular application of the degenerate electron
gas properties to the structure of white dwarf stars.[1] More recent developments include the
application of the One-Component Plasma model (OCP) and the OCP freezing transition to the
evolution of white dwarfs, pressure metallization of hydrogen in planets, the determination of
the helium abundance in the Sun via the observed oscillation spectrum, and phase separation
in white dwarfs, giant planets and brown dwarfs. Because of the extreme conditions to which
matter is subjected in many astrophysical contexts, astrophysics is a great "consumer" of the
more exotic theories of dense matter. On the other hand, stars in general often provide the only
sites for application and for testing these theories. In this paper, we illustrate the interaction
between these two fields by discussing one example.

---

* Equipe associée au CNRS

"

We have recently computed new models for the interior of the giant planets Jupiter and Saturn.[2] The structure and global composition of these planets can only be inferred by indirect means. Both have fluid interiors and are noticeably distorted by rapid rotation (rotation period $\approx$ 10 hours). The reaction of the planet to rotation, as measured by departures from spherical symmetry in the gravitational field provides integral constraints on the density profile throughout the planet. This is currently the most powerful probe of the interior of Jupiter and Saturn. Other methods include the response to tidal perturbations caused by satellites and global oscillations. The latter method enjoys great success in stellar astrophysics but until recently, pulsations in giant planets had remained undetected.[3] Confirmation of the detection of oscillations in Jupiter, an accurate determination of their frequency spectrum and subsequent modeling of the pulsations would constitute a major breakthrough in the study of its interior.

Our effort was motivated by the availability of a new hydrogen equation of state (EOS) which naturally predicts a first order molecular-metallic phase transition, or plasma phase transition (PPT), in the regime of pressure ionization at finite temperatures.[4] While the existence of a molecular-metallic phase transition in hydrogen at zero temperature is well established on theoretical grounds, it is still uncertain whether it occurs at temperatures of a few thousand degrees. Currently, diamond anvil experiments are revealing the richness of the phase diagram of hydrogen below room temperature, but the observed phenomena are only marginally relevant to the PPT where thermal effects play an important role. Other theoretical estimates of the PPT[5] rely on a few crude assumptions and do not provide the necessary details along the coexistence curve to use them in astrophysical applications.

Jupiter and Saturn are good testing grounds for theories of pressure ionization of hydrogen: they are the only sites where this phenomenon is known to occur with certainty, and a large fraction of their mass lies in the particularly difficult regime where $0.5 < \rho(\mathrm{g/cm}^3) < 5$ and $10^3 < T(\mathrm{K}) < 10^4$. Bodies less massive than Saturn do not reach high enough pressures to experience pressure ionization. In objects more massive than Jupiter, such as the elusive brown dwarfs, the PPT occurs very close to the surface and pressure-ionization affects only a minute fraction of their mass. While the PPT alters their thermal structure, the mechanical structure of brown dwarfs remains virtually unchanged.

Our application of this new EOS to interior models of giant planets has two purposes: 1) we wish to verify that the new EOS (and the PPT in particular) is compatible with our knowledge of giant planets and leads to acceptable models, and 2) determine the astrophysical consequences of applying this EOS to giant planets. Previous studies of similar scope have not treated hydrogen metallization in a rigorous, thermodynamically consistent fashion. Some treatments assumed that thermodynamic quantities can be smoothly interpolated between the low-density regions where individual atoms and molecules can be validly assumed, and the high-density regions where the thermodynamics is dominated by free electrons in nearly plane-wave states.[6,7] Other treatments assumed that the transition between these regions occur at an abrupt phase boundary, but the boundary was calculated by extrapolating solid-state models of the phases on either side, which could not correctly reveal the necessary termination at a critical point at high temperature.[8] Here we present the first models of Jupiter and Saturn based on a detailed description of the PPT, arising from a single free energy model which addresses the problems of pressure ionization directly.

The Equation of State

In a first approximation, the envelopes of Jupiter and Saturn share the same composition as the Sun: roughly 90% hydrogen and 10% helium by number with heavier elements contributing less than 1%. The calculation of accurate thermodynamics for a hydrogen and helium mixture under the range of conditions found in the interiors of giant planets presents substantial difficulties.

This is particularly true of the regime of pressure ionization, at densities of ≈ 1 g/cm^3. In view of the current lack of a satisfactory theoretical understanding of this problem, we have adopted a H/He equation of state based on a compositional interpolation between a pure hydrogen EOS and a pure helium EOS. This greatly simplifies the calculation of the EOS for any mixing ratio of H and He.

The hydrogen EOS was designed for applications to relatively cool and dense objects such as giant planets and substellar brown dwarfs and details can be found in Refs. 9 and 10. The Helmholtz free energy model is based on the "chemical picture", in the sense that we assume the existence of independent, bound configurations such as H atoms and H$_2$ molecules, interacting with pair potentials. Our model for an interacting system of H$_2$, H, H$^+$ and e can be seen as consisting of a "neutral" (H$_2$ + H) and a fully ionized (H$^+$ + e) model, which represent, respectively, the low-density, low-temperature and the high-density and/or high-temperature limits of the general model.

The thermodynamics of the fully ionized plasma are computed in the Screened One-Component Plasma model using the hypernetted chain theory for a temperature and density dependent screened Coulomb potential. Agreement with Monte Carlo simulations is excellent.[11] In the "neutral" model, interactions are described with the WCA fluid perturbation theory[12] using realistic potentials. The configuration pressure and internal energy of the binary mixture agree with Monte Carlo simulations to better than 3%. All known bound states of H$_2$ and H are included in the internal partition function. The effect of near-neighbor interactions are included in the internal partition function sum with an occupation probability formalism[13] which provides a reasonable description of pressure dissociation and ionization.[9]

The two models are combined in a *single* expression for the free energy for H$_2$, H, H$^+$ and e where charged-neutral interactions are included in the form of a polarization potential.[14] The chemical equilibrium of this four component mixture is obtained by numerically minimizing the total free energy.

This free energy model is thermodynamically unstable in the regime of pressure ionization of hydrogen, where it predicts a first order phase transition between an essentially molecular, low-density phase and a partially ionized "metallic" high-density phase.[4] The transition occurs at pressures of ≈ 1 Mbar and ends at a critical point located at $T_c = 15300$ K, $P_c = 0.614$ Mbar and $\rho_c = 0.35$ g/cm^3. The entropy of the partially ionized phase is *higher* than the entropy of the molecular phase by about $0.5\,k_B$ per proton.

The free energy model for the EOS of helium is much simpler than for hydrogen. It is also not as reliable, but this is acceptable since helium typically contributes only 10% of all particles present in most astrophysical applications. The phase diagram of helium is also separated in two regimes.

In the high-density region ($\rho > 10$ g/cm^3), helium is assumed to be fully ionized. The thermodynamics of the plasma is based on the Monte Carlo simulations of Hubbard and DeWitt.[15] At densities below 1 g/cm^3, temperature ionization of helium is described by the ideal Saha equations. Interactions between helium atoms are included *ex post facto*, using the soft sphere variational fluid theory[16] and a potential derived from shock tube experiments.[17]

Pressure ionization of helium may also occur discontinuously through a plasma phase transition. However, our low density model for helium is too crude to address this issue and we smoothly interpolate thermodynamic quantities across the intermediate density range. This is a reasonable approach since the pressures found in the deep envelopes of giant planets are not thought to be sufficient to induce pressure ionization of helium.

Given pure hydrogen and helium equations of state, we can evaluate the thermodynamics of H/He mixtures with a suitable interpolation in composition. The additive volume rule is found to be a reliable interpolation method.[18] For a system at pressure P and temperature T, we have:

$$\frac{1}{\rho(P,T)} = \frac{1-Y}{\rho_H(P,T)} + \frac{Y}{\rho_{He}(P,T)},$$

(1)

where Y is the helium mass fraction and ρ is the mass density. The entropy S (an extensive thermodynamical variable) is interpolated by replacing $1/\rho(P,T)$ by $S(P,T)$ in Eq. 1. and by adding the ideal entropy of mixing, to recover the correct ideal gas limit.

Even though it has been shown to work quite well, the additive volume rule does not address the question of interactions between helium and hydrogen. This becomes particularly important in the pressure ionization regime where a small admixture of helium may have a large effect on the EOS. The PPT of hydrogen may be strongly affected by the presence of helium, and it may vanish altogether. In addition, a self-consistent solution of the PPT phase equilibrium in a mixture of hydrogen and helium would result in a discontinuity in the helium mass fraction across the coexistence curve. This is a complex problem of high-pressure physics which is well outside the scope of the present study. We assume that the presence of helium does not affect the hydrogen PPT and that the helium mass fraction is constant in each phase.

Interior Models of Giant Planets

For a self-gravitating fluid body in rotation, the equation of hydrostatic equilibrium

$$\nabla P = \rho \nabla (V + Q),$$

where

$$V(r) = G \int \frac{dr' \rho(r')}{|\vec{r} - \vec{r'}|}$$

is the gravitational potential (G is the gravitational constant) and

$$Q(\xi) = \int_0^\xi \omega^2(\xi')\xi' \, d\xi'$$

is the centrifugal potential, and the equation of state $P(\rho)$ form a closed set. This two dimensional equation can be solved with a Legendre polynomial expansion to obtain the dimensionless gravitational moments

$$J_n = \frac{-2\pi}{Ma^n} \int_{-1}^1 d\mu' \int_0^a r'^{n+2} P_n(\mu')\rho(r',\mu') \, dr',$$

where M is the mass of the planet, a is the equatorial radius at a pressure of 1 bar, μ' the cosine of the colatitude, and P_n the Legendre polynomial of order n. Note that as n increases, J_n is more heavily weighted towards the surface of the planet. Also, for a fluid planet, $J_n = 0$ for odd values of n.

The primary observational constraints on a given EOS for Jupiter and Saturn is that the pressure-density relation $P(\rho)$ must, when coupled with the hydrostatic equilibrium equation and the planetary rotation period P_{ROT}, agree with the observed mass M, equatorial radius a and the gravitational zonal harmonics J_2, J_4 and J_6. Higher order harmonics have not been measured and will not be very useful as constraints on the EOS since they are affected by atmospheric dynamics and differential rotation.

Because of the integral nature of the constraints, a simple inversion procedure is not possible and a number of assumptions must be made to obtain an acceptable model. Based on previous studies of the interior of Jupiter and Saturn, we assume a basic model consisting of a dense rock core (magnesium-silicates and iron in solar proportions) surrounded by an "ice" mantle (a solar mixture of oxygen, carbon and nitrogen which form the volatile molecular species H_2O, NH_3 and CH_4 at lower pressures) obeying $P(\rho)$ relations given in Ref. 7. The bulk of each planet consists

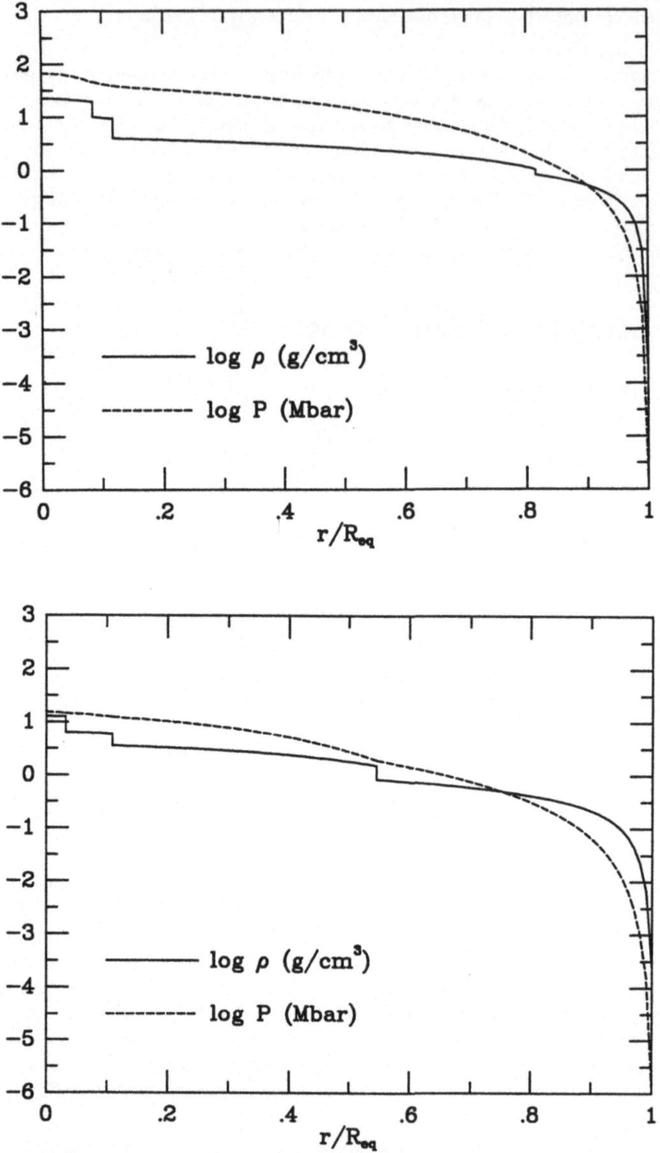

Fig. 1. Pressure and density profiles of optimized models of Jupiter (top panel) and Saturn (bottom panel), plotted as a function of mean radius. Discontinuities in the density clearly mark the boundaries of the four layers of the models: rocky core, ice mantle, metallic and molecular parts of the envelope.

of a thick envelope of hydrogen and helium. The helium abundance is assumed to be constant within a given phase of hydrogen, but is allowed to change discontinuously across the PPT. Such behavior is expected according to the Gibbs phase rule, quite apart from the possibility of H-He immiscibility predicted by Stevenson.[19]

This relatively simple model has four free parameters, the core mass M_c (rock core and ice mantle), the mass of the ice mantle M_{ice}, and the helium mass fractions in each of the phases of hydrogen, Y_I and Y_{II}, where index I refers to the molecular part of the envelope and II to the metallic part. These parameters are optimized to fit the observational constraints.

There is an additional constraint related to the composition of the envelope. Because there are no significant fractionation processes for H and He at work in Jupiter and Saturn, we expect that the overall abundance of helium relative to hydrogen must be the same as in the Sun, $Y = Y_\odot = 0.28$. We will see that all computed models require a helium mass fraction significantly in excess of that value, a result which is interpreted as due to the presence of elements heavier than helium distributed throughout the envelope. While the simple model described above does not explicitly allow for this situation, it is possible to compute the abundance of heavy elements *ex post facto*, based on the difference between the optimized Y_I and Y_{II} values and $Y_\odot$.

Optimized Models of Jupiter and Saturn

Our most satisfactory models, which fit all constraints, are shown in Fig. 1. Table I gives their most interesting parameters.

It is particularly interesting that models computed with an EOS which is smoothly interpolated across the pressure ionization regime (*i.e.* without a PPT) do not lead to satisfactory models of Jupiter if the interpolation range is as large as a decade in density.[2] This shows that even if the integral nature of the constraints render this method insensitive to details of the EOS, it certainly allows us to distinguish between extreme descriptions of pressure ionization. This suggests that pressure ionization of hydrogen occurs rather suddenly, if not through a first order transition. In the case of Saturn, it is not possible to obtain a satisfactory model without invoking a discontinuity in Y in the envelope. This is actually interpreted as a strong indication for a large helium abundance gradient in the envelope of Saturn.

Our models are consistent with the general picture which emerged over the last decade.[20,6-8] Both Jupiter and Saturn have a small core of heavy elements (rock + ice) surrounded by a thick H/He envelope which is significantly enriched in heavy elements compared to solar abundances (see below). Considering the variety of EOS models used in these investigations, this general picture of the structure of Jupiter and Saturn is secure. However, apparently small differences in $P(\rho)$ in the pressure ionization regime (See Fig 2. of Ref. 2) lead to important variations in the core mass, core composition, and the abundance of heavy elements. These in turn affect the astrophysical interpretation and the nature of the processes which lead to the inferred structure.

The comparison of helium abundances in each part of the envelope, Y_I and Y_{II} is particularly interesting. In Jupiter, we find that the discontinuity in Y ($\Delta Y = 0.036$) in the envelope is small and not very significant in the light of other uncertainties in our calculation. On the other hand, $\Delta Y = 0.48$ for Saturn. Even when allowance is made for the presence of heavy elements, this Saturn model shows a large discontinuity in the helium abundance. Because of the rather simple structure of our assumed models, this result really indicates the existence of a large helium abundance gradient in the envelope of Saturn. Such differentiation in the envelope of Saturn can develop from the insolubility of helium in a hydrogen plasma, which was originally suggested to account for the observed excess luminosity of the planet.[21] As the planet cools, part of the envelope becomes subcritical and helium rich bubbles form and sink towards the center of the planet since they are denser than the surrounding medium. This releases gravitational

Table I Optimized models of Jupiter and Saturn

	JUPITER	SATURN
M ($\oplus$)	317.7	95.1
M_c ($\oplus$)	5	1
M_{ice}/M_c	0.50	0.95
P_c (Mbar)	67.4	15.5
T_c (K)	22600	11900
P_{PPT} (Mbar)	1.71	1.93
T_{PPT} (K)	6880	6070
Y_I	0.29	0.25
Y_{II}	0.326	0.73

Table II Heavy element abundances

	JUPITER	SATURN
Y_I'	0.18 ± 0.04	0.06 ± 0.05
Y_{II}'	0.297 ∓ 0.007	0.46 ∓ 0.04
Z_I	0.079 ∓ 0.027	0.12 ∓ 0.02
$M_{Z,I}(\oplus)$	3.6 ∓ 1.2	5.0 ∓ 1.1
Z_{II}	0.034 ± 0.008	0.44 ± 0.03
$M_{Z,II}(\oplus)$	9.3 ± 1.9	22.5 ± 1.7

Subscript I and II refer to the molecular and metallic parts of the envelope, respectively. Error estimates are based on the uncertainty on the observed atmospheric mass fraction of helium, Y_I'. The heavy element mass fractions, accounting for the excess density of the fitted adiabat compared to H/He adiabats with the given Y' values, are Z_I and Z_{II}. The corresponding masses are $M_{Z,I}$ and $M_{Z,II}$.

energy which is radiated at the surface of the planet. Calculations show that this mechanism can account for the excess luminosity of Saturn and that it started about 2×10^9 years ago.[22] The luminosity of Jupiter can be explained in terms of the release of primordial heat and gravitational contraction alone and it is not believed that He/H immiscibility plays any role in that planet. Our models strongly support this picture of the role of immiscibility of He in pressure-ionized H in giant planets: The small ΔY found in Jupiter indicates little He/H differentiation, while the large value of ΔY obtained for Saturn shows that this process is well under way. Detailed evolutionary calculations taking this process into account have not been performed to this date. This result bears on the phase diagram of He/H mixtures. Most calculations find that the critical temperature for a solar mixture of helium and hydrogen of about $T_c = 8\,000$ K.[15,19] This value is consistent with our understanding of He/H separation in the evolution of Jupiter and Saturn.[22] Recently, this problem was studied with a different approach.[23] Based on very detailed zero temperature calculations for He/H alloys and a simple prescription to extend their model to finite temperature, they find $T_c = 15\,000$ K$\pm 3\,000$ K for a mixture containing 7% of helium (roughly the solar abundance). With such a high value of T_c, He/H phase separation would start much earlier in the evolution. We expect that differentiation would be complete in Saturn, with no energy source left to explain the present excess luminosity. Conversely, we would be facing the problem of too high a luminosity in the case of Jupiter, where the phase separation would be currently underway. It appears that a value of T_c much higher than 8000 K is inconsistent with 1) the present day luminosities of Jupiter and Saturn, 2) the magnitude of helium abundance gradients in the envelope, as obtained from gravitational constraints.

As mentioned above, we find that both Jupiter and Saturn have hydrogen-rich envelopes which contain a substantial fraction of material in addition to helium. That is, if we assume that the pressure-density relation in the envelopes of both planets follows the $P(\rho)$ relation for pure hydrogen and helium, with helium in strict proportion to hydrogen, and that the observed atmospheric abundance fixes the helium abundance in the metallic phase region by helium conservation, then this assumption is not confirmed by our models. The envelopes of both planets are everywhere too dense. The density excess is too large to be accounted for by including the effect of solar proportions of heavy elements, which would not exceed 0.02 of the total mass.

The mass fraction of heavy elements Z is determined as follows. We first assume that the observed atmospheric helium abundance (Y_I') prevails throughout the molecular part of the envelope (phase I). The helium mass fraction Y_{II}' in the metallic phase (phase II) is fixed by helium mass conservation, where the global He/H mass ratio must be solar, *i.e.* $Y = 0.28$. Given those helium mass fractions (which are lower than those determined by the fitting procedure, see Table II), an EOS for the heavy elements and using the additive volume rule to obtain an EOS for a mixture of hydrogen, helium and heavy elements, one can obtain Z at each pressure level in the planet (see Ref. 2 for details). The global Z value and total mass of heavy elements for each phase is given in Table II.

In both planets, we find more heavy elements distributed throughout the envelope than in the core. This suggests that these substances are mostly soluble in H/He mixtures under the conditions found in giant planets since they have not significantly segregated into the core since their formation, 4.5×10^9 years ago.

Adding up the masses of Z-component by referring to Tables I and II, we obtain 18 $M_\oplus$ in total for Jupiter (total mass 318 $M_\oplus$) and 29 $M_\oplus$ for Saturn (total mass 95 $M_\oplus$), where $M_\oplus$ is the mass of the Earth. This represents a threefold enhancement of heavy elements over the solar abundance for Jupiter and a factor of 15 for Saturn. On the other hand, Saturn has a smaller core than Jupiter, and it is nearly all "ice" with very little "rock" component. These parameters bear directly on the formation process of the two planets and on the physical and chemical conditions in the solar nebula at the distance where they formed. Models for the formation of the giant planets are based exclusively on the concept of "nucleated instability," in which accretion of rock and ice induces rapid accretion of a gaseous envelope around the forming core. In the nebula from which

Jupiter and Saturn formed, hydrogen and helium are found essentially fully in the gas phase. Water will be largely condensed out as ice, though in the Jupiter zone there may be epochs with less water ice during which the accretion rate of the nebula is higher. The origin of the nitrogen and carbon in Jupiter and Saturn is complicated, as they form CO and N_2, both highly volatile species. Much of the enhancement in these species may come from planetesimals formed further out in the solar system which were gravitationally scattered to intersect Jupiter and Saturn.[24] The carbon and nitrogen trapped in this cold material will be a mixture of reduced and oxydized species.[25] Thus the particular mix of ices which form the heavy element enhancement in these planets is complex, but may ultimately be constrained in part by the kind of modeling described above.

Conclusion

Since we were able to abtain satisfactory models of Jupiter and Saturn, the EOS presented here, and the PPT of hydrogen in particular, have passed their first experimental test. This EOS is *not* ruled out by observational constraints from the giant planets Jupiter and Saturn, and the thermodynamic model for the metallization of hydrogen remains viable.

We confirm the general conclusions of previous work on the structure of these two planets: they both have a central core of heavy elements surrounded by a massive hydrogen-rich envelope which is significantly enriched in heavy elements when compared to solar abundances. However, our models have much smaller central cores and larger amounts of heavy elements distributed throughout the envelope.

The application of thermodynamic descriptions of H and He to giant planet models which fit the mechanical constraints provided by the gravitational moments of the planet is presently the only means to gain insight into the helium abundance in the planet, the possible differentiation of the H/He envelope, the abundance of heavy elements (C, N, O, Mg, Fe ...), as well as their internal structure and the nature of the formation process. Conversely, giant planets are natural sites for testing theories of dense matter and provide hints as to the behavior of matter in the regime of pressure ionization of hydrogen. For example, this work indicates that the critical temperature for a He/H mixture of solar proportions cannot be much higher than about 8000 K. Also, it appears that heavy elements are largely soluble in He/H mixtures under the conditions found in Jupiter and Saturn.

It should be clear that our knowledge of the interior of giant planets and of the processes that lead to their formation and those at work during their evolution relies heavily on theories of dense matter. Further progress of our understanding of the largest planets of the solar system as well as the interpretation of the data returned by space probes Galileo (Jupiter) and Cassini (Saturn) will require more detailed investigations of dense matter problems such as 1) the pressure-ionization of hydrogen, with emphasis on the PPT, 2) the pressure-ionization of helium; does it also have a PPT? 3) how is the hydrogen PPT affected by the presence of helium, and 4) a reliable evaluation of the partitioning of helium and of other species across the PPT. This is particularly relevant to the interpretation of the measured atmospheric abundances of various elements.

This research was supported in part by NSF grant AST-8910780. D. S. gratefully acknowledges a postdoctoral fellowship from the Natural Sciences and Engineering Research Council of Canada.

References

[1] R. H. Fowler, Mon. Not. Roy. Ast. Soc. **87**, 114 (1926)
[2] G. Chabrier, D. Saumon, W. B. Hubbard, and J. I. Lunine, submitted to Astrophys. J. (1992)
[3] F.-X. Schmider, E. Fossat, and B. Mosser, Astron. Astrophys. in press, 1991

[4] D. Saumon and G. Chabrier, Phys. Rev. Lett. **62**, 2397 (1989)

[5] M. Robnik and W. Kundt, Astron. Astrophys. **120**, 227 (1983); W. Ebeling and W. Richert, Phys. Lett. **108A**, 80 (1985)

[6] W. B. Hubbard and G. P. Horedt, Icarus **54**, 456 (1983)

[7] W. B. Hubbard and M. S. Marley, Icarus **78**, 102 (1988)

[8] D. J. Stevenson and E. E. Salpeter, in *Jupiter*, T. Gehrels, Ed. (University of Arizona Press, Tucson, 1976), p. 85; V. N. Zharkov and V. P. Trubitsyn, *ibid*, p. 133

[9] D. Saumon and G. Chabrier, Phys. Rev. A **44**, 5122 (1991)

[10] D. Saumon and G. Chabrier, submitted to Phys. Rev. A (1992)

[11] G. Chabrier, J. Phys. France **51**, 1607 (1990)

[12] H. C. Andersen and D. Chandler, J. Chem. Phys. **53**, 547 (1970); J. D. Weeks, D. Chandler, and H. C. Andersen, *ibid.* **54**, 5237 (1971); *ibid.* **55**, 5422 (1971)

[13] D. G. Hummer and D. Mihalas, Astrophys. J. **331**, 794 (1988)

[14] W. D. Kraeft, D. Kremp, W. Ebeling, and G. Röpke, *Quantum Statistics of Charged Particle Systems*, (Plenum: New York, 1986)

[15] W. B. Hubbard and H. E. DeWitt, Astrophys. J. **290**, 388. (1985)

[16] M. Ross, J. Chem. Phys. **71**, 1567 (1979)

[17] W. J. Nellis, N. C. Holmes, A. C. Mitchell, R. J. Trainor, G. K. Governo, M. Ross, and D. A. Young, Phys. Rev. Lett. **53**, 1248 (1984)

[18] G. Fontaine, H. C. Graboske, Jr., and H. M. Van Horn, Astrophys. J. Supp. **35**, 293 (1977)

[19] D. J. Stevenson, Phys. Rev. B. **12**, 3999 (1975)

[20] A. S. Grossman, J. B. Pollack, R. T. Reynolds, A. L. Summers, and H. C. Graboske, Icarus **42**, 358 (1980)

[21] R. Smoluchowski, Nature **215**, 691 (1967); E. E. Salpeter, Astrophys. J. Lett. **181**, L83 (1973)

[22] D. J. Stevenson and E. E. Salpeter, Astrophys. J. Supp. **35**, 239 (1977); D. J. Stevenson, Ann. Rev. Earth Planet. Sci. **10**, 257 (1982)

[23] J. E. Klepeis, K. J. Schafer, T. W. Barbee III, and M. Ross, Science **254**, 986 (1991)

[24] D. J. Stevenson, Lunar Planet. Sci. Conf. **14**, 770 (1983)

[25] J. I. Lunine, in *Titan*, ESA-SP, in press (1992)

PARTICLE DRIVEN INERTIAL FUSION THROUGH CLUSTER ION BEAM

C. DEUTSCH and N.A. TAHIR[+]

L.P.G.P.[*] and GDR-CNRS 918 Bât. 212, Université Paris XI

91405 ORSAY Cedex, France

Abstract

Cluster ion beam with energy in the several tens of keV/a.m.u. range are considered as a novel direct driven for a simple fusion pellet made of Deuterium + Tritium fuel surrounded by a Lithium pusher. The driven-pellet interaction is calculated through the hypothesis of maximum multifragmentation followed by highly correlated ion debris motion. One thus gets enhanced stopping and ablation pressure in the hundreds Mbar range. An implosion is then completed in 5 nsec.

Introduction

In the field of particle-driven inertial confinement fusion (ICF), one is now witnessing a persistent interest in very heavy drivers with the smallest possible charge-to-mass ratio [1],[2]. Up to now, the corresponding mass range has extended over a large scale: from heavy atomic ions up to macroparticles (containing 10^{22} atoms) driven to hypervelocities ($\sim$ 50-1000 kms).

The purpose of this paper is to speculate for the first time about driving potentialities afforded by intense cluster ion beams (hereafter referred to as CIB). Today it is well known [3] that cluster ions containing nearly any number of constituent atomic ions may be easily prepared, identified, selected, and even transported.

As far as direct drive compression is concerned, a straightforward manipulation of scaling laws based upon a Bethe-like stopping formula displays almost at once the obvious payoffs of playing with a heavy projectile.

CIB for ICF

The straightforward argument runs as follows [4]. A spherical shell, of radius r, thickness Δr, and density ρ, is assumed to stop an ion beam of ion mass M_i, stripped charge state Z_i, and energy E_i. Since

---

[+] on leave of absence from ICTP, B.O. Box 586, 34100 - Trieste, Italy.

[*] Associé au C.N.R.S.

the ions lose their energy predominantly by Coulomb interactions with the electrons in the ragion Δr, we can write for the ions the following:

$$\frac{dE}{dx} \approx - \frac{4\pi N Z_i^2 e^4 M_i}{2mE_i} \ln \Omega \tag{1}$$

where m is the electron mass and N is the electron number density.

Integrating equation (1), by assuming approximate constancy of the Coulomb logarithm $\ln \Omega$, gives the approximate ion range λ as:

$$\lambda \approx \mu E_i^2 / 4\pi N Z_i^2 e^4 M_i \ln\Omega = \Delta r. \tag{2}$$

The specific energy deposition ε_d, obtained by assuming that most of the target mass is in the shell Δr, is:

$$\varepsilon_d = n N_i E_i / 4\pi r^2 \lambda \rho \approx \text{const.}, \tag{3}$$

where N_i is the total number of ions of energy E_i in n beams entering and stopping in the spherical annulus in a pulse length Δt. For a current I in each beam, $N_i = I\Delta t/Z'_i e$, where Z'_i is the ion charge state in the beam. Putting $\rho \propto N$ and using equation (2) in equation (3) gives the following approximate scaling law:

$$n I M_i Z_i^2 \Delta r / r^2 E_i Z'_i \approx \text{const.} \tag{4}$$

An enhanced M_i, for instance, could allow for a smaller beam intensity I or a larger neck radius r.

Among a number of additional and intriguing possibilities, CIB would permit direct drive through momentum directly imparted to a pure DT fuel hollow target.

This highlights the momentum rich beam (MRB) [2] concepts. One can thus expect a smoother compression with a lower energy threshold, of the order of 0.2-0.5 MJ requested for ignition. Maschke speculated [2] on an ~ 10 μsec pulse length, for a 100 kA/cm^2 Cs$^+$ beam, accelerated to few hundred keV. Obviously, CIB offer attractive and more flexible alternatives for achieving similar goals.

Fragmentation and stopping

An advantage of clusters with narrow mass distributions is that they can be accelerated in linacs and other multistage acceleration devices. Single-stage accelerators can be used with any distribution of masses, but are limited in the energy they can provide.

At Orsay, we are currently [5] investigating the possibility of accelerating from the terminal of a tandem accelerator Au clusters built on 2 up to 7 atoms, with a positive charge equal to 1 or 2.

The considered energy range will be $35 < E/A < 380$ keV. Another ion source aims at producing organic clusters containing several thousands of atoms with one positive charge per one thousand atomic mass. We thus expect to accelerate linearly clusters $A = 50000$ and $Z = 50$. Finally, it should be appreciated that the crucial fragmentation issues are simplified here by taking for granted the so-called maximum entropy principle (hereafter referred as MEP). This implies that the given cluster projectile will break under impact into the largest number of its smallest building blocks.

Namely, the ionized atoms.

Atomic units (a.u.) are used in the sequel.

In most cases of interest, i.e., when the projectile kinetic energy per amu is larger than 10 keV, the stopping processes of cluster ions matter are likely to be preceded by a fragmentating event. Such an occurrence is highly dependent on the beam-target pair interaction.

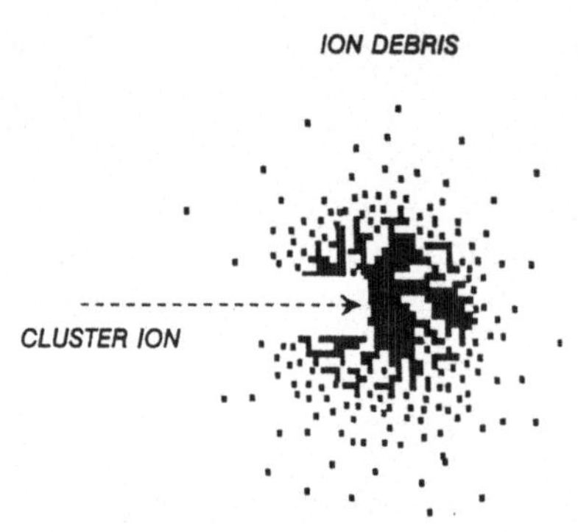

Fig. 1 - Cluster ion multifragmentation in Target

In the given energy range the cluster projectile is highly likely to experience a partial coulomb explosion. This means that the resulting debris will be at least once ionized. Moreover, their relative velocity is expected to be small compared to the projectile one over most of its quasi-linear range within the target. Therefore, these debris are expected to fly in a highly correlated motion with relative distances of order of a Bohr radius a_0 , as illustrated on Fig. 1.

The given Coulomb explosion takes place on a femto second scale length, which supports the MEP model.

Other fragmentation scenarios based essentially on combinatorial arguments might also be considered. However, they do retain as an exact asymptotic limit, the MEP scenario.

This corresponds to a sudden projectile-target interaction with a maximum produced disorder compatible with initial conditions.

A crucial simplification concerning the subsequent ranges calculations is afforded by the velocities ratio of the ion debris to the initial CIB one.

Two such debris, are expected to experience, at most, when located within a a.u. distance, a Coulomb repulsion ∼ 1 Rydberg (13.6 eV). So, an initial CIB with a kinetic energy ∼ a few tens of keV/a.m.u. will impart a nearly unchanged velocity to the resulting charged debris.

The repulsion velocity of the latter being nearly two orders of magnitude smaller than the initial CIB.

The target is likely to display a rather simple structure (Fig. 4a). It could be made essentially of roughly 4 mg of D+T fuel, surrounded by an outer/Li/shell of pusher material in close analogy with momentum reach beam (MRB) targets proposed recently [2],[8].

The main point we want to emphasize here is the tremendous flexibility introduced into the accelerator constraints by using CIB as heavy ion drivers.

Taking for instance, a given plausible CIB with M_i = 50000, Z'_i = 50 and $E_i/M_i \sim$ 10 keV, we see from Eq. (1) that a complete fragmentation into ion debris with unit charge is equivalent to a standard direct drive (HIBALL) heavy ion beam (HIB), provided $nI \sim$ 1 ampere.

Other parameters featuring Eq. (1) are expected to retain their HIB values. Such a modest requested beam intensity, 4 orders of magnitude below its HIB counterpart, seems to lies within the reach of existing linear accelerating structures. Radio frequency quadrupole (RFQ) are obvious candidates.

As far as target compression is concerned, the present MRB concept introduces the possibility of a directly imposed external pressure (hammer effect) on the pellet.

Let us consider, for instance, $Au_{N_C}^+$ metallic clusters with N_C = 2-7, and assume that we have enough sources for bombarbing uniformly the target surface. This is quite feasible if each ion source is followed by a tandem like accelerating structure [2],[6].

Now, we turn to the most plausible CIB stopping scenario. The debris resulting from the cluster impact on pellet are expected to fly in a highly correlated relative motion. The given target are modeled presently by a fully degenerate electron jellium with a T = 0 Fermi temperature. Such a homogeneous medium is characterized by a dimensionless parameter $r_S = (\frac{4}{3} \pi n_e)^{-1/3} a_o^{-1}$, in terms of the electron number density n_e.

Then, we consider a given cloud of cluster ion debris in target, as a 2-body superposition. The stopping analysis then puts emphasis on the individual ion contribution, and on the correlated one, as well.

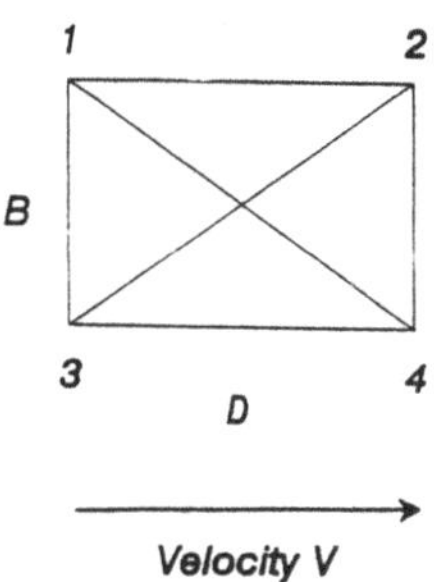

Fig. 2- 4-cluster stopping in jellium target

The latter is treated here as a superposition of dicluster contribution [9],[10], according to a model due to Basbas and Ritchie [7].

We start our analysis by considering a frozen (polarized) configuration of ion debris flying in target, closely to the initial CIB trajectory. This picture stems from the fragmentation scenario outlined, above.

Obviouly, many geometries can be taken into account. Fig. 2 thus features a polarized four-atom cluster structure with respect to its velocity in target.

Suitable structural averages allowing a more flexible description will be considered elsewhere.

According to the 2-body superposition principle advocated above, the stopping power of a N-cluster is straight forwardly given by [8]

$$S_c = (\sum_i Z_i^2)\, S_p + 2 \sum_{1 \le i < j \le N} Z_i Z_j\, S_v(B_{ij}, D_{ij}) \quad , \tag{5}$$

$$\equiv \text{POINT} + \text{CORR} \quad ,$$

in terms of (m = electron mass) the pointlike contribution

$$S_p = \frac{e^2 \omega_p^2}{v^2} \ln\left[\frac{2mv^2}{\hbar \omega_p}\right] \quad , \quad \text{with } \omega_p = \frac{3^{1/2}}{r_s^{3/2}} \text{ a.u.} \quad , \tag{6}$$

and the correlated one $S_v(B,D)$ worked previously for the cylindrical geometry depicted in Fig. 2.

The present jellium modelling is particulary well adapted to the stopping behaviour of a Li target. This latter has a density d = 0.534 g/cm^3 , so r_s = 3.27. Moreover, Li exhibits the smallest first energy of ionization ($\sim$ 3 eV), and it is very easily turned into a jellium of degenerate free electrons under heavy ion impact. The Bethelike formalism developed above is valid provided the ion debris velocity, nearly colinear to the initial CIB one, remain larger than $(\frac{\omega_p}{2})^{1/2}$. For the Li shell target, at hand, this implies v $\ge$ 0.39 a.u.

As a significant result, we evaluate on Fig. 3, the total stopping in a Li target including correlation, of 4 charges disposed on a tetrahedron geometry polarized along velocity $\vec{v}$.

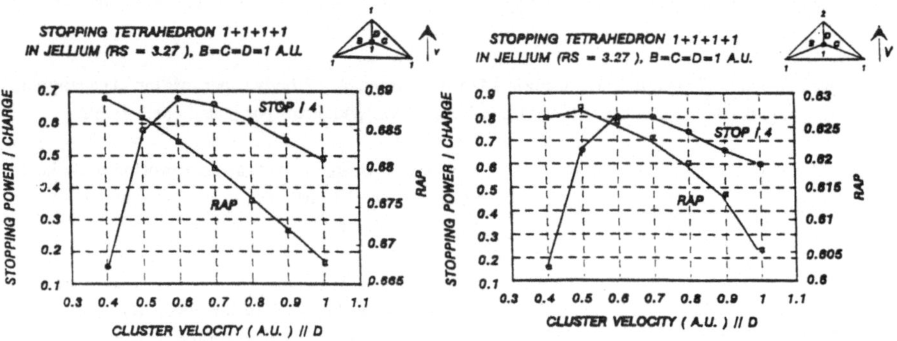

Fig. 3 - Stopping of rectangular tetrahedrons (B = C = D = 1 a.u.). B,C$\perp$ velocity $\vec{v}$. $\vec{D}//\vec{v}$ with various charges distributions. — a/ 1+1+1+1 , b/ 1+2+1+1.

In agreement with the quadratic form (5), the plotted ratio RAP = CØRR/Sc is the largest for equal charges case. Corresponding ranges are given in Table 1 for different ion debris repartitions and velocities.

According to the nearly isotropic fragmentation processes advocated above, it appears plausible to evaluate an ablation pressure resulting from a homogeneous distribution of ion debris within a volume $(N_c^{1/3}\, a_0)^2\, R$. R is the linear average cluster range in target.

Table 1- Cluster stopping in pellet

Tetrahedrons 4-clusters and cubiclike 8-clusters stopping in the Li outer shell (d = 0.534 g/cm, r_s = 3.27) featured by ranges in µm and ablation pressures in 100 MBars. Ions charges in clusters are given at vertices. The vertex-vertex distances projected on cartesian axis parallel and orthogonal to velocity are all taken equal to 1 a.u. (B = C = D = 1). In agreement with the averaged stopping estimates displayed on Figs. 3.

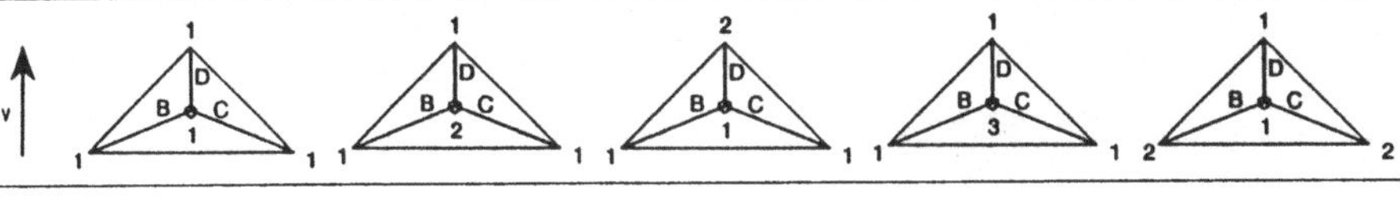

$$\frac{E}{A} = 10.228 \text{ KeV/a.m.u.} \quad (v = 0.8 \text{ a.u.})$$

Range (µm)	10.61	12.69	4.28	7.20	4.55
Abl.Pr. (100 MB)	1.69	2.23	4.20	2.50	3.95

$$\frac{E}{A} = 15.982 \text{ Kev/a.m.u.} \quad (v = 1 \text{ a.u.})$$

Range (µm)	18.45	4.78	3.18	12.04	3.61
Abl.Pr. (100 MB)	1.53	5.91	8.87	2.34	7.83

Pellet compression

In terms of the projectile kinetic energy E, the ablation pressure thus reads as (Energy/volume)

$$P(100 \text{ Mbar}) = \frac{N_c^{1/3} \times E(\text{keV/a.m.u.}) \times 0.0565 \, M_i}{R(\mu m)} \quad , \tag{7}$$

Eq. (7) is illustrated in Table 1 for a Li target. Huge pressures may indeed be achieved at reasonable projectile energies. The atom distribution within cluster is taken homogeneous in space.

We suppose a solid density of cluster material all around pellet.

The above stopping results allow for a one dimensional calculation of the direct drive compression for the target given in Fig. 4a. We thus use a three temperatures (ion, electron and radiation) code [9] to unravel the time evolution of several radial distributions for temperatures, matter density and pressure.

Thanks to the huge avalaible ablation pressure (Table 1), the implosion is completed within 5 nsec. Such a behaviour it at variance with the 35 nsec duration of standard heavy ion scenarios. Above all, these calculations highlight the considerable potential interest of a dynamic tamper arising from synchronous clusters impact on pellet.

These latter do not behave solely as a driver bringing energy and momentum into target. They also provide radiative shielding and hydrodynamic tampering. Fig. 4c pertains to a mid-compression time evolution. The dissambly phase starts after 4.5 nsec.

Summarizing, we have disclosed a novel scheme for particle-driven compression of a simple inertial fusion pellet. It makes use of a cluster ion beam as a driven. Cluster fragmentation in target results in a highly correlated and enhanced stopping in the Lithium pusher. One thus gets an ablation pressure in the hundreds Mbar range.

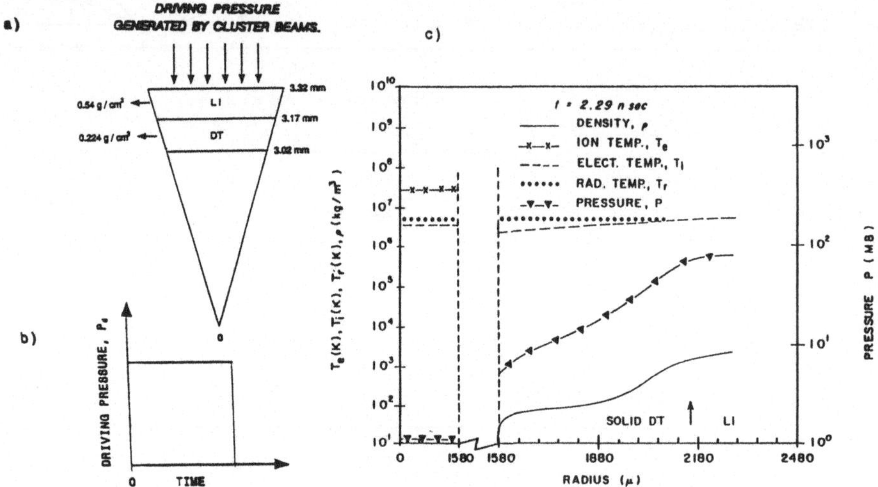

Fig. 4 - a/ Target initial conditions; b/ Time variation of input pressure pulse.
　　　　 c/ Matter temperature, T; radiation temperature, Tr; density, ρ, and pressure P.
　　　　 vs target radius at t = 2.29 nsec.

References

[1]　F. WINTERBERG, *Z. Phys.* **A 296**, 3 (1980) and also
　　　T. YABE and T. MOCHIZUKI, *Jpn. J. Appl. Phys.*, **22**, L262 (1983)

[2]　A.W., MASCHKE, *Proceed HIF 84* (INS-Tokyo) p. 168 (1984)

[3]　M.A. DUNCAN and D.H. ROUVRAY, *Sci. Am.*, **261**, 60 (1989)

[4]　T.D. BEYNON, *Phil. Trans. R. Soc. London*, **A 300**, 613 (1981)

[5]　S. DELLA-NEGRA, D. GARDES and Y. LE BEYEC, Private communication.(1989) and
　　　C. DEUTSCH, *Laser and Part. Beam* **8**, 541 (1990)

[6]　D.C. WILSON and A.W. MASCHKE, LANL Preprint (1990) (unpublished)

[7]　G. BASBAS. and RITCHIE, *Phys. Rev.*, **A25**, 1943 (1982)

[8]　C. DEUTSCH, *Laser and Part. Beam* (1991) and
　　　also *Trans. Am. Nucl. Sci* (1991)

[9]　N.A. TAHIR, K.A. LONG and E.W. LAING, *J. Appl. Phys.*, **60**, 899 (1986)

STRONGLY COUPLED PLASMA IN LASER-TARGET EXPERIMENTS

A. Nowak-Goroszczenko, W. Mróz, J. Wołowski and E. Woryna

Institute of Plasma Physics and Laser Microfusion,
00-908 Warsaw 49, P.O. Box 49, Poland

1. INTRODUCTION

Properties of plasma are controlled by both the hydrodynamical parameters and the Coulomb type interaction of charged particles in plasma. The ideal plasma is called to be such a plasma in which the Coulomb coupling energy is neglegible compared with the kinetic energy of particles in the plasma. The measure of discrepancy from the ideality is accepted to be the coupling constant of plasma thought as the ratio of the mean interaction energy of neighbouring particles to their mean kinetic energy. The coupling constant of an ideal plasma is considerably less than one. The plasma, the coupling constant of which appears to be greater than one is referred to as the non-ideal or strongly coupled plasma. The coupling constant of the plasma in which a classical statistics (the Boltzmann statistics) holds, can be expressed in the following way [1]:

$$\gamma = (ze)^2/(akT) = 2.32 \times 10^{-7} (n_i [cm^{-3}])^{1/3}/T(eV) \tag{1}$$

where ze is the ion charge, T — the plasma temperature, n_i — the ion concentration, $a = [3/(4\pi n_i)]^{1/3}$, is the ion-sphere radius — so called Wigner-Seitz radius.

In the presented work the non-ideality strenght of different types of plasma created by intensive laser radiation has been estimated. The targets of different atomic number — from polystyrene to gold have been investigated. Subjected to analysis were the hot plasma heated directly by the laser radiation in the focus area, the plasma heated by X-ray radiation generated in the hot plasma in the case of targets of atomic number Z > 13 [2] and the plasma generated at the back surface of thin foil.

The strenght of plasma coupling was estimated on the basis of the plasma parameters obtained from investigations performed by using the corpuscular diagnostic methods.

2. EXPERIMENTAL SET-UP

The investigations were performed in the IPPLM by using the Nd laser system ($\lambda = 1.06 \ \mu m$) of parameters: the laser energy $E_L < 10$ J, the pulse duration $\tau_L \cong 1$ ns, the radiation power intensity at the target

$q_0 = 10^{13} - 10^{14}$ W/cm^2. In investigations the solid plane targets of $(C_8H_8)_n$, SiO$_2$, Al, Cu Ta and Au and foils of Al and Au were used. Set-up of diagnostics shown in Fig. 1 enabled investigation of the plasma expanding in direction of incident laser radiation (in the case of solid targets and foils) as well as the plasma generated at the back surface of the target (in the case of foil targets).

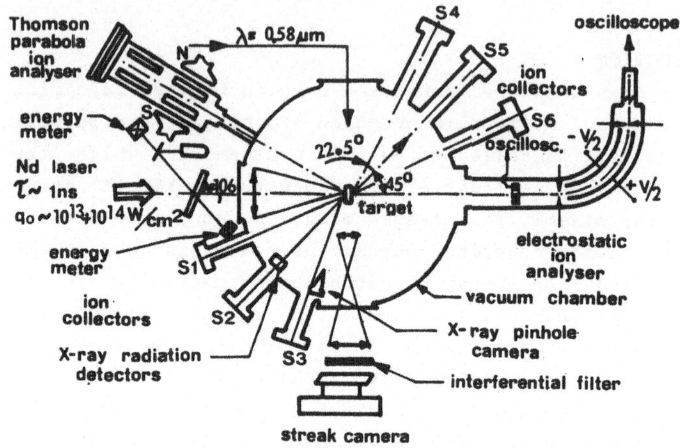

Fig. 1. Experimental set-up.

Presented investigations were carried out with the use of ion collectors, Thomson mass spectrograph and electrostatic ion analyser [3].

3. THE METHOD FOR DETERMINATION OF PLASMA PARAMETERS

The electron temperature T_e of plasma was determined from relation between mean ion energy $\overline{E}_i$, temperature T_e and the mean ion charge z_0 of plasma. For one-dimensional adiabatic expansion [4]:

$$\overline{E}_i = 3.33(\overline{z}_0 + 1)T_e \tag{2}$$

The mean ion energy $\overline{E}_i$ was found from relation:

$$\overline{E}_i = \int (dQ/dE_i)E_i dE_i / \int (dQ/dE_i)dE_i \tag{3}$$

where: $dQ/dE_i = d(N_i e\overline{z})/dE_i$ is the ion charges distribution as a function of ion energy, which was obtained from transformation of ion current pulses measured by ion collector, $\overline{z}$ is the mean ion charge in the place where ion collector is localized.

The mean ion charge $\overline{z}_0$ in the hot region of plasma was calculated with the use of modified Shearer-Barnes formula [5]:

$$\overline{z}_0 = B\left\{T_e(keV)/[1 + (B/Z)^2 T_e(keV)]\right\}^{1/2} \tag{4}$$

where: Z is the atomic number of the target material, B – the constant dependent on the kind of target material (Table 1).

Table 1. Values of constant B for selected target materials.

Target	C	SiO$_2$	Al	Cu	Ta	Au
B	35	37	40	45	47	50

The mean ion charge $\bar{z}$ of plasma in the place of localization of the
ion collector was calculated on the basis of an analytical solution of
recombination problem presented in the paper [5].

Justification of possibility of application of the presented method
for the ideal plasma in the range of $\gamma < 1$, for the non-ideal plasma
in the range $\gamma > 1$, was confirmed by the results of numerical calculations [6].

4. THE MEASUREMENT RESULTS

4.1. The plasma created at the front surface of target

As a result of performed investigations it is found, that the ion current pulses from the ion collectors for the target materials of Z $\leqslant$ 13
(polystyrene, glass and aluminium) essentially differ from the pulses
obtained for targets of Z > 13 (copper, tantalum and gold), which is
illustrated in Figs 2 and 3. These ion pulses were normalized to their
maxima. Ion collector were located on the time-of-flight base L =
= 44.2 cm at the angle 22.5°. In the case of targets of Z $\leqslant$ 13 one can
distinguish two ion groups on the ion pulse: the thermal ions of the
basic material of the target and the light ions contaminations,
hydrogen mainly. In the case of targets with Z > 13 one can distinguish three groups of ions on the ion pulse: the thermal ions of the

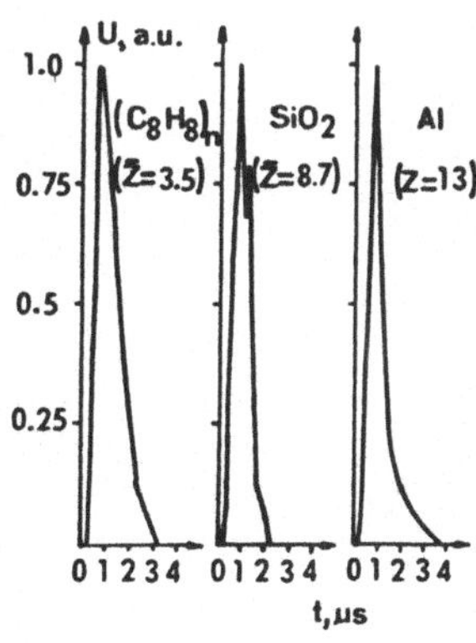

Fig. 2. Ion collector oscillograms
obtained from low-Z targets.

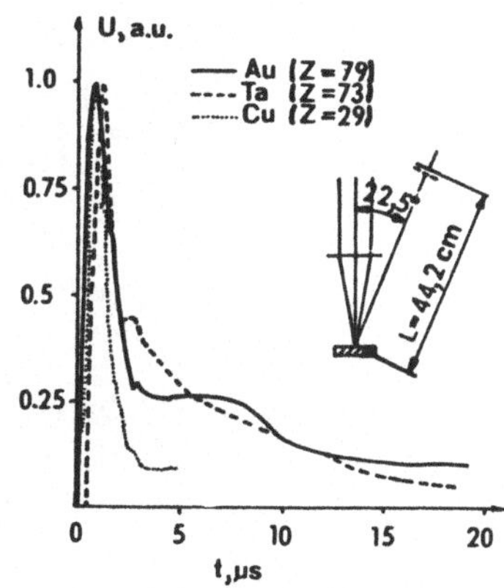

Fig. 3. Ion collector oscilograms
obtained from high-Z targets.

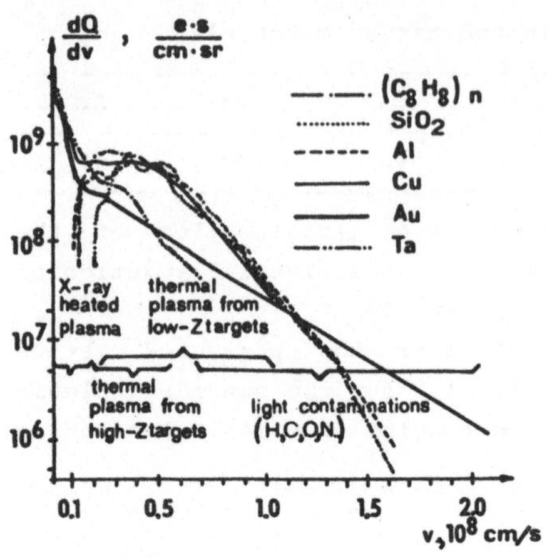

Fig.4. Ion velocity distributions.

basic material of the target, the ions slower than thermal ions from the plasma of the basic material generated as a result of heating the target by X-ray radiation and the light ion contaminations [2,7] – H, C, N and O mainly. The range of velocities of the above mentioned ion groups are shown on the velocity distributions presented in Fig. 4. In order to identify the ion kind, the measurement results obtained from the Thomson mass spectrograph (Fig. 5) were used.

In Figs 6 – 9 are presented dependences of the energy carried by ions, of the number of ions, of the mean ion energy and of plasma temperature on the atomic Z number of the target material for the thermal ions and the ions of plasma heated by X-ray radiation. These quantities are normalized to the unity of the solid angle.

Above mentioned parameters of the thermal as well as heated by X-ray radiation plasmas are shown in Table 2.

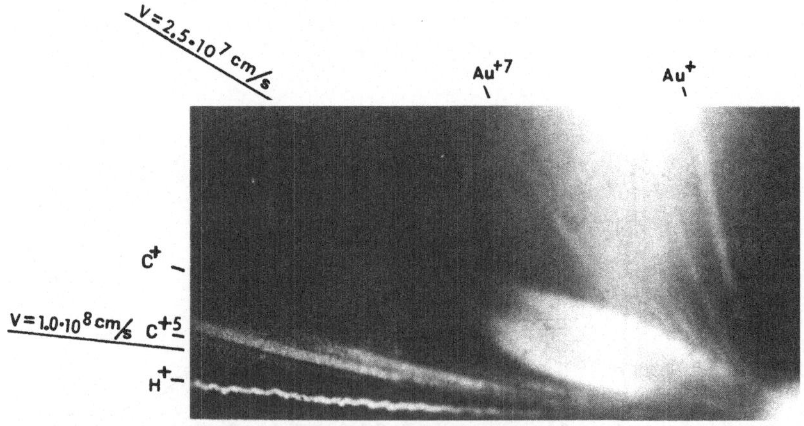

Fig.5. Thomson parabola obtained from Au target.

4.2. Plasma created at the back surface of target

In Fig.10 is shown the typical oscillogram of the ion current from 6 μm thick Al foil from back surface of the target.

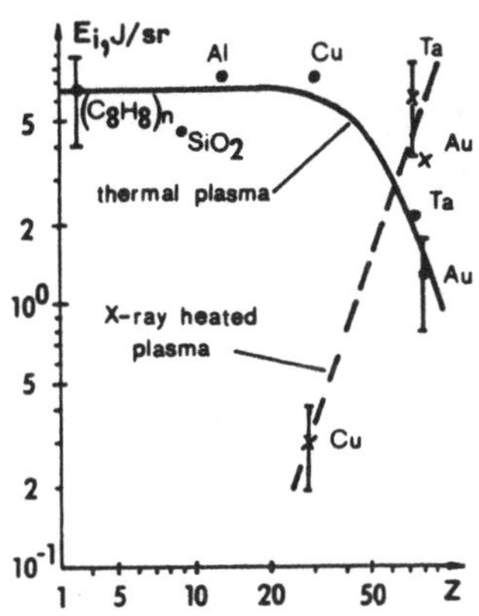

Fig.6. Enery carried by ions as a function of target Z-number for thermal and X-ray heated plasmas.

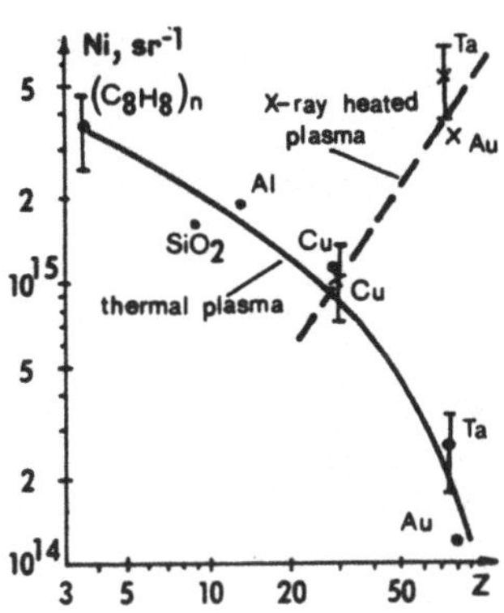

Fig.7. The number of ions as a function of target Z-number for thermal and X-ray heated plasmas.

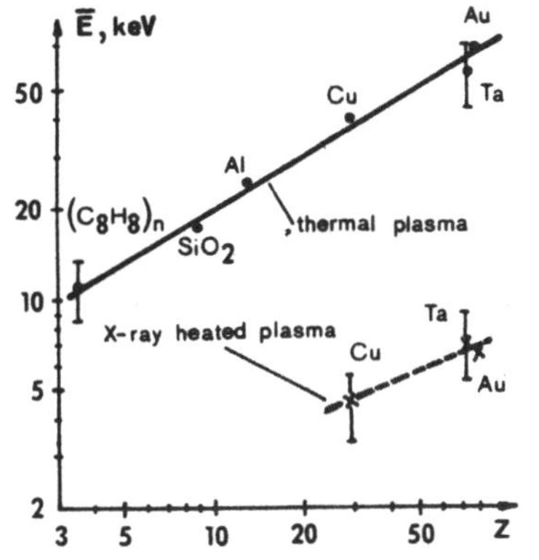

Fig. 8. Mean ion energy as a function of Z-number for thermal and X-ray heated plasmas.

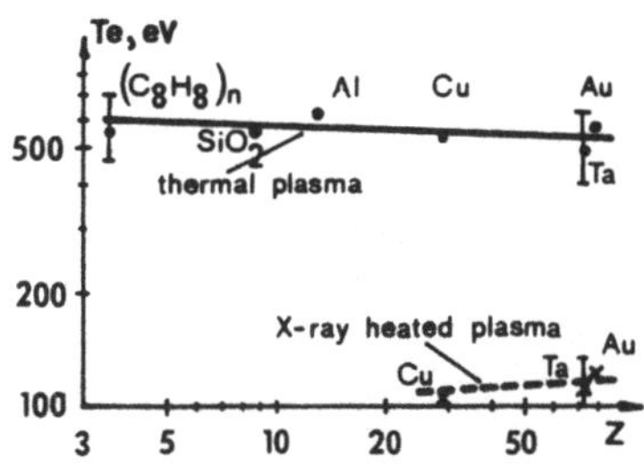

Fig. 9. Electron temperature as a function of Z-number for the thermal and X-ray heated plasmas.

The characteristic feature of the oscillograms of the ion current, recorded from the back surface of the target is the occurrance of several maxima. In Table 3 are shown the plasma parameters calculated using the method described in the point 3 at assumption that maxima of the ion current origin from the ions of the basic material of the target expanding from the plasma with concentration of $n_i = 10^{22}$ cm^{-3}.

Table 2. Parameters of thermal and X-ray heated plasmas.

Target	$\bar{z}_0$		$\bar{E}$, keV		T_e, eV	
	thermal plasma	X-ray heated plasma	thermal plasma	X-ray heated plasma	thermal plasma	X-ray heated plasma
Au	34.47	16.61	67.2	6.7	585	119
Ta	30.15	15.51	57.0	6.9	496	114
Cu	21.89	13.03	39.9	4.5	547	104
Al	12.05	–	24.6	–	614	–
SiO_2	9.44	–	17.3	–	550	–
$(C_8H_8)_n$	3.50	–	10.9	–	555	–

Target	$\bar{z}$		N_i, sr^{-1}		E_i, J/sr	
	thermal plasma	X-ray heated plasma	thermal plasma	X-ray heated plasma	thermal plasma	X-ray heated plasma
Au	2.78	0.65	1.2×10^{14}	3.3×10^{15}	1.3	3.5
Ta	2.56	0.51	2.6×10^{14}	5.4×10^{15}	2.1	5.9
Cu	2.06	0.48	1.1×10^{15}	1.0×10^{15}	7.2	0.3
Al	1.41	–	1.9×10^{15}	–	7.5	–
SiO_2	1.09	–	1.6×10^{15}	–	4.5	–
$(C_8H_8)_n$	0.90	–	3.6×10^{15}	–	6.4	–

Tabele 3. Parameters of Al plasma created at the back surface of Al target (Fig.10).

No. of peak	v, cm/s	T_e, eV	$\bar{E}_i$, eV	$\bar{z}_0$
1	1.2×10^7	76	2 130	8.5
2	6.0×10^6	26	500	5.8
3	1.4×10^6	3.5	27	2.4

5. ESTIMATION OF THE PLASMA COUPLING CONSTANT

The coupling constant of the plasma was calculated from relation (1) [1], assuming $z = \bar{z}_0 = \sum_j (z_j n_{i,j}) / \sum_j n_{i,j}$ (j = 1 to z_{max}), on the basis of the plasma parameters shown in Tables 2 and 3. The calculation

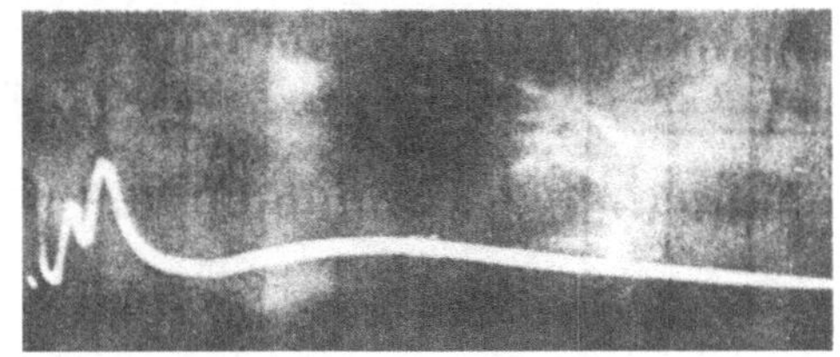

Fig.10. Ion current oscillogram from back surface of 6 μm thick Al
target.

Table 4. Results of calculation of the coupling constant of different types of the laser plasma (for thermal plasma $n_{e,cr} = n_i\bar{z}_0 = 10^{21}$ cm^{-3} and for plasma heated by X-ray radiation, $n_i = 10^{22}$ cm^{-3}).

Target	Thermal plasma	X-ray heated plasma
$(C_8H_8)_n$	0.01	–
SiO_2	0.04	–
Al	0.05	–
Cu	0.09	8.15
Ta	0.14	10.53
Au	0.14	11.81

For the back surface of aluminium target it was obtained the following values for the coupling constant for each maximum of the ion current shown in Fig. 10, resulted by the ions emitted from the plasma regions of parameters presented in Table 3:

$$\gamma_1 = 4.8; \qquad \gamma_2 = 6.5; \qquad \gamma_3 = 8.2$$

For examination is subjected the physical nature of appearance of the ion current peaks in the plasma created at the back surface of the target.

6. CONCLUSIONS

As a result of interaction of the laser radiation with targets of Z >
> 13, due to the conversion of the laser radiation into X-ray radiation, a diffusion of X-ray radiation into the target bulk occurred. As a result of absorption of this radiation is generated the strongly coupled plasma ($\gamma \sim 10$) of concentration as in a solid state, of temperature $T_e \sim 100$ eV and of large mean ion charge ($\bar{z}_0 \sim$
~ 15).

In the case of thin foil targets with thickness of order of micrometers, as a result of energy dissipation of the shock wave, onto the back surface of the target, generated is non-ideal plasma of low mean ion charge, with solid state and even higher concentration due to the material compression resulted by the shock wave, of temperatures $T_e \sim$
~ 10 eV and $\gamma \sim 3 - 5$.

The investigation results, presented in this paper, show that corpuscular diagnostics can be applied for determination of the parameters of non-ideal plasma.

REFERENCES

1. S. Ichimaru, Strongly Coupled Plasmas: High-Density Classical Plasmas and Degenerate Electron Liquids, Rev. Modern Physics **54**, 4, 1017 (1982).

2. W. Mróz, A. Nowak-Goroszczenko, J. Wołowski, E. Woryna, Investigations of Laser Interaction with High-Z Plasmas, Proceedings of 21th European Conference on Laser Interaction with Matter, October 21-25, 1991, Warsaw, Poland (submitted to Laser and Particle Beams, 1992).

3. J. Wołowski, E. Woryna, S. Denus, A.A. Erokhin, Yu.A. Zakharenkov, W. Mróz, G.V. Sklizkov, J. Farny, A.S. Shikanov, Mass-Spektrometricheskie Issledovania Plotnoj Plazmy na Ustanovke "Kalmar", Trudy FIAN **149**, 125 (1984).

4. H. Puell, Heating of Laser Produced Plasmas Generated at Plane Solid Targets, Z. Naturforsch. **25**, 1807 (1970).

5. J. Farny, Phil. Tesis, Military Technical Academy, Warsaw 1985

6. B. Kärcher, private communication.

7. W.Mróz, J. Farny, M. Kolanowski, A. Nowak-Goroszczenko, Investigations of Absorbed Laser Energy Redistributions in High-Z Plasmas, IPPLM Report No. 1/91, Warsaw 1991 (submitted to Laser and Particle Beams, 1992).

THOMAS-FERMI CALCULATION OF LASER BEAM AND
PARTICLE BEAM RANGES IN STRONGLY COUPLED PLASMA.

A.Ya. Polishchuk and V.E. Fortov
Institute for High Temperatures, USSR Academy of Sciences
Izhorskaya str. 13/19 , Moscow 127412 , USSR

Abstract

A novel approach suggested to calculate the photon and ion ranges in plasma at extreme conditions is presented. The proposed approach is based on the approximation that many-electron ions imbedded into free electron gas are considered as plasma clots which are characterized by collective character of excitation spectrum . This approach contains new physics and differs considerably from usual quantum-mechanical models using one-particle approximation for excitation spectrum. The present results are compared with experimental data and other calculations for hot dense   plasma.

1. EFFECTIVE PHOTON AND FAST ION RANGES IN PLASMA.

Spectral absorption coefficient $\eta(\omega)$ determines the characteristic range length $l(\omega)$ and frequency-dependent opacity $x(\omega)$:

$$l(\omega) = 1/\eta(\omega) ; \qquad x(\omega) = \eta(\omega)/\rho \tag{1}$$

where ρ is a density of matter.

Effective photon ranges in optically thick plasma l_R - Rosseland and optically thin plasma l_P- Planck are defined by well-known relations [1] :

$$l_R = \int_0^\infty l(\omega)\ G_R(u)\ du$$

$$l_P = \int_0^\infty l^{-1}(\omega)\ G_P(u)\ du \tag{2}$$

where $u = \omega/T$, T - temperature, and

$$G_R(u)=\frac{15}{4\pi^4}\ \frac{u^4 e^{-u}}{(1-e^{-u})^2}\ ;\qquad\qquad G_P(u)=\frac{15}{\pi^4}\ \frac{u^3}{e^u-1}\qquad\qquad(3)$$

Spectral absorption coefficient $\eta(\omega)$ of electromagnetic radiation at the frequency ω is determined as a ratio of energy loss rate per unit volume Q to Poiting's vector S and may be expressed through the macroscopic dielectric function $\varepsilon(\omega)$:

$$\eta(\omega)=\frac{\omega^2}{c^2}\ \frac{\text{Im }\varepsilon(\omega)}{\text{Re }k(\omega)}\qquad\qquad(4)$$

where c - speed of light in vacuum, $k(\omega)$ - wave vector connected to dielectric function by the relation :

$$k(\omega)=\frac{\omega}{c}\ (\varepsilon(\omega))^{1/2}\qquad\qquad(5)$$

Dielectric function $\varepsilon(\omega)$ also allows to calculate stopping power of matter for fast charged particles moving with the velocity V.

$$F=\frac{4\ \pi}{V^2}\ \langle n_e\rangle\ \ln\frac{2\ V^2}{\langle\omega\rangle}\qquad\qquad(6)$$

where $\langle\omega\rangle$ is the average excitation energy determined by dielectric function

$$\ln\ \langle\omega\rangle=\frac{1}{2\pi^2\langle n_e\rangle}\ \int\limits_o^\infty\omega\ \text{Im}\ \frac{1}{\varepsilon(\omega)}\ \ln\ \omega\ \ d\omega\qquad\qquad(7)$$

with $\langle n_e\rangle$ being the average numerical electron density.

The principal value in considered model is the electron density distribution in Wigner-Seitz cell. The nucleus with charge Z is placed at cell center . Quantum-statistical electron density functional theory at finite temperatures in its gradient formulation [2] is used to calculate the electron distribution profile n(r) at different temperatures and densities of matter.

The macroscopic dielectric function of matter can be calculated by definition via the relation

$$D_a(\omega)=\varepsilon(\omega)\ E_a(\omega)\qquad\qquad(8)$$

where E_a and D_a are space averaged strength of the electric field and the electrical induction. In the Wigner-Seitz cell model the averaging procedure is restricted by the cell volume.

2. LONG-WAVELENGTH APPROXIMATION FOR CALCULATING INTRA CELL ELECTROMAGNETIC FIELD.

Long-wavelength approximation for calculating the interaction between the electromagnetic radiation with wavelength of λ and the plasma clot of characteristic length of a fulfills the condition of $\lambda << a$. Characteristic length of heavy atom (ion) with nucleus charge Z may be estimated as a $\approx Z$ in accordance with Thomas-Fermi model. Then, the inequality $\lambda >> a$ comes to the following limitation on the incident radiation frequency: $\omega << c\ Z^{1/3} \approx 10$ keV Thus, long-wavelength approximation for interaction between radiation and heavy atom (' ion) is true up to X-ray region.

Quasi-stationary approximation to Maxwell equations is true for long-wavelength limit $\omega << c\ Z^{1/3}$. Intra cell electrical field is described by potential V (r) which obeys the equation :

$$\text{div } (\varepsilon(r,\omega)\ \nabla\ V(r,\omega))=0 \tag{9}$$

where $\varepsilon(r,\omega)$- dielectric function distribution in the cell.

Consider external electric field directed in Z- axis and introduce the respective polar coordinate frame with the origin in the cell centre. Then the boundary conditions to equation (9) can be formulated as finiteness condition at the origin V(0) << 0 and continuity conditions imposed at the cell boundary on normal component of electrical field E and tangent component of electrical induction $\varepsilon(r,\omega)$ E

$$\frac{\partial V(r,\theta,\phi)}{\partial r}\Big|_{r=R} = \frac{-\partial V(r,\pi-\theta,\phi+\pi)}{\partial r}\Big|_{r=R}$$

$$\tag{10}$$

$$\frac{\partial V(r,\theta,\phi)}{\partial \theta}\Big|_{r=R} = \frac{-\partial V(r,\pi-\theta,\phi+\pi)}{\partial \theta}\Big|_{r=R}$$

These conditions can be satisfied by setting

$$V(r)=V(r)\ \cos\ \theta \tag{11}$$

the equation for $V(r)$ being

$$V''(r) + \left(\frac{2}{r} + \frac{\varepsilon'(r,\omega)}{\varepsilon(r,\omega)} \right) V'(r) - \frac{2}{r^2} V(r) = 0 \qquad (12)$$

Let us define constants $V_1(\omega)$ and $V_2(\omega)$ by joining expression

$$V(r) = -V_1 r + V_2/r^3 \qquad (13)$$

and its derivative to the solution of (10) at $r=R$.

We use the simplest approximation for local dielectric function. This leads nevertheless to sufficiently sound results :

$$\varepsilon(r,\omega) = 1 - \frac{4\pi\, n(r)}{\omega^2} + i0 \qquad (14)$$

When $\operatorname{Im} \varepsilon(r,\omega) = 0$, equation (10) has a singularity connected with the possibility of the function $\varepsilon(r,\omega)$ to turn into zero at a certain point $r=r_0$. Then, in the vicinity of singularity, the equation (10) is presented in the form :

$$V''(r) + \left(\frac{2}{r} + \frac{1}{r-r_0} \right) V'(r) - \frac{2}{r^2} V(r) = 0 \qquad (15)$$

General solution of equation (15) is written in the form :

$$V(r) = C_1 \phi_1(\rho) + C_2(\phi_1(\rho) \ln \rho + \phi_2(\rho)) \qquad (16)$$

Here $\rho = r-r_0$, C_1 and C_2 - complex constants, $\phi_1(\rho)$ and $\phi_2(\rho)$ regular functions in the vicinity of the point $\rho = 0$, while $\phi_1(0) = \phi_2(0) = 1$

Note that potential $V(r)$ diverges logarithmically at the point $r=r_0$. To elucidate the connection between the values of potential (16) on the different sides from branching point $\rho = 0$, remember the existence of infinitesimal imaginary addition to dielectric function $\operatorname{Im} \varepsilon(r,\omega)$. Picking out the regular branch of nonanalytical function $\ln \rho$ in accordance with positive sign of this imaginary addition gives

$$\ln \rho = \ln |\, r-r_0| - i\pi\, \theta(r-r_0) \qquad (17)$$

where θ being the step-function. Therefore this analysis shows the route of the numerical calculation of macroscopic dielectric function using Eq.(8).

3. COMPARISON WITH EXPERIMENTAL RESULTS AND OTHER CALCULATIONS.

Let us compare the results based on the model described above with the experimental data and other calculation methods.

Fig. 1 shows the comparison between the calculations of Rosseland photon range in aluminum plasma at density 0.5 g/cm^3 and temperatures from 300 K to 10^6K using the present PC model and HFS method taking different ion populations into account [3].

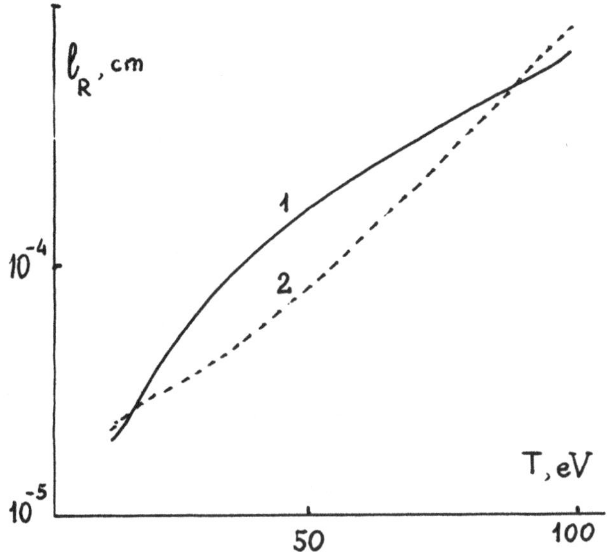

Fig. 1. Rosseland photon range versus temperature for *Al* at ρ = 0.5 *g/cm*   1 – PC model,   2 – ionic HFS method

Fig.2 shows the comparison of PC model calculations of Rosseland mean opacities versus temperature for gold plasma with calculations of [4]. This model presents a semi-empirical version of average atom model fitted to quantum mechanical calculations [5]. The maximum value of Rosseland mean opacity is given by the maximum opacity theorem [6]. For comparison, this limit is also shown.

Fig.3 represents the comparison of calculations based on PC model with experimental results on stopping of 0.35 TW/cm^2 proton beam in 6.2μm aluminum foil obtained on KALIF [7]

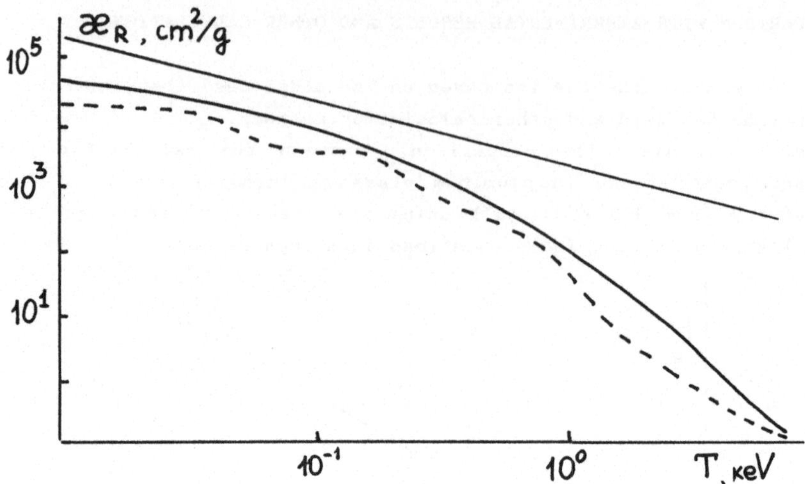

Fig. 2.   Rosseland   mean opacity   versus temperature for *Au* at ρ = 0.1 *g/cm* . Calculations by PC model are indicated   by solid line. Dashed line represents results obtained   by [4] , straight line is the limit   given by the maximum opacity theorem.

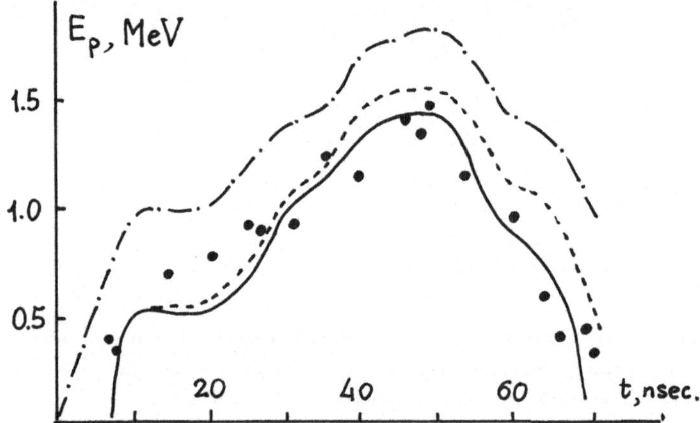

Fig.3. Energy loss of 0.35 TW/cm^2 proton beam in 6.2 Al foil versus the time. Dashed-dotted line represents the input energy,   measured output energy. PC model calculations : solid line- for heated and expanded target, dashed line - for cold solid density target.

REFERENCES

[1] Ya.B. Zel'dovich and Yu.P.Raizer Physics of Shock Waves & High Temperature Hydrodynamics Phenomena (New York:Academic Press,1966)

[2] A.Ya. Polishchuk 1987 Solid State Commun. 61, 193 (1987)

[3] A.F. Nikiforov , Preprint No 114, Institute of Applied Mathematics (Moscow,1982)

[4] G.D. Tsakiris and K. Eidmann , J.Quant.Spectr.Rad.Transf.38,353 (1987)

[5] SESAME'83: Report on the Los Alamos equation-of-state Library,T4 - Group,Report No LALP-83-4 (Los-Alamos, N.M.)

[6] B.H. Armstrong and R.W. Nicholls , Emission, Absorption and Transfer of Radiation in Heated Atmospheres (Oxford: Pergamon Press,1972)

[7] W. Bauer , H. Bluhm , B. Goel Institut fuer Neutronenphysik und reactorentechnik preprint, INR 1601 (1988).

Teubner - Texte zur Physik

Vol. 13: Progress and Trends in Applied Optical
Spectroscopy. Proceedings of the Fourth Sympo-
sium Optical Spectroscopy (SOS 86) held in
Reinhardsbrunn, Germany, 1986
Ed. Fassler/Feller/Wilhelmi. 291 pages.
Bound DM 39,-. ISBN 3-322-00444-9

Vol. 15: Münchow/Reif, **Recent Developments in
the Nuclear Many-Body Problem. Vol. II**
180 pages. Bound DM 24,-. ISBN 3-322-00313-2

Vol. 16: Localization in Disordered Systems
Proceedings of the International Seminar held
in Bad Schandau-Ostrau, Germany, 1986
Ed. Weller/Ziesche. 232 pages.
Bound DM 31,-. ISBN 3-322-00512-7

Vol. 17: Ulbricht u.a., **Thermodynamics of Finite
Systems and the Kinetics of First-Order Phase
Transitions**
208 pages. Bound DM 28,-. ISBN 3-322-00491-0

Vol. 19: Plath, **Diskrete Physik molekularer
Umlagerungen**
248 pages. Bound DM 33,-. ISBN 3-322-00492-9

Vol. 22: Anishchenko, **Dynamical Chaos in Physical Systems - Experimental Investigation of
Self-Oscillating Circuits**
211 pages. Bound DM 28,50. ISBN 3-322-00713-8

Vol. 23: Irreversible Processes and Selforga-
nization. Proceedings of the 4th International
Conference IPSO, Rostock, Germany, February 1989
Ed. Ebeling/Ulbricht. 255 pages.
Bound DM 34,50. ISBN 3-322-00735-9

Vol. 24: Akhmanov/Emel'yanov/Koroteev, **Interaction of Strong Laser Radiation with Solids and
Nonlinear Optical Diagnostics of Surfaces**
204 pages. Bound DM 28,-. ISBN 3-322-00488-0

Preisänderungen vorbehalten.

B.G. Teubner Verlagsgesellschaft
Stuttgart · Leipzig

Teubner - Texte zur Physik

Ebeling/Förster/Fortov/
Gryaznov/Polishchuk

Thermophysical Properties of Hot Dense Plasmas

By W. Ebeling, A. Förster, both Berlin,
V.E. Fortov, V.K. Gryaznov and
A.Y. Polishchuk, all Moscow

1991. 315 pages. 14,5 x 21,5 cm. (Vol. 25).
Bound DM 48,-
ISBN 3-8154-3010-0

New challenging technologies based upon laser
and particle beams, plasma light sources, X-ray
lasers, and controlled fusion, the arc physics
and astrophysical research focus theoretical
investigations and applied research on strongly
coupled plasmas. The book provides new results
on the plasma composition, the equation of state
and other thermodynamic properties for plasmas
with higher charges and for mixed plasmas.
Using different theoretical models also energy
level shifts, ionization and recombination
kinetics, electrical and thermal conductivity,
light absorption and plasma stopping power are
discussed. Besides a review of the theory,
useful semiempirical formulae are derived.
Tables of thermophysical properties of plasmas
which are especially relevant for practical
applications close the book.

B.G. Teubner Verlagsgesellschaft
Stuttgart · Leipzig

Diese Reihe wurde geschaffen, um eine schnellere Veröffentlichung physikalischer Forschungsergebnisse und eine weitere Verbreitung von physikalischen Spezialvorlesungen zu erreichen. TEUBNER-TEXTE erscheinen in deutscher oder in englischer Sprache. Um Aktualität der Reihe zu erhalten, werden die TEUBNER-TEXTE im Manuskriptdruck hergestellt, da so die geringeren drucktechnischen Ansprüche eine raschere Herstellung ermöglichen. Autoren von TEUBNER-TEXTEN liefern an den Verlag ein reproduktionsfähiges Manuskript. Nähere Auskünfte darüber erhalten die Autoren vom Verlag.

This series has been initiated with a view to quicker publication of the results of physical research-work and a widespread circulation of special lectures on physics. TEUBNER-TEXTE will be published in German or English. In order to keep this series constantly up to date and to assure a quick distribution, the copies of these texts are produced by a photographic process (smal-offset printing) because its technical simplicity is ideally suited to this type of publication. Authors supply the publishers with a manuscript ready for reproduction in accordance with the latter's instructions.

Cette série de textes a été créée pour obtenir une publication plus rapide de résultats de recherches physiques et de conférences sur des problèmes physiques speciaux. Les TEUBNER-TEXTE seront publiés en langues allemande ou anglaise. L'actualité des TEUBNER-TEXTE est assurée par un procédé photographique (impression offset). Les auteurs des TEUBNER-TEXTE sont priés de fournir à notre maison d'édition un manuscrit prêt a être reproduit. Des renseignements plus précis sur ia forme du manuscrit leur sont donnés par notre maison.

Эта серия была создана для обеспечения более быстрого публикования результатов физических исследований и более широкого распространения физических лекций на специальные темы. Издания серии ТОЙБНЕР-ТЕКСТЕ будут публиковаться на немецком или английском языках. Для обеспечения актуальности серии. ее издания будут изготовляться фотомеханическом способом. Таким образом. более скромные требования к полиграфическому оформлению обеспечат более быстрое появление в свет. Авторы изданий серии ТОЙБНЕР-ТЕКСТЕ будут предоставлять издательству рукописи. удовлетворящие требованиям фотомеханического печатания. Более подробные сведения авторы получат от издательства.

B. G. Teubner Verlagsgesellschaft Stuttgart · Leipzig
O - 7010 Leipzig, Postfach 930